MAFF

Pesticides, Cereal Farming and the Environment

THE BOXWORTH PROJECT

Edited by

Peter Greig-Smith

Geoff Frampton

Tony Hardy

London : HMSO

ISBN 0 11 242876 2

Contents

Foreword

The environment is today in everyone's thoughts, no more so than the issue of modern farming practices and their integration with the protection of the countryside. MAFF has a major responsibility to ensure that farming methods, especially the use of chemical pesticides, are environmentally safe. Indeed, MAFF is spending some £50 million to fund research and development (R&D) on environmental topics in 1991/92. One of the foremost recent examples of successful MAFF R&D is the Boxworth Project, the results of which are described fully and clearly in this book.

At the time when the Boxworth Project was conceived, in the late 1970s, environmental concerns were less topical than they are today. It was all the more commendable, therefore, that a group of forward-thinking scientists and agronomists within MAFF's Agricultural Development Advisory Service (ADAS) should plan a major project to address the possibility of long-term cumulative effects of the complete package of pesticides used by cereal farmers. MAFF was alert to the importance of such effects, and had the foresight to back the proposal with substantial funding over a ten-year period.

Though its importance may not have been widely recognised at the outset, the Boxworth Project certainly achieved a high public profile as it developed. The wealth of information in this book, on a broad sweep of environmental topics, shows the rapid progress in this area of science in the 1980s. The Boxworth Project has paved the way for several new projects, including two large-scale research studies being carried out by ADAS, again on MAFF's Experimental Farms.

I greatly enjoyed my visit to Boxworth, which gave me the opportunity to appreciate the valuable work which has been undertaken there. I welcome the production of this book, which gives a fascinating account of the complex ways in which pesticide use can affect wildlife and natural enemies of pests. It is a fitting culmination to a major effort by scientists within MAFF, the Agricultural and Food Research Council, the Natural Environment Research Council, and Universities. The results endorse the wisdom of carefully managed, selective use of pesticides as the best way to achieve cost-effective crop protection for cereals. The Project is also a clear demonstration of MAFF's commitment to environmental R&D.

The Baroness Trumpington
Minister of State
Ministry of Agriculture, Fisheries & Food

List of Principal Authors' Addresses

Dr A J Burn National Farmers Union, Agriculture House, Knightsbridge, London SW1X 7NJ

Mr M R Fletcher Central Science Laboratory, Ministry of Agriculture, Fisheries and Food, London Road, Slough, Berks SL3 7HJ

Dr G K Frampton Dept of Biology, University of Southampton, Biomedical Sciences Building, Bassett Crescent East, Southampton SO9 3TU

Dr P W Greig-Smith Central Science Laboratory, Ministry of Agriculture, Fisheries and Food, Tangley Place, Worplesdon, Surrey GU3 3LQ

Mrs M Hancock ADAS, Ministry of Agriculture, Fisheries and Food, Block C, Government Buildings, Brooklands Avenue, Cambridge CG2 2DR

Dr A D M Hart Central Science Laboratory, Ministry of Agriculture, Fisheries and Food, Tangley Place, Worplesdon, Surrey GU3 3LQ

Mr R H Jarvis C/O Boxworth Experimental Husbandry Farm, Boxworth, Cambridge CB3 8NN

Dr I P Johnson C/O Dr J R Flowerdew, Department of Zoology, Downing Street, Cambridge CB2 3EJ

Dr E J P Marshall Long Ashton Research Station, Long Ashton, Bristol, BS18 9AF

Mr K A Tarrant Central Science Laboratory, Ministry of Agriculture, Fisheries and Food, London Road, Slough, Berks SL3 7HJ

Dr H M Thompson Central Science Laboratory, Ministry of Agriculture, Fisheries and Food, London Road, Slough, Berks SL3 7HJ

Dr G P Vickerman Department of Biology, University of Southampton, Biomedical Sciences Building, Bassett Crescent East, Southampton, SO9 3TU

Mr D J Yarham ADAS, Ministry of Agriculture, Fisheries and Food, Block C, Government Buildings, Brooklands Avenue, Cambridge, CB2 2DR

Origins and aims of the Boxworth Project

Peter W Greig-Smith (MAFF Central Science Laboratory, Worplesdon)

Introduction

Scientific research is often conducted in a sheltered atmosphere, at least until its results are complete and published. From time to time, however, certain studies achieve a greater contemporary significance, at least within their own field, by virtue of novel approaches that bear promise of breaking new ground. The Boxworth Project was one such study. It has become a byword within the circles of science and agriculture that are concerned with the ecological side-effects of intensive farming practices and with the development of more environmentally benign approaches to crop production.

Such topics are now a commonplace part of the public concerns that attend the interface between man's activities and his environment. However, that has not always been the case. It was not long ago that the imperative to increase agricultural production led to high inputs of agrochemicals on many crops, with few doubts about the potentially damaging side-effects they may have. These products were all screened for safety before approval for their use was granted, so that in theory any risks of environmental effects should have been identified before use became widespread.

The Boxworth Project was conceived at a time when such assumptions were being questioned. Results from long-term monitoring in cereal-growing areas of southern England by the Game Conservancy were beginning to reveal drastic reductions in invertebrate populations over a period in which pesticide usage had increased (Potts, 1986). At the same time, the wisdom of over-using chemicals as an insurance against crop losses was being challenged from the point of view of efficient and cost-effective crop protection. The Ministry of Agriculture, Fisheries & Food's Agricultural Development Advisory Service (ADAS) and others were developing approaches to 'managed' crop protection, in which pesticides would be used only when necessary, and targetted closely at relevant pest, weed or disease problems (Anon., 1984).

Such was the debate in the late 1970's, when the Boxworth Project had its origins. The Project was therefore timely, but it had added importance because it broke new ground in the way that studies of farming systems are approached.

With one or two exceptions (eg El Titi, 1991), most previous studies of the environmental effects of agrochemicals had either involved small plot trials, following standard experimental designs directed at measurement of particular predicted effects, or consisted of monitoring in areas of farmland with minimal interference. The first approach inevitably restricts investigations to single agrochemical uses, whereas the latter suffers a lack of experimental control, so that effects may be confounded, and misunderstood as a result. In the Boxworth Project, these limitations were discarded, in an attempt to carry out an experimental study on a large scale, approaching that of real farm use. This was not free from problems, however, (see Chapter 2), but it brought the major benefit of allowing the combined, cumulative effects of an entire annual package of pesticides to be examined over several years.

Unique at the time of its inception, the Boxworth Project has had a great influence on other studies that have subsequently taken a 'whole-farm' approach to the effects of agricultural pesticides (and other practices) on the environment.

History of the Boxworth Project

The Project grew from ideas and discussions held by ADAS in the mid-1970's. In particular, consultations between Dr Peter Bunyan and Dr Peter Stanley, of the Tolworth Laboratory, and members of the Institute of Terrestrial Ecology (ITE) and the Game Conservancy served to affirm the need for a new approach to examine the potential long-term, combined effects of pesticide regimes on wildlife.

This led to a discussion document produced in mid 1978, which identified the concept of a large-scale study that would address these issues. The specific notion of comparing the contrasting extremes of high and low pesticide input systems was developed further later that year, in a proposal by Mr Stanley Evans, ADAS Regional Agronomist at Leeds, to establish an experimental study.

From that point, the planning gathered momentum. A series of meetings during 1979 involved debates about the scale of the study, its location, and the principles of the pesticide regimes that should be examined. It was agreed to concentrate on cereal crop protection, because of the large UK acreage, the relatively high levels of pesticide use on cereals at the time, and the emergence of treatment thresholds for managed chemical control.

The possibility of comparing a no-pesticide regime to high and low inputs was abandoned, in view of difficulties in maintaining a viable crop without any pesticides. For practical reasons, it was considered too difficult to obtain well-matched, very large blocks of land of a size sufficient to permit population studies of birds and mammals. Instead, plans were developed to take a site under moderate pesticide use, study it for a baseline period, and then switch parts to the chosen experimental regimes in order to determine the ecological consequences.

On this basis, a series of research proposals, for work over a seven-year period, was presented to the ADAS Central Research & Development Committee (CRDC) in late 1979. With CRDC agreement, the technical aspects were further refined, and Boxworth Experimental Husbandry Farm was chosen as a suitable location. A plan for complementary studies by ADAS staff (from Boxworth, Cambridge, and Tolworth Laboratory), the Weed Research Organisation, and scien-

tists to be funded by MAFF research fellowships at Universities, was drawn up. Funding was agreed in 1981, in time for the Project to start at Boxworth in the autumn of that year. Coordination of the many elements of the Project was consigned to a Steering Group, initially under the chairmanship of the ADAS Eastern Regional Agricultural Scientist, Mr Peter Wiggell, with Dr Tony Hardy (Tolworth Laboratory) as its secretary and coordinator of the research studies. This group gathered for the first time in November 1981, and subsequently met to review progress and plans, on 22 occasions up to the end of the study.

Both in the early planning, and in the course of the Project itself, discussions were held with numerous other organisations, formally and informally. Research at the Game Conservancy had been instrumental in stimulating the Project, and close liaison was maintained over methodology and ideas connected with invertebrate studies. The Institute of Terrestrial Ecology (ITE) at Monks Wood Experimental Station was consulted, particularly about vertebrates, and at an early stage it was agreed that ITE would participate in studies of birds. However, because the size of the experimental blocks was eventually smaller than originally intended, this possibility was not pursued. Discussions were also held with the Nature Conservancy Council, the Agricultural and Food Research Council, the Natural Environment Research Council, University researchers, and the agrochemicals industry, about the concept, design and methodology of the Project. Support for the Project's aims was given in the recommendations of the 7th report of the Royal Commission on Environmental Pollution (1979).

With this background, the Project began in late 1981, following a programme that is explained in Chapter 2. After several years, criticisms of the Project's design were raised by those who felt that a lack of replication invalidated the aims of the work (see Chapter 2). These were taken up by the House of Lords Science and Technology Committee in 1984, whose report suggested that the design should be reviewed, to assess whether any changes were necessary or feasible. The resulting review involved further discussions with leading experts in the major interested organisations. It concluded that any change mid-way through the Project would be disruptive, and also would jeopardize certain aspects of the research. The original programme was therefore endorsed, and proceeded as planned until 1988, the end of the seven-year period. As that time approached, there was strong pressure from several quarters to extend the Project and continue the comparisons between high and low input strategies. In the event, this was resisted, in favour of starting new projects elsewhere, but conducting a smaller continuation study at Boxworth with different aims (see Chapter 2). The latter was complete by the autumn of 1991, ten years after the start of the Project itself.

Throughout the Project, the growing interest in its findings led to frequent visits by individuals and groups from government research institutes, conservation bodies, the farming and agrochemicals industries, policy makers and others. The Project was also featured on radio, television, in articles in the Press, and in numerous lecture presentations to interested groups. These many outlets ensured that the aims of the work, and particularly the questions it was *not* designed to address, became widely known, both nationally and throughout Europe.

Aims of the Project

There were three linked objectives addressed by the Boxworth Project. The principal aim was to examine and compare the environmental side-effects of contrasting pesticide regimes. These were chosen to represent a high chemical input, designed to insure against the occurrence of any pest, weed or disease problems, and realistic reduced-input systems, relying on selective chemical control when required and modifications of husbandry to avoid problems arising.

Second, it was considered important to monitor the economics of crop production under these regimes, in order to establish the relevance of

the results and give preliminary indicators of whether reduced-input farming is commercially viable.

Third, the Project sought to identify any difficulties that might be entailed in the practical operation of reduced pesticide systems.

The ways in which these aims were addressed are described in detail in Chapter 2.

Contents of the book

As the Project proceeded, interest in its scientific results was met by the publication of a series of Annual Reports (Hardy, 1983–86; Greig-Smith, 1987–89a). These served to outline the studies being conducted, and to highlight significant provisional findings. Though useful as interim statements of progress, these reports did not include the depth of scientific detail necessary for a full interpretation of the data. That will be covered by papers published in scientific journals, many of which have already appeared (Appendix V), although some aspects have not yet been fully analysed.

A need was also felt for a less technical account of the work, particularly in view of the broad range of studies involved. Accordingly, a series of general articles was produced (Hardy, 1986; Jarvis, 1988; Greig-Smith, 1989b, 1990, 1991), leading to this book, which provides a final overview of the entire Project. The aim of the book is to explain, without a wealth of detail, the reasons for the Project and its design, the range of studies carried out, and the principal implications of the results.

Chapter 2 provides a summary of the decisions reached during the planning of the Project, and outlines the main features of the design which they generated, including the size and layout of the experimental areas and the choice of pesticide programmes. It is followed by three chapters which deal in turn with the crop protection problems against which these programmes were employed. The major pests (Chapter 3), weeds

(Chapter 4) and diseases (Chapter 5) occurring at Boxworth are listed, the proposed use of pesticides for insurance and supervised control is explained, and the results of monitoring these problems during the project are presented. Appendices I and II provide a complete catalogue of the pesticides included in the planned crop protection programmes, and list those that were actually applied in each year of the Project.

Chapter 6 describes the performance of crops in the Project fields, including yields and grain quality, and the costs of production under the three experimental regimes.

Throughout most of the Project, there was no systematic, regular monitoring of the distribution and persistence of pesticides after they had been applied to the crops. However, a number of short studies of specific pesticide applications were undertaken. These included an investigation of spray drift of summer insecticides. The results are described in Chapter 7, along with a survey of pesticide residues in soil water from certain fields, which was carried out at the end of the Project to determine whether there had been any accumulation of residues under the high-input regime.

Chapter 8 describes work carried out on plants found in the wheat fields and their margins. This extended beyond the surveys needed to assess weed infestations, and considered the variety and density of plant species other than the pernicious weeds.

The next three chapters concern research studies of invertebrates. An intensive programme of regular sampling provided information about changes in the populations of pests, beneficial species, and others in the crop (Chapter 9). This was complemented by a specific investigation of the interactions between pests and the 'natural enemies' which may help to keep them in check (Chapter 10). A separate study was carried out to monitor invertebrates living in the soil (Chapter 11).

Possible effects of pesticides on mammals were also investigated. A programme of trapping revealed patterns in the occurrence of mice, voles and shrews in fields and adjacent hedgerows

(Chapter 12). As well as looking for trends through the course of the Project, this work addressed the question of short-term changes at the time of particular pesticide applications. A brief study of rabbits was undertaken when it became clear that these animals were present in the wheat fields at the time of summer aphicide use (Chapter 13).

Three chapters deal with studies of birds. Chapter 14 describes the results of monitoring the populations of songbirds breeding in and around the Project fields, and includes an assessment of effects on their breeding performance. Chapter 15 focuses on one use of pesticides—spraying of insecticide for summer aphid control—that seemed particularly likely to affect breeding birds. This involved research on a single bird species, the tree sparrow, and Chapter 16 presents the results of three other short studies on particular species that may be affected by insecticide use.

Finally, Chapter 17 summarises the findings from all aspects of the Boxworth Project, and discusses their implications for understanding the ecology of cereal fields, for revealing the environmental consequences of high-input, insurance use of pesticides, and for the development of less damaging, lower-input systems.

Chapter 17 also includes an outline of the further studies that have been planned as a result of the Boxworth Project. One of these is a continued comparison of the pesticide regimes at Boxworth, but on a smaller scale, and with a specific aim to examine the recovery of invertebrate populations in the high-input fields once pesticide use is returned to a lower level. This work continued for three years after the end of the main Project, and its results will be described in two progress reports, to be produced after the second and third continuation years.

Several other research studies have arisen as a result of the Boxworth Project. These include investigations of birds in orchards, and two experimental projects concerned with the design and ecological effects of low-input systems (Cooper, 1990). As these new areas of research progress, the significance of the Boxworth Project for the development of more environmentally benign farming systems will become apparent.

Acknowledgements

The Boxworth Project has involved a very large number of people, all of whom have helped to ensure its success. Appendix VI lists those who have been practically involved in planning, managing and carrying out the study. In addition, Miss P Hill and Miss K Bowyer (MAFF Central Publications Unit) and Mr R Middleton and Mr M Haskins (HMSO) have put in much effort to ensure the production of this book.

The following experts kindly reviewed the main chapters of the book, and their comments have been valuable in guiding the final versions: Dr R G McKinlay (Pests), Dr V W L Jordan (Diseases), Mr R J Chancellor (Weeds, and Flora), Mr E S Carter (Crop Performance), Dr J P Dempster (Invertebrates), Dr J Gurnell (Mammals), Dr J J D Greenwood (Birds), Dr N W Moore (Birds), and Dr P J Bunyan (Summary and Recommendations).

References

Anon. (1984). 'Managed disease control: winter wheat.' *Ministry of Agriculture, Fisheries & Food Leaflet No. 831*. MAFF (Publications), Alnwick.

Cooper, D A (1990). 'Development of an experimental programme to pursue the results of the Boxworth Project'. *Proceedings of the 1990 Brighton Crop Protection Conference—Pests and Diseases*, vol. 1, pp. 153–162.

El Titi, A (1991). 'Twelve years experience of integrated wheat production at the commercial farm of Lautenbach, South West Germany.' In: Firbank L G, Carter N, Darbyshire J F and Potts G R (Eds). *The Ecology of Temperate Cereal Fields*. Blackwell Scientific Publications, Oxford pp. 399–411.

Greig-Smith, P W (Ed.) (1987). *The Boxworth Project: 1986 Annual Report*. ADAS, Tolworth.

Greig-Smith, P W (Ed.) (1988). *The Boxworth Project: 1987 Annual Report*. ADAS, Tolworth.

Greig-Smith, P W (Ed.) (1989a). *The Boxworth Project: 1988 Annual Report*. ADAS, Tolworth.

Greig-Smith, P W (1989b). 'The Boxworth Project—environmental effects of cereal pesticides.' *Journal of the Royal Agricultural Society of England*, **150**: 171–187.

Greig-Smith, P W (1990). 'The Boxworth Project'. *Pesticide Outlook*, **3**: 16–19.

Greig-Smith, P W (1990). 'The Boxworth Project'. *Pesticide Outlook*, **3**: 16–19.

Greig-Smith, P W (1991). 'The Boxworth experience: effects of pesticides on the fauna and flora of cereal fields'. In: Firbank L G, Carter N, Darbyshire J F and Potts G R (Eds). *The Ecology of Temperate Cereal Fields*. Blackwell Scientific Publications, Oxford pp. 333–371.

Hardy, A R (Ed.) (1983). *The Boxworth Project: 1982 Annual Report*. ADAS, Tolworth.

Hardy, A R (Ed.) (1984). *The Boxworth Project: 1983 Annual Report*. ADAS, Tolworth.

Hardy, A R (Ed.) (1986). *The Boxworth Project: 1985 Annual Report*. ADAS, Tolworth.

Hardy, A R (1986). 'The Boxworth Project—a progress report.' *Proceedings of the 1986 British Crop Protection Conference, Pests and Diseases*, **3**: 1215–1224.

Jarvis, R H (1988). 'The Boxworth Project'. In: Harding D J L (Ed) *Britain since Silent Spring*. Institute of Biology, London pp. 46–55.

Potts, G R (1986). *The Partridge: Pesticides, Predation and Conservation*. Collins, London.

Royal Commission on Environmental Pollution (1979). *Seventh Report. Agriculture and Pollution*. HMSO, London.

Design and management of the Boxworth Project

Peter W Greig-Smith * and **Anthony R Hardy** # (MAFF Central Science Laboratory, Worplesdon* and Slough#)

Introduction

The aim of the Boxworth Project was to address the possible large-scale, long-term effects of pesticides under conditions as close as possible to real farm use. This had not previously been done experimentally, and there was therefore no ready-made and tested model for the design of such a project. Although apparently quite simple in concept, this objective raised a number of considerable practical difficulties in planning, establishing and operating the experiment.

The first consideration was the choice of a suitable site, or sites, on which to carry out the work. Because of limits on resources, each contrasting approach to pesticide use could not be carried out on a whole working farm. That approach would also have led to problems in obtaining appropriately matched farms as 'control' comparisons. A compromise was therefore reached, to undertake the Project on the largest scale possible within the confines of a single farm.

Because the work was to be experimental, following a fixed programme of pesticide use rather than monitoring a farm under unplanned, flexible management, it was important to be sure that full control could be exercised over pesticide applications and all other aspects of husbandry. A detailed record of past cropping and inputs was also needed, to allow full interpretation of results obtained during the Project. For these reasons, it was decided to site the work on a MAFF Experimental Husbandry Farm. Boxworth was chosen as the most suitable within an intensive cereal-growing area, where there was scope to divert a large part of the farm's 300 hectares to the Project. Table 2.1 lists the principal aspects of the design and management of the study. Each of these is discussed in turn below.

Approaches to pesticide use

Originally the intention was to compare a high-input prophylactic approach to crop protection ('Full Insurance') with a lower, managed input, based on assessing the need for each treatment by monitoring pest, weed and disease levels ('Supervised'). This would have provided a contrast between the high levels of input practised by a substantial number of the most intensive cereal producers in the eastern counties of England in the late 1970s, against the more cautious approach then being developed by ADAS. At the time, there was also growing interest in other means of reducing pesticide inputs, including the choice of disease-resistant varieties, and the pros and cons of spring versus autumn sowing. Such aspects affect the need for chemical inputs, with economic and environmental implications. Therefore it was decided that a third approach to pesticide use should be tested as part of the Boxworth Project. Described as 'Integrated', this was intended to incorporate as many modifications to cropping,

Table 2.1 Aspects involved in the design and management of the Boxworth Project.

Experimental Design	Husbandry
Replication	Approaches to pesticide use
Layout of treatment areas	Pesticide programmes
Matched triplets of fields	Use of fertilizers
Intensive study fields	Cropping
Timetable	
Replicated plot trial	

Management of the Project	Studies Undertaken
Steering Group	Yield and economic assessments
Operational aspects	Monitoring of pests, weeds and diseases
	Research studies

husbandry, and pest control as were possible within the constraints imposed by the experimental design described below.

In practice, many aspects of the Supervised and Integrated programmes were so similar that they can be regarded together, as a single comparison against the Full Insurance approach.

An important decision taken at the outset was to keep the three pesticide programmes the same throughout the life of the Project, despite any changes in common farming practice that might occur during the period. This was essential to enable the identification of any long-term effects, without confounding differences due to other causes. Inevitably, by the end of the Project, there was a risk that one or other of the pesticide programmes might seem unrealistic in the light of current practice. This was indeed the case. The Full Insurance approach now seems much less representative of even the most intensive cereal farming than it did a decade earlier. However, its experimental importance, as a maximum realistic contrast to the managed approach, is undiminished.

Replication

The desire for large study areas on a single farm, without limitless resources, raised a dilemma. At one extreme, each of the treatment areas could be sited as a single large block of land, thereby reducing edge effects and preserving the idea of a 'whole-farm' treatment as far as possible. In contrast, there are obvious advantages in establishing a series of properly replicated areas under the three styles of pesticide use, to allow the application of the statistical tests normally used in agricultural trials. Conventionally, such plots are laid out in randomized blocks, Latin squares, or other standard arrangements, and sited in uniform fields where any differences can be attributed to the treatments applied to them, using Analysis of Variance (Cochran & Cox, 1957). This approach is clearly incompatible with experimental units on a field-scale, let alone on a farm-scale. Nevertheless,

in principle it would have been possible to arrange the field treatments in such a way that some of the confounding variation between fields could be accounted for statistically. Thus, if the farm comprised several soil types, or areas with different amounts of non-crop habitats, fields in each area could be divided among the three pesticide treatments, and examined as a group, rather than having all the Full Insurance fields in one soil type, for example. This would permit a satisfactory, if not ideal, form of replicated statistical testing.

However, even this compromise carries difficulties. If the areas under different treatments were sufficiently small and close together, there would be a risk of contamination by spray drift, and mixing of wild populations from areas under different forms of management. Most importantly, the isolation of individual fields would deprive the experiment of the power to investigate those large-scale effects that were the prime reason for its establishment.

The solution adopted was to retain each treatment as a single large block, maximizing the chance of identifying large-scale effects, at the expense of not being able to use the conventional approach to testing replicated differences. This is justified for several additional reasons:

First, the nature of the environmental effects that might be observed was not known at the beginning of the Project. Accordingly, the work was largely an exploration to reveal these effects, rather than an experiment to measure the size of an already-specified effect (which is the basis of all replicated trials). The Boxworth Project was thus partly a 'test-bed' for ideas, to be investigated further in subsequent studies.

Second, as indicated above, the greatest priority was to assess the largest and longest scale effects, that had not previously been addressed by testing of individual agrochemical products.

Third, there are alternative approaches to testing whether any effects observed were likely to be due to the pesticide treatments. For example, measuring the typical range of variation in the density of an animal population provides a basis for judging whether a change occurring after a pesticide application is larger than would be expected for other reasons. Confidence in the interpretation would be strengthened if such changes occurred consistently in different fields or in the same field at different times, within a single treatment area. This cannot prove cause and effect, of course, but it is able to provide a weight of evidence that may strongly suggest the existence of a real pesticide-related effect. In some cases, such indications were then pursued in more detail by small-scale experiments within the overall Project.

Finally, a replicated trial of small plots was set up in one of the Project fields, to provide a stronger basis on which to interpret certain changes at the field scale.

Figure 2.1   Map of Boxworth Experimental Husbandry Farm, showing the location of the fields included in the Project.

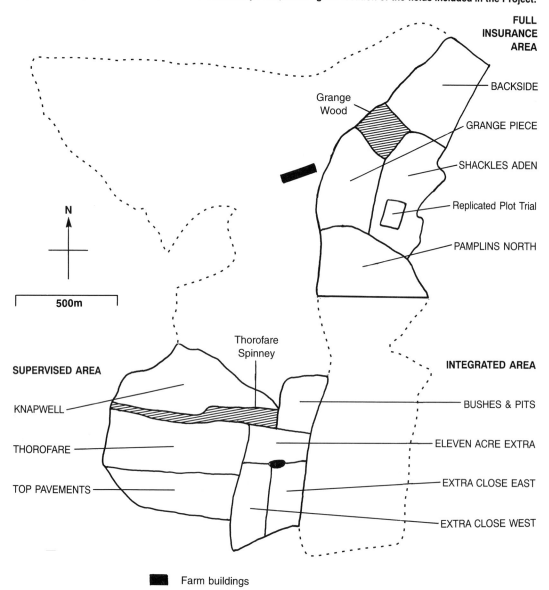

FULL INSURANCE AREA

BACKSIDE

Grange Wood

GRANGE PIECE

SHACKLES ADEN

Replicated Plot Trial

PAMPLINS NORTH

N

500m

Thorofare Spinney

SUPERVISED AREA

INTEGRATED AREA

KNAPWELL

BUSHES & PITS

THOROFARE

ELEVEN ACRE EXTRA

TOP PAVEMENTS

EXTRA CLOSE EAST

EXTRA CLOSE WEST

▮ Farm buildings

On balance, therefore, it was felt that the large areas chosen offered sufficient opportunity to gauge the likely reality of the effects observed. What is more, none of the critics who emphasized the shortcomings of this approach were able to devise a preferable experimental design to address the primary aims of the Project.

Layout of treatment areas

The main priority was to match as closely as possible the Full Insurance fields with those in the Supervised and Integrated areas, in order to minimize confounding variation. This involved selecting similar numbers of fields of approximately the same size in each area, checking that the underlying soils were not markedly different, and ensuring that the amount and distribution of non-crop habitats (woods, hedges, etc.) were approximately similar.

Another consideration was to keep the high- and lower-input areas separate from each other, and to select parts of the farm where the surround-ing land was managed in a compatible way. Thus the Full Insurance area should ideally be bordered by intensively-farmed fields, and the Supervised and Integrated areas by a lower-input style of farming.

These needs were met by the choice of eleven fields, in two blocks at opposite ends of the farm (Figure 2.1). The Full Insurance area comprised four fields (53 ha), the Supervised contained three (46 ha) and the Integrated had four (22 ha), two of which were operated as a single unit. These fields are listed in Table 2.2. Before the start of the Project, the fields had been used for growing cereals, other arable crops, and grass for cattle, but there were no consistent differences in productivity among the groups of fields chosen to form each experimental area. Similarly, average previous use of pesticides had been consistent in all fields.

Once selected for the Boxworth Project, these fields were managed solely according to the needs of the Project. They were used for other trials only occasionally, when small plots could be placed

Table 2.2 Fields used in the Boxworth Project, showing their size and the crops grown in each (W = winter wheat, Beans = Winter beans, OSR = oilseed rape).

		Baseline years		Treatment phase					Continuation Study	
	Size (ha)	1982	1983	1984	1985	1986	1987	1988	1989	1990
Full Insurance area										
Grange Piece	12.0	Beans	W	W	W	W	W	W	W	W
Shackles Aden	13.0	W	W	W	W	W	W	W	–	–
Pamplins North	12.6	OSR	W	W	W	W	OSR	W	W	W
Backside	15.6	W	W	W	OSR	W	W	W	W	W
Supervised area										
Knapwell	16.9	W	W	W	W	W	W	W	W	W
Top Pavements	11.5	W	W	W	W	W	OSR	W	W	W
Thorofare	17.3	W	W	W	OSR	W	W	W	–	–
Integrated area										
Extra Close East	5.7	W	W	W	W	W	W	W	–	–
Extra Close West	6.2	W	W	W	W	W	OSR	W	–	–
Eleven Acre Extra/ Bushes & Pits	10.7	W	W	W	OSR	W	W	W	–	–

in certain fields without jeopardising the comparisons between areas.

Matched triplets of fields

Some of the major variations in crop growth, and in populations of pests and wildlife in fields, are due to the timing of agricultural operations, particularly sowing date. For some pesticide uses, there is quite a broad 'window' within which application is possible, and effects may be very different if they occur early in this period, rather than at its end. Accordingly, it was important to remove this factor in comparing pesticide effects.

In order to do so, fields were assigned to triplets, matching individual fields to partners in the other two areas. All husbandry operations (cultivations, sowing, applications of fertilizers and pesticides, harvesting and straw disposal) were carried out as near as possible in time on each of the three fields in each triplet group. These groups were as follows:

	Triplet I	Triplet II	Triplet III
Full Insurance fields:	Backside	Pamplins North	Grange Piece
Supervised fields:	Thorofare	Top Pavements	Knapwell
Integrated fields:	Bushes & Pits + Eleven Acre Extra	Extra Close West	Extra Close East

This left one Full Insurance field (Shackles Aden) that was not matched with fields in the other areas, but was generally treated along with Grange Piece.

Overall, this system was very successful in standardizing the times at which operations were carried out, although it did require a different approach to management, and was less efficient than the usual practice of treating adjacent fields at the same time. However, inevitably there were some occasions on which weather conditions prevented all three fields of a triplet group being treated at the same time. Serious delays of this kind occurred only rarely during the five treatment years of the Project, and the majority of agrochemicals were applied at the intended time. However, there were a few occasions on which applications to all three fields in a triplet, or only to the Full Insurance fields, were delayed by bad weather, or problems of access on wet ground. These differences in timing may have had a substantial influence on some animal populations (Chapter 9).

Intensive study fields

The area of the Project was too great to allow research studies to be carried out in equal depth on all fields. Therefore, one triplet group of fields (Grange Piece, Knapwell and Extra Close East) was selected for intensive study. This involved more detailed sampling and observations than on the other fields, where data were gathered to strengthen the interpretation of results from these three intensive study fields.

Timetable

Once the plan for the Project had been decided, the chosen fields had to be brought from their previous cropping and levels of input to the appropriate experimental programmes. Thus, some fields in which there had been long runs of cereals needed a break under a different arable crop, to provide an opportunity to control grass weeds and volunteer cereals, before entering another period of cereals. This switch was achieved during a preliminary two-year period before the start of the main, experimental phase.

A second reason for the two-year 'baseline' period was to allow researchers to examine the populations of wild animals and plants present in the fields and their surroundings at the start of the work. This was essential to provide a background against which to interpret changes observed later, and to reveal inherent differences between fields. Also, the researchers needed an opportunity to try out the monitoring and experimental techniques they proposed to use, and to determine the scale and frequency with which samples should be taken.

Accordingly, from just after harvest in 1981 to harvest 1983, the fields were farmed under similar, moderate pesticide inputs. In autumn 1982, all fields were sown with winter wheat, to standardize their entry into the experimental phase of the Project. Following the harvest of 1983, there was a five-year period of contrasting Full Insurance, Supervised and Integrated treatments (Table 2.2).

This meant that the Project was scheduled to finish at the harvest of 1988. As that time approached, there was much debate, and comment from other organizations that the Project should continue in the same form, in order to trace any further population changes that might occur. While this might have added valuable information, it was felt that the resources necessary to continue operating the Project and monitoring populations could be better spent in new research, pursuing the implications of the Project's results with slightly different approaches, taking account of the lessons learnt at Boxworth (see Chapter 17).

The main part of the Project was completed, as planned, in autumn 1988. However, it was decided to continue some work at Boxworth, in order to see whether certain invertebrate populations, which had been adversely affected by the Full Insurance regime, would recover once pesticide use returned to a more moderate level. Accordingly, a three-year continuation period was initiated, in which the set of fields involved, and the pesticides applied to some of them, were changed. Five of the original ten fields were kept for this work. Two of the Full Insurance fields (Grange Piece and Pamplins North) were switched to the Supervised approach, and were matched with their partners from the triplet groups in the Supervised area (Knapwell and Top Pavements). Backside field was

retained under a Full Insurance regime, to provide a high-input standard against which to assess any recovery that might occur in the other fields. The findings of the continuation study are not included in this book, but will be described in a later report.

Cropping

The aim of the Project was to investigate the consequences of contrasting pesticide inputs in cereal production. Monoculture of cereals, especially winter wheat, is now rare, as it leads to increased problems with soil-borne diseases. It also entails high workloads at sowing and harvest. Most farmers intersperse crops of field beans or oilseed rape to overcome these problems and at the same time to boost yields in the subsequent wheat crops.

For the purposes of the Boxworth Project, attention was focused on crop protection chemicals applied to winter wheat. In order to isolate the effects of pesticides, cropping had to be standardized as far as possible, without losing the credibility of a realistic farming system. The compromise reached was to grow winter wheat in all fields, subject to a break crop of oilseed rape at 5-yearly intervals (Table 2.2). The break was staggered among triplet groups, so that there were only three fields sown to rape in two of the five years. The three intensive study fields were kept in continuous winter wheat throughout, to allow the maximum long-term information to be gathered.

The varieties of winter wheat grown each year were varied according to the current recommended range (Table 2.3). The wheats grown in the Full Insurance area were selected from the National Institute of Agricultural Botany (NIAB) list, based on their high-yielding characteristics. Thus, there was a change from 'Norman' in the early years of the Project (a high-yielding feed variety) to the quality milling wheats 'Avalon' and 'Mercia' in the later years. The same varieties were grown in the corresponding fields in the Supervised area. However, variety choice was one of the extra

Table 2.3 Varieties of winter wheat, and of oilseed rape (in brackets), grown in each of the Project fields during the five-year treatment period.

	1983–84	1984–85	1985–86	1986–87	1987–88
Full Insurance area					
Grange Piece	Norman	Norman	Norman	Avalon	Mercia
Shackles Aden	Norman	Norman	Norman	Avalon	Mercia
Pamplins North	Norman	Fenman	Fenman	(Bienvenu)	Brock
Backside	Norman	(Jet Neuf)	Brock	Galahad	Galahad
Supervised area					
Knapwell	Norman	Norman	Norman	Avalon	Mercia
Top Pavements	Norman	Fenman	Fenman	(Bienvenu)	Brock
Thorofare	Norman	(Jet Neuf)	Brock	Galahad	Galahad
Integrated area					
Extra Close East	Norman	Norman	Blend*	Rendezvous	Rendezvous
Extra Close West	Fenman	Blend*	Blend*	(Mikado)	Rendezvous
Eleven Acre Extra/ Bushes & Pits	Aquila	(Jet Neuf)	Blend*	Rendezvous	Rendezvous

*Blend of Brock, Brimstone and Norman

modifications possible in the Integrated fields, where varieties were chosen primarily for resistance to disease. Consequently, the varieties grown in the Integrated fields were not necessarily the same as in the rest of their matched triplets.

Selection of disease-resistant varieties is not straightforward, for good resistance to one pathogen does not generally indicate equal protection against all diseases, and a balance must be struck according to the problems considered likely to predominate locally. The selection of varieties, and its consequences, are considered in Chapter 4.

Another approach to disease control is to grow different varieties in adjacent fields, so that any one disease will not affect all crops to the same extent (Wolfe et al., 1981). This approach was tried in the Integrated area in the 1983–84 season, with the varieties Norman, Fenman and Aquila each grown in one field. In the following two years, the principle was extended by growing a blend of three varieties in each of the three fields (Table 2.3). This was not entirely successful in reducing inputs (Chapter 4), and was abandoned in the final years

of the Project, in favour of a return to the use of a single variety, the eyespot-resistant Rendezvous.

Varieties of oilseed rape, in the two years that the crop was grown, were chosen according to similar principles of high yield for the Full Insurance and Supervised areas, and disease-resistance for the Integrated area.

Pesticide programmes

The philosophy of the Full Insurance programme was to identify all those pests, weeds and diseases likely to occur at Boxworth, and to apply a pesticide in anticipation of the need for control. Most of the crop problems, and the treatments used to combat them, were typical of what might apply to cereal-growing in most parts of the UK, although there were some exceptions due to the particular local conditions at Boxworth. For example, wheat bulb fly *Delia coarctata* is rarely a problem at Boxworth, and therefore no routine seed treatment against this pest was incorporated, as it might have been on many farms in eastern England. However, there were treatments against most of the post-emergence pest problems that can affect wheat crops, including slugs, aphids and stem-boring flies. Similarly, the herbicide and fungicide programmes were chosen to control the weeds and pathogens liable to occur at Boxworth, but were typical of a heavy crop protection approach that might be applied to cereals elsewhere. Chapters 3–5 explain the pesticide programmes in more detail, and the planned treatments are listed in Appendix I.

For the reasons explained earlier, it was decided to keep these programmes constant throughout the Project. This was necessary to provide experimental rigour, but it meant a loss of flexibility to respond to changes in current practice. However, when a preferred new product or formulation became available, it was considered permissable to substitute it for the chemical previously used for that crop protection treatment. The rigidity of the pesticide programmes raised some problems. For example, in 1986 government approval for use of triazophos against yellow cereal fly was restricted to the period after 31 March, which would have prevented its use as planned in the Full

Insurance programme. The problem was overcome by applying for an experimental permit to cover the use of triazophos outside the approved period.

Reduced rates of herbicides were used in the Supervised and Integrated areas, if it was considered that weed control could be achieved without a full-rate application. Similarly, part-field treatments were allowed in the Integrated area if weed problems did not affect the whole field. Both strategies helped to reduce pesticide inputs to the minimum necessary for crop protection.

When several pesticides were needed in a field, and the chemicals could be safely mixed, they were applied together in a single spraying operation.

In practice, the full suite of planned applications in the Full Insurance area was not always realised, chiefly because of problems of access to fields in inclement weather. Table 2.4 shows the average number of annual treatments, compared to the planned Full Insurance programme. For insecticides, herbicides and fungicides, three-quarters of the planned applications were made, although in some years the full programme was achieved (see Appendix II). This level of input was recognised to be artificially high for Boxworth, where pest problems are not usually serious, but it

Table 2.4 Pesticide use in the fields of the three treatment areas, during the baseline years and in the five-year experimental phase of the Project. Figures are the average numbers of annual applications. The planned Full Insurance programme is also shown.

	Insecticides	Herbicides	Fungicides	Total
Full Insurance area				
average 1982–83	0.8	6.2	2.2	9.0
planned programme	7	7	5	19
average 1984–88	5.2	5.2	4.0	14.4
Supervised area				
average 1982–83	1.0	4.6	2.4	8.0
average 1984–88	0.8	3.4	2.6	6.8
Integrated area				
average 1982–83	1.0	3.6	2.6	7.2
average 1984–88	0.9	2.8	2.4	6.1

was representative of pesticide use by the most intensive cereal growers when the Project started.

The Supervised and Integrated approaches both allowed pesticide use to be reduced to about the same extent, which was on average less than half of the applications made to Full Insurance fields. This was particularly marked for insecticide use, which was only one-sixth of that in the high-input area (Table 2.4).

Both these reduced-input programmes gave a level of pesticide use that was comparable to the husbandry of the fields before the start of the Project, which is reflected in the inputs in the two baseline years (Table 2.4). This confirms that the Full Insurance area experienced a sudden rise from moderate to high inputs, followed by a return to a moderate level during the continuation phase, whereas the Supervised and Integrated programmes did not differ from previous conditions, and therefore would not be expected to have produced any major ecological changes.

Use of fertilizers

In keeping with the aim of standardizing all agricultural operations other than pesticide use, the levels of nitrogen and phosphate fertilizers added to fields were similar in all treatment areas. However, fertilizer inputs varied between triplet groups according to the needs of the crop. Phosphate was applied every third year, at a rate of 450 kg of 46% P_2O_5 per ha, and nitrogen was used annually in the spring, according to ADAS recommendations of 230 kg per hectare on oilseed rape, and 200–220 kg per hectare on wheat crops, reduced to 150 kg per hectare in the first year after oilseed rape.

Replicated plot trial

Although most of the interest in the Project was directed at the study of field-scale effects, for which large treatment areas were necessary (see above), it was recognized that certain aspects were amenable to testing by a small-scale replicated plot trial. For example, yield differences and disease prob-

lems can be studied in small plots, and some less mobile animals, such as springtails, are likely to show effects on a small scale.

Therefore it was agreed that a replicated plot trial should be included in the Project, to complement information obtained from field-scale monitoring and to give an indication of the validity of conclusions drawn from the field-scale results.

Originally, the plan was to site this trial outside the fields being used for the Project. However, that was changed in favour of a location in one of the Full Insurance fields. The reason for this decision is that the plots would be less affected by their surroundings in a high-input field, where pests, diseases and wild animal populations were reduced, than where there might be substantial spread into plots, confounding the results.

The replicated plot trial was established in the autumn of 1983 in Shackles Aden field (which was not matched as part of a triplet group). It consisted of 24 plots, each measuring 24 m × 12 m, arranged as shown in Figure 2.2. This design was selected to provide sufficient replication to identify differences between the treatments. The size and spacing of plots were considered to be great enough that spread of diseases or of soil fauna would not outweigh any treatment differences. They were also based on convenience of access by tractor, passing down a tramline in the centre of each plot, thereby facilitating the application of agrochemicals, and other operations. The set of plots was monitored for yield and grain quality, plant diseases, and soil fauna, until the end of the Project.

It was straightforward to reproduce the Full Insurance and Supervised programmes within the plots, since the same variety of wheat was grown in both these areas. However, the Integrated approach, involving a different, disease-resistant variety, could not be readily incorporated in the plots. As an alternative, it was replaced by a 'Minimum Input' treatment, in which no insecticides or fungicides were used and herbicides were avoided unless absolutely necessary to prevent unacceptable spread of weeds to other plots.

Figure 2.2 Layout of the replicated plot trial sited in Shackles Aden field from 1983 to 1988.

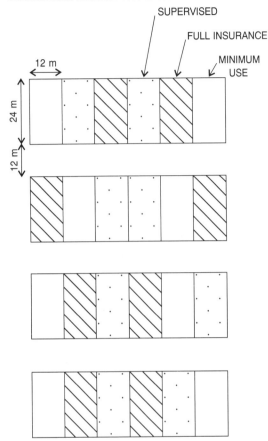

SUPERVISED

FULL INSURANCE

MINIMUM USE

12 m

24 m

12 m

Yield and economic assessments

Alongside the ecological assessment of environmental side-effects of pesticides, their economic implications were also addressed, by recording the performance of the crops, and the costs of production, under each of the three pesticide regimes.

At harvest, field yields were measured separately, and assessments of several aspects of grain quality were made. This information was combined with records of input costs such as seed and chemicals, and the value of the crop, to indicate the financial margins of each approach to production. The results of this work are described in Chapter 6, which compares the three areas, taking account of inherent differences in yield

between fields, which were revealed in their long-term cropping performance.

Monitoring of pests, weeds and diseases

Regular monitoring was carried out in order to provide information on the build-up of problems in the crops, which formed the basis of decisions on whether or not pesticide treatments were required in the Supervised and Integrated fields. Entomologists and plant pathologists from ADAS in Cambridge visited the farm frequently to make assessments of the status of each pest and disease. Recommendations for treatment were then made, according to ADAS thresholds for effective crop protection (MAFF, 1982). The criteria remained largely consistent throughout the Project, although there was a change in the ADAS threshold for summer aphid numbers in 1986 (Chapter 3), which is unlikely to have materially affected the results or their interpretation.

Monitoring of weeds was slightly different. For most of the Project, assessments of weed densities were made three times each year by staff from the Agricultural and Food Research Council (AFRC) at Long Ashton Research Station. Recommendations were based on ideas being developed for the use of a 'crop equivalents' method to measure the economic importance of weed infestations (see Chapter 5). This approach, which was more detailed than would be employed on most farms, was based on weed assessments made at a time well before treatments would be carried out. Later in the Project (in 1987 and 1988) crop-walking visits were also made by agronomists from ADAS at Cambridge to supplement this predictive approach with direct measurements of current weed infestations, to allow immediate, more realistic, decisions on the need for herbicide applications.

For all pesticide use in the Supervised and Integrated areas, and for treatments to Full Insurance fields when there was doubt about whether an application could be completed successfully, decisions were the responsibility of the Farm Director. This meant that his local knowledge and previous experience of the farm could be combined effectively with the information from monitoring.

Studies of weeds and diseases were extended well beyond the minimum monitoring necessary to generate recommendations for treatment. Sampling within fields was more extensive, and was carried out more frequently, than would be the case on a farm where the sole aim was to assess infestations for the purposes of control. In addition, monitoring was undertaken in a similar way in the Full Insurance fields, in order to establish how well, or how poorly, the Full Insurance programme matched real crop protection needs.

Research studies

In order to obtain a broad ecological insight from the Project, a wide range of research studies was instituted. These included work on birds, for which the treatment areas were not ideally large, and investigations of much less wide-ranging wildlife, such as the soil fauna, which can be studied on a much smaller scale. There are obvious practical difficulties in making consistent comparisons between treatment areas for all these diverse groups. However, the research programme was designed to integrate information where appropriate on issues such as interactions between predators and prey.

Birds were studied by staff from the MAFF Central Science Laboratory, who surveyed the population densities and breeding success of the common birds on the farm. They also investigated the consequences of exposure to particular pesticides for selected species, in collaboration with Reading University (see Chapters 14–16).

MAFF awarded funds for three studies carried out throughout the entire Project by postdoctoral researchers at universities. One of these was for investigations of small mammal populations, particularly those of the wood mouse *Apodemus sylvaticus*, in and around the cereal fields. The results of this work, by Cambridge University, are described in Chapter 12.

Southampton University was commissioned to survey populations of invertebrates in the crop, including pests, their predators and parasites, and other groups. This was not only to identify effects on the populations of these species for their own importance, but also to provide information relevant to other research, for example on animals that feed on invertebrates. Therefore the results of this work were used in the studies of birds and small mammals, and in a second university study of invertebrates, based at Cambridge University. This concerned the relationships between the main crop pests (aphids, slugs and cereal flies) and their natural enemies (the predatory and parasitic crop invertebrates).

The soil fauna, which includes different species to those above ground, was also investigated, through regular sampling by staff from the MAFF Central Science Laboratory at Tolworth, to complement the other invertebrate studies. The three invertebrate studies are reported in Chapters 9–11.

Studies of the flora in fields by staff from Long Ashton Research Station were linked to assessments of weeds for monitoring treatment thresholds, and extended to detailed surveys of the distribution patterns of plants in fields and their boundaries. Their results are presented in Chapter 8.

As well as the direct ecological effects of pesticides, research examined the fate of pesticides applied to fields (Chapter 7). Several exercises to measure the amounts of certain chemicals deposited in fields and at their margins, and their persistence, were carried out by staff from the MAFF Central Science Laboratory based at Tolworth and Harpenden. Researchers from the Institute of Terrestrial Ecology at Monks Wood Experimental Station also took the opportunity in 1986 to conduct an investigation of spray drift in some of the Project fields. In the final year of the Project, samples of drain water from two fields were analysed for the presence of pesticide residues at the ADAS National Pesticide Residues Unit in Cambridge, to determine whether there were any major differences between the treatment areas.

In addition to these major studies, a number of small research projects were undertaken. These included assessments of the exposure of rabbits to pesticides (Chapter 13), and surveys of soil microbial organisms during the continuation phase of the Project.

Steering group

All aspects of the Project, including operational matters on the farm, monitoring, and research studies, were coordinated by a Steering Group. This consisted of representatives from ADAS (at Boxworth, Eastern Region, and the Central Science Laboratory), AFRC at Long Ashton Research Station, and the MAFF Chief Scientist's Group in London, as well as the principal researchers involved. From the beginning of the Project until 1985, the group met under the chairmanship of Mr P Wiggell, and thereafter of Dr M J Griffin.

The Steering Group was charged with reviewing the progress of the work, deciding on changes to be made, and dealing with any problems or compromises that arose. This approach to management was a very successful way of overseeing a large and complex project, requiring the integration of diverse work by a large number of people. Despite the potential for conflict between the requirements of different researchers, and interference between those wanting to do different things in the same place, there were no serious difficulties of this kind.

Operational aspects

Although the Project was designed to reveal what effects would occur on a typical working farm, the experimental nature of the study entailed many changes to normal working practice:

Facilities required A new low-ground-pressure, high-clearance field sprayer was purchased for the Project, to ensure access to fields and avoid excessive damage during winter crop treatments. The researchers needed space for equipment, accommodation (in caravans), and the use of power and other services. The necessity for separate measurements of yield and grain quality for each field, and the replicated trial plots, meant that facilities for grain handling and storage, and for quality assessment, had to be provided.

Disruption of other farm activities Although the Project fields were kept discrete from the rest of the farm, the special husbandry needs of the Project took time from other activities, as well as increasing the cost of farm operations.

Visitors As the Project developed, there were frequent visits by interested groups, requiring facilities for lectures, refreshments and transport around the farm.

Communication Effective liaison between farm staff, researchers, and personnel involved in monitoring was essential. It often involved consultation at short notice in order to make rapid decisions about treatments. Detailed up-to-date records also had to be available on request.

Liaison with neighbouring farms For full interpretation of the results obtained at Boxworth, there was a need for information on pesticide use, and other husbandry, in the fields next to the treatment areas. Also, agreement had to be reached on the management of joint boundaries, so as not to interfere with some of the research studies.

Bringing the fields into the correct condition to start the Project was not straightforward. Some were in a cropping sequence that was not appropriate to lead into a series of consecutive wheat crops, but the two-year baseline period allowed this to be adjusted (for example, winter beans were grown in Grange Piece in 1981–82 before it entered a sequence of six consecutive wheat crops). A more difficult problem was the need to make the treatment areas as uniform as possible before the main part of the Project. Hedgerow management practices were standardized, and several specific measures were taken to remove unusual features of fields. A new drainage scheme was installed in Knapwell field in September 1983, to replace the previous unsatisfactory drain system that had been constrained by an earlier boundary dividing the field, now removed. In Thorofare field, there was a heavy infestation of brome grass, which was reduced by a combination of ploughing and herbicide applications.

Several problems arose during the 5-year treatment period, making it difficult to adhere to the planned prophylactic programme and to recommendations based on monitoring. For winter application, in particular, there was always a risk of delay because of adverse weather or wet ground conditions. In 1986, patchy crop damage occurred in Backside field when the herbicide isoproturon was applied in mid-March following frost heave. Such events are liable to distort the comparisons of field yields, for example. Another unpredictable occurrence was the accidental burning of one of the hedges bordering part of the Supervised and Integrated areas in August 1983, when straw burning on a neighbouring farm got out of control.

Finally, the operation of the Full Insurance regime itself entailed some confounding changes not directly related to pesticides. The greater number of sprays meant that extra crop damage was likely to be caused by frequent passes of the tractor. This was particularly noticeable for oilseed rape crops in 1985. However, it is not likely to have seriously biased any results other than yield measurements.

References

Cochran, W G & Cox, G M (1957). *Experimental designs*. 2nd Edition. Wiley, New York.

MAFF (1982). *Cereal Pests*. MAFF Reference Book 186. HMSO, London.

Wolfe, M S, Barratt J A & Jenkins, J E E (1981). 'The use of cultivar mixtures for disease control.' In: J F Jenkyn and R T Plumb (Eds). *Strategies for the control of cereal disease*. Blackwell Scientific Publications, Oxford. pp 73–80.

Pests

Mary Hancock (ADAS, Cambridge)

Introduction

Boxworth EHF is typical of heavy land farms in eastern England, and its pest problems are typical of the area. Slugs and aphids are the pests most frequently needing treatment on cereals. All the pests of oilseed rape occur on the farm, but rarely at damaging levels. A full monitoring and control programme for both crops was devised, to cover all the major pests that could be expected to occur in eastern England.

Pest control programmes for winter wheat and oilseed rape were drawn up at the start of the Project. At the same time, monitoring schemes and thresholds similar to those normally recommended by ADAS in advisory work on commercial farms were established. A few modifications were made later to take account of changes in thresholds (eg the treatment threshold for cereal aphids changed from numbers of aphids per ear to percentage of tillers infested) and new chemicals (eg the substitution of cypermethrin in place of permethrin for the control of barley yellow dwarf virus).

All monitoring was done by ADAS entomologists based at Cambridge. Recommendations for action were made to the Director of Boxworth EHF, who decided what to do based on the control programmes drawn up at the start of the Project. Occasionally, a recommended treatment could not be applied at the most effective time because of the weather. Sometimes this prevented an application taking place at all.

Monitoring of pests

Winter wheat

The ADAS monitoring and risk assessment techniques are described in ADAS Pamphlet P858, and are outlined briefly below.

Each pest was monitored in the Supervised and Integrated fields, using the techniques that a grower or his agronomist would use, and treatment recommendations were made according to ADAS practice. Each field was dealt with as an individual unit and treatments were only applied if pest numbers or risk in that field warranted it. This meant that on some occasions, only one field in the Supervised area was treated.

Slug activity was monitored with lines of baited traps placed across each field from September (post-harvest) to about November (crop establishment). Molluscicidal pellets were applied before drilling if an average of four or more slugs per trap was found in a field over a three-day trapping interval.

Some species of cereal aphids, especially the bird cherry-oat aphid (*Rhopalosiphum padi*), transmit barley yellow dwarf virus (BYDV). The risk can be reduced by killing any aphids in the crop in autumn when the plants are most vulnerable. The incidence of BYDV depends on several factors—the proportion of aphids carrying the virus, the number of migrating aphids, the weather during the migration period and the availability of host plants. During the Project, the risk of BYDV was assessed using the general risk for that year (usually based on an infectivity index) and the particular risk for the field (based on drilling date and the incidence of volunteer (non-crop) cereal plants or grass weeds). Where necessary, plants were examined in the field to determine infestation levels. Treatment was applied in late October or November, if necessary.

Besides the effect they may have by transmitting BYDV, cereal aphids can cause yield loss in the summer by feeding on a crop and reducing the growth of the grain. Two species are likely to be a problem at this time—the grain aphid (*Sitobion avenae*) and the rose-grain aphid (*Metopolophium dirhodum*). Aphid populations were monitored during June and July by visual counts of the numbers of aphids on ears and tillers. The threshold for action at flowering (Growth Stages 60 to 69) was changed during the Project, from five aphids per ear to 66% of ears infested. A threshold for treatment (50% tillers infested) at an earlier growth stage, when the developing ear was enveloped in leaf (Growth Stage 45: 'booting'), was introduced towards the end of the study period. There was no treatment threshold for use after flowering, when recommendations, if necessary, were based on experience.

The larvae of wheat bulb fly (*Delia coarctata*) and yellow cereal fly (*Opomyza florum*) attack the central shoot of cereal tillers, causing the shoot to yellow and eventually die—the 'deadheart' symptom. The eggs of both species are laid in the soil in the summer/early autumn (wheat bulb fly) or late autumn/early winter (yellow cereal fly) and the risk of attack can be assessed from the number of eggs laid. Soil samples were taken in September/ October for wheat bulb fly and in December for yellow cereal fly. Where warranted, treatments were applied in January (egg hatch) or March/April (deadheart) for wheat bulb fly and in January— March (egg hatch to plant invasion) for yellow cereal fly.

The larvae of frit fly (*Oscinella frit*) can attack cereals in autumn if eggs are laid on the young cereal plants, or on grass or cereal volunteers before the crop is drilled. The risk depends on the locality of the crop, the incidence of frit flies in late summer and early autumn and the availability of host plants (the crop, weeds or volunteers) for eggs. Where the risk is significant, inspection of the crop for 'deadheart' damage is recommended at the 1–2 leaf stage. If necessary, treatment was usually applied in October, soon after crop emergence.

The approach to wheat bulb fly control needs explanation. This pest has seldom been a problem at Boxworth because the farm is not in the major area of incidence of the pest (the Fens, and western Suffolk and Norfolk). Also, the current and proposed rotations did not include the crops most

attractive to the pest. However, it was not clear how oilseed rape might affect the incidence of the pest in the following wheat crops. Therefore, an option for treatment against wheat bulb fly in the Full Insurance area was retained, to be used if treatment proved necessary in either or both of the other two areas.

Oilseed rape

Four pests were monitored on oilseed rape. Slugs were assessed prior to drilling using baited traps. There is no specific threshold for treatment, and molluscicidal treatments were applied when the risk of damage was considered to justify it.

Samples of plants were dissected in October—December and in February—March for the larvae of the cabbage stem flea beetle (*Psylliodes chrysocephala*). Treatment was recommended in the autumn if there were five or more larvae per plant. Larger numbers are needed to justify treatment in the spring.

Blossom beetle adults (*Meligethes* spp.) can be a problem on winter-sown crops but usually invade after flowering has started, when the plants are no longer vulnerable. The numbers of beetles were monitored by beating plants before buds opened; treatment was recommended and applied if an average of 15 or more beetles per plant was recorded.

The larvae of cabbage seed weevil (*Ceutorhynchus assimilis*) develop from eggs laid inside pods and reduce yield by eating the seed. The holes made by female weevils when laying eggs enable female brassica pod midges (*Dasineura brassicae*) access into pods to lay their eggs. Midge larvae also feed on rape seed. Control of the midge depends on the control of the weevil. The need for treatment was based on an assessment of the numbers of adult weevils made by beating plants. Treatment was advised if the average cabbage seed weevil population across the field exceeded one weevil per plant on one sampling occasion, or 0.5 weevils per plant on two or more occasions where at least three assessment visits were made. Treatment was applied at about the end of flowering (Growth Stage 69).

Choice of treatment

The pesticides used in the Full Insurance and Supervised areas were chosen because they were recommended or approved for use against that particular pest and represented those commonly chosen by farmers. For many pests there were several alternatives to choose from. In those cases decisions were influenced by the most effective chemical or application technique for the job, and what the intensive arable grower was most likely to use. All the chemicals were broad-spectrum in action, affecting many different types of invertebrate.

In the planned control programme for the Integrated area, it was intended that specific chemicals, which significantly affect only the intended target, would be substituted for the products used in the other two areas. Unfortunately, there was only one such specific chemical available for commercial use (pirimicarb, an aphicide), and no non-chemical control techniques were possible. Delaying drilling, to avoid BYDV for example, would have been too great a change because the crops would not have been at a comparable stage of development to those in the other areas. In addition, a change in husbandry that favours the control of one problem may well mitigate the control of another. Throughout the Project, pest control was constrained by what was acceptable commercially.

A list of pesticides used in the three areas is given in Appendix II.

Pest control decisions

1983–1984

The main features were relatively high numbers of slugs recorded in the autumn and of aphids in the summer. Maximum numbers detected on the surface in each field ranged from 4.0 to 8.8 slugs per trap per 3 nights; all were the grey field slug (*Deroceras reticulatum*). Cereal aphid numbers in the Supervised and Integrated fields built up rapidly in the summer, to reach a maximum that ranged from 5.4 to 12.3 aphids per ear (all above the threshold), and demeton-S-methyl or pirimicarb were applied to fields in these areas. The commonest species was the grain aphid although rose-grain aphid was also present. At the same time numbers in the Full Insurance fields, where demeton-S-methyl had been applied, were less than one aphid per ear.

1984–1985

All pest levels were low. The maximum number of slugs trapped in any field up to drilling was only 1.7 slugs per trap per 3 days. Cereal aphids were relatively scarce during flowering but increased in numbers rapidly thereafter in favourable weather while the crops ripened only slowly. Again, the grain aphid was the commonest species. At the milky ripe/early dough stages (Growth Stages 79 to 83), 2.5–6.3 aphids per ear were present. A decision was taken not to spray in the Supervised and Integrated areas, based on the results obtained in ADAS trials of treatments applied at these growth stages.

1985–1986

Again, no pests exceeded thresholds in any of the Supervised or Integrated fields and, therefore, treatments were not recommended. The maximum number of slugs recorded in any monitored field was 2.0 slugs per trap per 3 days. The crops were examined for cereal aphids three times during the summer (at the end of flowering, milky-ripe and early dough: Growth Stages 69, 79 and 83). Population levels remained low until the last assessment, when numbers present ranged from 2.3 to 18.9 aphids per ear. At this stage most of the plants had senesced, with green leaf on the flag leaf only, and no spray recommendation was made.

1986–1987

The year was characterised by a lack of pests. No thresholds were exceeded and no treatments recommended in the Supervised and Integrated areas. The maximum number of slugs caught in any monitored field was less than 1.0 per trap per 3 days. Similarly, cereal aphid numbers in the summer were very low with a maximum in any field of 2.0 aphids per ear, recorded at the milky-ripe growth stage.

1987–1988

Several pests were abundant. Slug activity was greater in the autumn than for several years with a maximum ranging from 0.5 to 22.3 slugs per trap per 3 days. There was no clear link between the numbers of slugs found and previous cropping with oilseed rape.

Numbers of wheat bulb fly eggs detected by soil sampling in the autumn were the highest recorded in the five years. The treatment threshold was reached in Extra Close West but it was decided not to apply a preventative treatment. Subsequent tiller damage in the spring did not justify action and confirmed that treatment at egg hatch would have been unnecessary.

The risk of BYDV was estimated to be low and no autumn treatments were applied. Small patches of infested plants were seen in the spring in the sheltered parts of Top Pavements and Thorofare where colonies of bird cherry-oat aphid and some grain aphid had survived from the autumn. These patches were not extensive enough to affect the field yields but probably acted as local sources of aphids in the spring and summer. Cereal aphid populations increased generally across four fields (all the Supervised fields and Extra Close West in the Integrated area) by early June before the start of flowering. In these fields from 64 to 88 per cent of tillers were infested, so treatment was recommended and applied. In two more Integrated fields (Extra Close East and Eleven Acre Extra) the threshold was exceeded during flowering and pirimicarb was applied.

Aphid populations in the remaining, unsprayed, Integrated field (Bushes & Pits) never increased sufficiently to warrant treatment and crashed after flowering, possibly because of higher numbers of natural parasites and predators and a spell of unsettled, cooler weather. Grain aphid was the most abundant species with isolated, relatively large populations of bird cherry-oat aphid in virus-infected patches in Top Pavements and Thorofare. Very few rose-grain aphids were seen, except in Eleven Acre Extra where about 5% of tillers were infested with this species.

Oilseed rape (1984–1985 & 1986–1987)

No treatments were advised in any of the fields in either year.

Costs of pest monitoring

The main cost was the time spent in examining crops for pests, sampling and extracting insect eggs from samples. The average time spent in gathering information for pest control decision-making was 7.25 hours for each field of wheat each year. This could be reduced to about three hours per field by relying more heavily on past experience and general intelligence for that season obtained from elsewhere (eg ADAS intelligence reports and the farming press) and less on field-specific information. In this Project, travelling to and from Boxworth from Cambridge cost additional money and time.

Discussion

No significant damage was caused by any pest that failed to reach its threshold (and therefore was not treated) except possibly in 1985–1986 when there was a late increase in aphid populations in the summer. Yields then were lower than expected and with hindsight it seems that treatment of the more heavily infested crops would have been worthwhile. Recent unpublished ADAS trials results suggest

Table 3.1 Summary of applications of insecticides and molluscicides to fields in the three treatment areas. Figures are the numbers of fields of winter wheat treated each year. In two years (1984–5 and 1986–7), oilseed rape was grown in one field in each area (*), or in one case, two fields (**).

Pest problem	Full Insurance area (4 fields)				
	1983–4	1984–5*	1985–6	1986–7*	1987–8
Aphids in autumn (BYDV)	4	3	0	0	3
Aphids in summer	4	3	4	3	4
Frit fly	4	0	0	3	3
Yellow cereal fly	4	3	4	3	4
Wheat bulb fly	0	0	0	0	0
Slugs	4	3	4	3	4

Pest problem	Supervised area (3 fields)				
	1983–4	1984–5*	1985–6	1986–7*	1987–8
Aphids in autumn (BYDV)	0	0	0	0	0
Aphids in summer	3	0	0	0	3
Frit fly	0	0	0	0	0
Yellow cereal fly	0	0	0	0	0
Wheat bulb fly	0	0	0	0	0
Slugs	3	0	0	0	2

Pest problem	Integrated area (4 fields)				
	1983–4	1984–5*	1985–6	1986–7*	1987–8
Aphids in autumn (BYDV)	0	0	0	0	0
Aphids in summer	4	0	0	0	3
Frit fly	0	0	0	0	0
Yellow cereal fly	0	0	0	0	0
Wheat bulb fly	0	0	0	0	0
Slugs	4	0	0	0	4

that yield increases of up to about 5% might have been obtained in that year. In other years when summer aphicides were not applied in the Supervised and Integrated area, yield losses would not have exceeded about 1%.

A few operational problems were experienced. It was often difficult to obtain comprehensive information about slugs because traps were damaged or the trapping period was interrupted by cultivations. The tenacious clay soil at Boxworth EHF made sampling the soil for insect eggs a frustrating, and occasionally an impossible task.

The recommendations made for cereal pest control are summarised in Table 3.1. The differences in pesticide usage between the Supervised and Integrated areas and the Full Insurance area were considerable (Figure 3.1). As shown in Figure 3.1, the cereal crops in the Full Insurance area usually received a total of six insecticidal or molluscicidal applications each year. In the Supervised and Integrated areas no such applications were made at all to the crops grown from 1984 to 1987. In 1983–1984, each cereal crop was treated twice and

Figure 3.1 Average number of insecticide/molluscicide applications made annually to each field during the treatment years of the Project (1984–88).

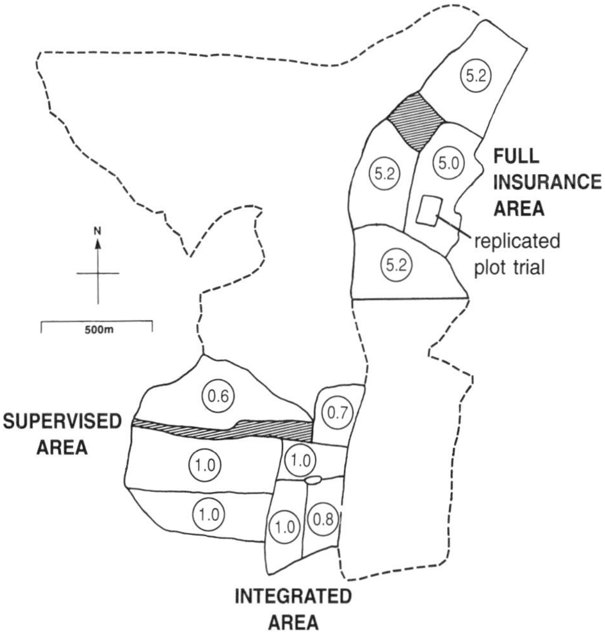

in 1987–1988 five crops (two in the Supervised area) were treated once. With such a contrast in pesticide use, it would not be surprising if large differences in non-target invertebrate species were seen between the Full Insurance and the other two areas. The contrast in number of treatments was similar for the oilseed rape crops. No applications were made to the oilseed rape crops grown in the Supervised and Integrated areas throughout the Project, whereas the crops grown in the Full Insurance area each received three insecticidal/molluscicidal treatments. However, the seed sown in all areas was of the usual commercial standard and as such was supplied already treated with gamma-HCH as a seed treatment.

Summary

Pest numbers were low or moderate on cereal crops in the three seasons between 1984 and 1987, but were higher in 1983–1984 and 1987–1988, when treatments were applied against slugs and aphids in the summer on most crops in the Supervised and Integrated areas. Pest levels on the oilseed rape crops were too low to warrant treatment at any stage of the Project.

Weeds

4

Jon Marshall (Long Ashton Research Station)

Introduction

Soils at Boxworth EHF are heavy alkaline boulder clays and the associated weed flora includes blackgrass (*Alopecurus myosuroides*), wild-oat (*Avena fatua*) and brome grasses (*Bromus* spp.) as well as a range of dicotyledonous species. Prominent among these are chickweed (*Stellaria media*), speedwell species (*Veronica hederifolia* and *V. persica*) and cleavers (*Galium aparine*). All weed species compete with the crop, though to varying extents and at different times (Tottman & Wilson, 1989). Among the most competitive species are the grass weeds and cleavers. Yield loss may be reduced or prevented by use of herbicides, and by cultural techniques such as ploughing, straw burning and altering the timing of sowing.

A weed control programme for the winter wheat crops was developed using a range of herbicides for particular weed species (Appendix I). The programme included a late-spring application of difenzoquat for the control of wild-oat. Glyphosate was applied before harvest for control of common couch (*Elymus repens*), a perennial grass, at a time when optimum herbicide uptake would be expected. Autumn applications of isoproturon were made for blackgrass control, and a sequence of tri-allate and metoxuron was used in autumn against the brome grasses (*Bromus sterilis* and *B. commutatus*), which are difficult to control with herbicides. Dicotyledonous weeds were controlled in autumn and spring with a range of herbicides, notably bromoxynil + ioxynil, mecoprop and fluroxypyr.

Decisions to apply herbicides in the Supervised and Integrated fields were made when thresholds of weed density associated with economically significant yield loss were exceeded. While predictive thresholds have been used for pests and diseases for some years, there had been little practical or theoretical consideration of spray thresholds for weeds at the outset of the Project, with the exception of work by Eggars & Niemann

(1980). A sampling programme was initiated to estimate weed populations within each field. If weed populations exceeded pre-determined thresholds (see p. 26), the Farm Director, with whom the final spray decision rested, was advised to apply herbicides.

Weed assessment methods

Weed populations were estimated at three times in the year. Autumn germination was assessed in November and an estimate of winter survivors and spring germination was made in March. Assessments were also made in early July, when early-flowering grass weeds were beginning to lodge and late-flowering species, such as wild-oat, were in flower. Grass populations in winter wheat crops were estimated in early July by counting inflorescences present in 0.25 m² quadrats, in a sampling grid based on tractor wheelings through the crop. Inflorescences were counted to estimate the amount of seed produced and thus the likely density of grass weeds in the following crop. Sampling intensity was six points per hectare with four random quadrats thrown away from the wheelings at each of the sample points, which were located every 50 paces along every third wheeling. Dicotyledonous plants were counted in November and March in 0.1 m² quadrats, on the same sampling grid used for assessing grasses in July. In addition, dicotyledonous species present in July quadrats were recorded. No weed assessments were made in the rape crops in July.

In order to assess the precision of weed estimates in quadrat counts, grass weeds in Top Pavements field were assessed in July 1984 using a much higher sampling intensity of 36 points per hectare. Variations in the distribution of weeds were examined, as well as the errors associated with different sampling procedures. A simple computer mapping programme was used to represent weed distributions.

Decision thresholds for herbicide applications

Thresholds for grass weed species

Cussans, Cousens & Wilson (1987) developed spray thresholds for use in the Boxworth Project, using information on the competitive effects and population dynamics of weeds. Data on the amount of viable seed to reach the soil, seed germination, and seedling survival, were used to predict grass weed density in the succeeding crop (Cussans & Moss, 1982). Treatment thresholds were based on grass inflorescence densities in the previous July, expressed as average inflorescence densities per m^2 for the whole field and for the weediest quarter of the field. If the mean inflorescence density in the weediest quarter exceeded 2.0 inflorescences per m^2 for any grass species (apart from wild-oat), the farm was advised to spray the following year. The threshold for wild-oat, a more competitive species, was set at 0.5 inflorescences per m^2. In the case of common couch, the advice given in early July was acted on before harvest the same year.

Thresholds for broad-leaved weed species

A different approach was required for dicotyledonous species because the variability of seed production and germination prevents reliable prediction of populations from one season to the next. In addition, the weed flora comprises a range of species that are not equally competitive. However, most broad-leaved weed herbicides affect most species. In order to accommodate the competitive effects of a mixed broad-leaved flora, Wilson (1986) developed the concept of 'crop equivalents', which give an indication of the relative competitive effects of different weed species, and set a threshold value for spraying at 5.0 crop equivalents per m^2. Crop equivalent values for different species ranged from 1.73 for cleavers, the

most competitive species, to 0.08 for parsley piert (*Aphanes arvensis*). The density of each broad-leaved weed species in November and March was multiplied by a crop equivalent value for that species. If the mean for summed crop equivalents for the weediest quarter of wheelings in autumn exceeded 5.0 crop equivalents per m^2, then the farm was advised to apply broad-leaved weed herbicides. More recent work by Cussans, Cousens & Wilson (1987) has indicated the need to modify some of the crop equivalents, particularly to increase the value for cleavers. However, crop equivalent values were kept the same throughout the Project.

At the outset of the experiment, it was agreed that the advice given on the basis of thresholds could be modified by the farm staff, If weed populations changed after the surveys were made.

Sampling precision and patchiness of weeds

An intensive study of Top Pavements field was made to examine plant distribution in more detail. This used quadrat counts at an intensity of 36 points per hectare, six times the usual intensity, which entailed sampling every 25 paces along every wheeling in the field. Grass weeds were patchily distributed across the field (Figure 4.1). It was also shown that the frequency distributions of density followed a negative binomial distribution, which indicates that the plants were clumped, with clumps occurring at random (Marshall, 1988).

The precision of estimated average field densities was affected by sampling intensity. Sampling at six points per hectare gave standard errors ranging from 30% to 80% of the average value, depending on the abundance of the species. Therefore, estimates of the density of weeds with field averages of less than 2.0 inflorescences per m^2 are likely to be imprecise. Although not practical in the Project, the results indicated that an intensity of sampling of 18 points per hectare would be required to estimate accurately the density of less

Figure 4.1   Computer-drawn map of inflorescence densities of barren brome
(*Bromus sterilis*) **in Top Pavements field at Boxworth.**

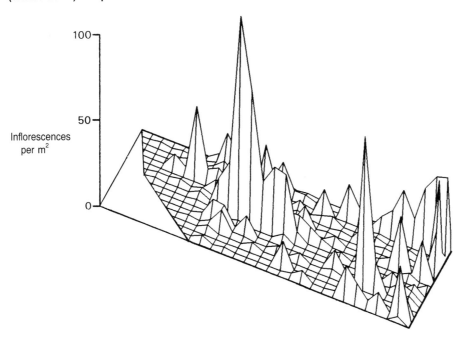

abundant species. The implications of these findings for the practical use of spray decision thresholds are discussed later.

Spray decisions for grass weed control

There were five common grass weed species at Boxworth: wild-oat, common couch, blackgrass, barren brome and meadow brome. Mean inflorescence densities in the weediest quarter of each field are shown in Figure 4.2. A particular feature of the data was the marked variation between fields within treatment areas and differences between areas at the outset of the Project. Blackgrass was prevalent in the Integrated area, while brome grasses were most abundant in Supervised area fields. In the Full Insurance area, which had the lowest grass weed populations, inflorescence densities were maintained below threshold levels after the two baseline years, with the exception of meadow brome.

During the baseline years, 1982 and 1983, difenzoquat was applied as a routine measure at half-rate (0.5 kg of active ingredient per ha) for the control of wild-oat. Thereafter, the decision to spray at a threshold of 0.5 inflorescences per m² was imposed in the Integrated and Supervised areas, and this resulted in many fewer applications compared with the Full Insurance area (Table 4.1) and the baseline years. After 1983, wild-oat herbicides were used on only five occasions in the Supervised and Integrated areas. Populations of wild-oat were negligible in the Full Insurance area and were generally low in the other two areas. In 1988, wild-oat populations increased in two Integrated area fields (Figure 4.2), almost certainly as a result of changing from tine cultivation and ploughing up dormant seed.

The amounts of couch in fields did not differ significantly between the areas, but there was more in 1983 than in other years. From 1985 onwards, the species was not recorded in the Full Insurance fields (Figure 4.2). Glyphosate was effective in maintaining low populations of couch for more than one season; applications to Extra Close West and Knapwell in 1983 were not required again until

Figure 4.2 Densities (per m²) of grass weed inflorescences in July of each year in the weediest quarter of each of the Boxworth Project fields. Spray decision threshold levels are shown by a dotted line.

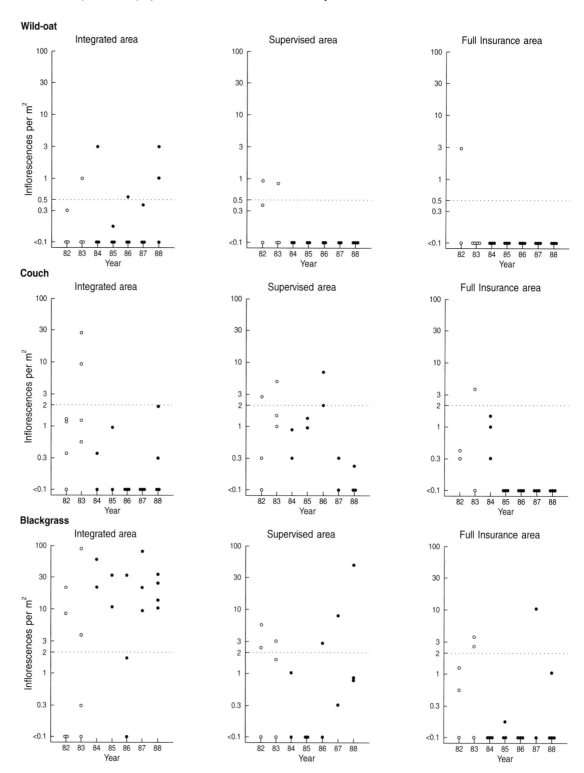

Figure 4.2—*continued*

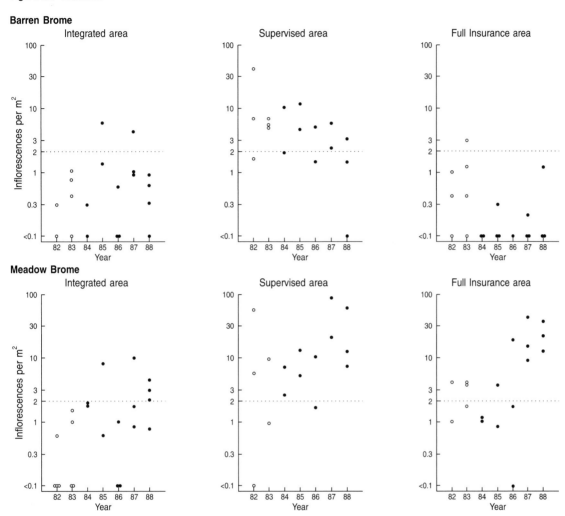

Barren Brome

Meadow Brome

three years later. A slow increase in populations was tolerated in several fields, before the threshold was reached and control was recommended. For example, average couch densities in Knapwell increased from 0.8 inflorescences per m^2 in 1984 to 6.8 inflorescences per m^2 in 1986. The field was then sprayed and by July 1987 only 0.3 inflorescences per m^2 were recorded.

Blackgrass inflorescence numbers fluctuated markedly (Figure 4.2). Overall, there were significantly higher densities in Integrated fields than in the other two areas, which had similar amounts. Over the treatment phase of the Project,

the Supervised area received three isoproturon sprays, compared to at least 12 in the other areas (Table 4.1). 1984, 1987 and 1988 showed blackgrass increases in most Integrated and Supervised area fields, following late spring applications of isoproturon. Large blackgrass plants are less susceptible to isoproturon, so that late applications were less effective. Increases were particularly marked in Thorofare field, where herbicide control in 1987 and 1988 was poor. Densities of blackgrass in 1986 in fields that followed oilseed rape were similar to those in the previous wheat crop. Metoxuron and tri-allate, which were used to control

Table 4.1   Summary of herbicide applications to fields in the three treatment areas. Figures indicate the number of fields of winter wheat treated each year. In two years (1984–5 and 1986–7), oilseed rape was grown in one field in each area (*), or in one case, two fields (**).

Weed problem	Full Insurance area (4 fields)				
	1983–4	1984–5*	1985–6	1986–7*	1987–8
Brome grass	4	3	4	3	3
Blackgrass	4	1	1	3	4
Wild-oat	4	3	4	3	4
Perennial weeds	4	0	4	1	0
Broad-leaved weeds	4	3	4	3	4

Weed problem	Supervised area (3 fields)				
	1983–4	1984–5*	1985–6	1986–7*	1987–8
Brome grass	3	2	2	1	2
Blackgrass	0	0	0	1	2
Wild-oat	1	1	0	0	0
Perennial weeds	0	0	3	0	2
Broad-leaved weeds	2	2	3	2	3

Weed problem	Integrated area (4 fields)				
	1983–4	1984–5**	1985–6	1986–7*	1987–8
Brome grass	2	0	1	0	0
Blackgrass	1	2	2	3	4
Wild-oat	1	1	0	0	1
Perennial weeds	1	0	2	0	2
Broad-leaved weeds	3	2	4	3	4

brome grasses, may also have given some control of blackgrass.

Barren brome (*Bromus sterilis*) and meadow brome (*B. commutatus*), were not always controlled well with the herbicide sequence of tri-allate followed by metoxuron. Meadow brome populations increased during the Project, while barren brome populations remained steady. Overall, there were significantly higher densities of barren brome in Supervised area fields than in Integrated or Full Insurance fields, and average densities of meadow brome were significantly greater in Supervised and Full Insurance fields than in the Integrated area (Figure 4.2). This result was unexpected in view of the lower herbicide inputs in the Integrated area. It exemplifies the initial field differences in weeds, which are likely to be linked with past management. Particularly large increases in meadow brome in Grange Piece, a Full Insurance field, resulted from high densities favoured by a farm experiment in part of the field.

Applications of metoxuron or tri-allate for brome grass control were made to the Supervised and Integrated fields on 13 occasions over the treatment phase of the Project (Table 4.1), which represents a saving of at least two sprays per field compared with the Full Insurance area. Iso-proturon, which is used to control blackgrass, may also have given some control of brome grasses.

Ploughing is an effective method for con-trolling barren brome because the seeds have little dormancy and, after burial, most germinate but do not successfully emerge from more than 13 cm deep (Froud-Williams *et al.*, 1980). However, despite frequent ploughing and herbicide appli-cations, barren brome populations in Knapwell field remained above threshold levels. In contrast, the lack of herbicides and ploughing on Thorofare field in 1986 and 1987 contributed to an explosion in the populations of meadow brome, but not of barren brome.

Spray decisions for broad-leaved weed control

Autumn and spring surveys of broad-leaved weeds

Five broad-leaved species were particularly common in the Project fields: cleavers (*Galium aparine*), charlock (*Sinapis arvensis*), chickweed (*Stellaria media*), field speedwell (*Veronica persica*) and ivy-leaved speedwell (*V. hederifolia*). A range of other species was found, including several uncommon arable weed species, such as dwarf spurge (*Euphorbia exigua*) and fluellen (*Kickxia spuria*). There were marked field-to-field differences and annual variations. For example, chickweed densities in spring 1987 varied between 1.7 and 19.3 plants per m² in fields within the Integrated treatment area. In 1988, densities in the field that had been weediest in 1987 reached only 0.1 chickweed plants per m².

Using plant counts made in November and March, the mean total crop equivalents per m² for the worst quarter of each field were used to advise

Figure 4.3 Total crop equivalents (per m²) for broad-leaved weeds in the weediest quarter of each field in autumn and spring. The threshold of 5.0 crop equivalents per m² is indicated by a dotted line.

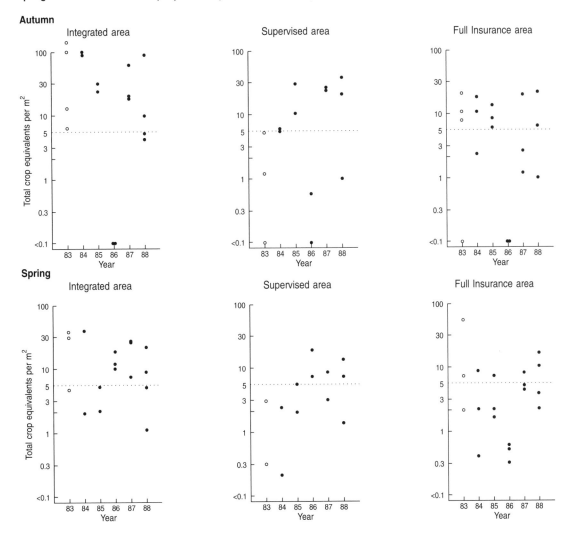

on the need for control measures (Figure 4.3). Note that spring herbicide applications were made after the assessment of weed populations in spring. There was marked variation in autumn weed populations. For example, in the autumn before the 1986 harvest, there was very little germination in any field. Typically, however, there were higher crop equivalent values per m^2 in autumn than in spring, with considerable winter mortality of weeds occurring in many fields, even without herbicide applications. Total crop equivalents per m^2 often increased in the season following an oilseed rape crop, reflecting increased amounts of weed seed produced in the crop.

Broad-leaved weeds in July surveys

The occurrence of broad-leaved weeds, measured as the percentage of sample locations at which weeds were recorded, is shown in Figure 4.4. During the baseline years and the first treatment year, many broad-leaved weeds survived to July and were recorded in quadrat assessments. Thereafter, occurrences were lower over all three treatment areas. There were no significant differences between areas (Figure 4.4), probably reflecting a similar input of broad-leaved weed herbicides.

The broad-leaved and other non-grass weed species present in wheat in early July, before harvest, are shown in Table 4.2. In 1982–1983 the number of species found in July was particularly low, but numbers were high in 1983–1984, though the reasons are obscure.

Spray recommendations for broad-leaved weed control

Densities of broad-leaved weeds were estimated in late November and again in March. Weather conditions often prevented autumn spray applications, so most broad-leaved weed herbicides were applied in April (Appendix II), after weeds were counted in March. However, the advice on whether or not to spray, which was given in autumn and spring on the basis of the crop equivalent thresholds, was overruled by the Farm Director on

12 out of 30 occasions. Herbicides were often applied to control late-germinating weeds or seedlings of 'single low' varieties of oilseed rape. As a result, there were only three times when a field remained unsprayed with herbicides throughout the season; Bushes & Pits and Thorofare in 1984, and Top Pavements in 1985 were unsprayed. Thus, it was not possible to assess the effectiveness of the crop equivalents threshold in reducing applications of herbicides or the build-up of weed populations.

Germination in the Project fields was variable, affected by autumn weather and the timing of sowing. For example, in 1985 little autumn germination of broad-leaved weeds was recorded. The experience of implementing the crop equivalents threshold at Boxworth indicates that the effects of weather conditions on germination and survival of both autumn and spring weed populations need to be taken into account. Further field tests of the technique are required.

Discussion

Advice on the control of grass weeds, provided on the basis of quadrat estimates of grass inflorescence densities and spray thresholds, was followed on the majority (86%) of occasions. As a result, significant reductions in numbers of herbicide applications for the control of grass weeds were achieved, compared to the Full Insurance programme, demonstrating the potential benefits of spray decision thresholds for weed control.

Counting part-field applications of glyphosate, Full Insurance fields received an average of 4.9 applications of herbicides for the control of grass and broad-leaved weeds each year between 1984 and 1988, compared with 3.3 in Supervised fields and 2.6 in Integrated fields.

Lowest weed populations were found in the Full Insurance fields. The threshold programme pursued in the Integrated and Supervised areas kept grass weeds under control with the exception of brome species (Marshall, 1987). Savings in herbicides were achieved in these areas in control-

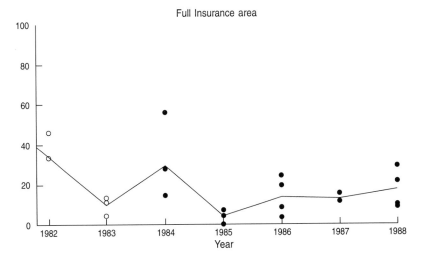

Figure 4.4   Percentage of field sampling locations at which broad-leaved weeds were recorded in July, between 1982 and 1988, in the three treatment areas. The lines indicate changes in the annual averages.

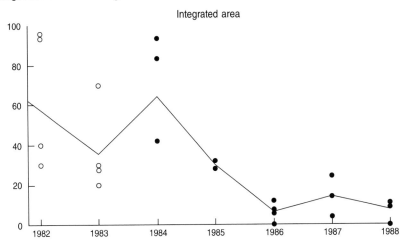

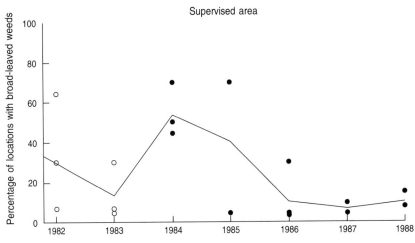

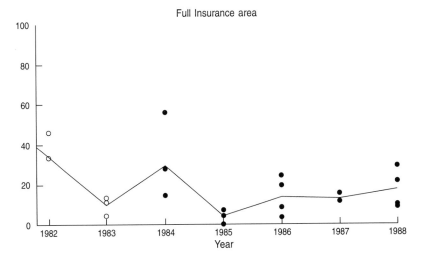

Table 4.2 Occurrences in July of the most common broad-leaved weeds in the three treatment areas during the baseline (1981–1983) and treatment (1984–1988) years of the Project.

	Full Insurance area							Supervised area							Integrated area						
	81–2	82–3	83–4	84–5	85–6	86–7	87–8	81–2	82–3	83–4	84–5	85–6	86–7	87–8	81–2	82–3	83–4	84–5	85–6	86–7	87–8
Black bindweed	+	+	+	+		+	+	+	+	+	+		+	+	+	+	+	+	+	+	+
Cleavers	+	+	+	+		+	+	+	+	+	+		+	+	+	+	+	+	+	+	+
Field bindweed	+	+	+	+	+	+	+	+	+	+	+	+	+	+	+	+	+	+	+	+	+
Chickweed				+	+	+	+	+		+	+	+	+	+	+		+				+
Charlock	+		+			+	+	+		+	+	+			+		+			+	
Fool's parsley	+		+			+	+			+	+				+	+	+				
Ivy-leaved speedwell			+	+		+				+											
Field speedwell		+	+		+	+				+				+	+		+	+	+	+	
Orache	+		+					+	+	+	+			+	+		+				+
Scarlet pimpernel				+		+	+	+	+	+		+		+							
Fat-hen	+		+				+	+	+	+	+							+			
Dwarf spurge					+			+		+		+	+								
Knotgrass			+					+		+											
Common hempnettle			+							+	+						+				
Prickly sow-thistle				+							+	+					+				
Round-leaved fluellen					+					+	+	+									
Shepherd's purse				+							+							+			
TOTAL SPECIES	9	5	13	4	6	11	12	11	6	20	15	11	7	8	7	4	14	7	5	7	5

ling wild-oat and couch. Poor control of blackgrass and brome species in some fields was a reflection of low herbicide activity. Blackgrass populations increased dramatically when herbicides were ineffective, reflecting the enormous potential for regeneration of the species (Moss, 1981).

Weeds were patchily distributed, both within and between fields. This presents some difficulties in designing sampling programmes to achieve precise estimates of weed populations. Six sampling points per hectare were used in the Project, but it was shown that a higher intensity would be required to estimate accurately the density of the less abundant weeds. However, estimates for more abundant species are more accurate and permit examination of gross population changes (Figure 4.2). It may be that estimates for operating decision thresholds need be less precise than those made to examine population trends (Cousens,1987). It is unlikely that a farmer or crop consultant would use quadrat estimates to decide on herbicide applications, because it is likely to be too time-consuming and therefore too expensive. An assessor would need to spend approximately 45 minutes per hectare, and even if this were done only once a year, the cost to a farmer would be considerable. While grass weeds may be accurately assessed in July, counts of broad-leaved weeds are less reliable. Some early-maturing species may be absent, and it is often impossible to count individual plants that intertwine. Traditional visual assessment of the whole field therefore remains the best practical method of deciding on weed control strategies.

Yield losses due to weeds were estimated by methods which are outlined in Appendix A. Annual losses for the three treatment areas are shown in Table 4.3, from which the averages were 4.7% in the Full Insurance area, 7.0% in the Supervised area, and 7.3% in the Integrated area.

Summary

At the start of the Project, the fields had differing weed problems. In the later years, broad-leaved weed herbicides were used in most fields (Table 4.1), so that surviving weed populations in the crop in July were uniformly low (Figure 4.4). Herbicide treatments maintained low populations of grass weeds in Full Insurance fields, though meadow brome was not controlled and increased in most fields (Figure 4.2). In the Supervised area, meadow brome was also poorly controlled by herbicides, though barren brome populations were restricted to near threshold levels. Blackgrass populations were poorly controlled in Integrated area fields, notably Bushes & Pits and Extra Close East, and from 1986 onwards the species increased markedly in Thorofare field. Spray thesholds worked well for the control of wild-oats and couch (Figure 4.2), demonstrating a potential method for deciding on the best use of herbicides.

Table 4.3 Yield losses of winter wheat due to grasses and broad-leaved weeds.

Percentage yield loss

	Full Insurance area		Supervised area		Integrated area	
	grasses	broad-leaved species	grasses	broad-leaved species	grasses	broad-leaved species
1983–4	0.6	1.4	1.8	0.3	3.5	7.3
1984–5	2.4	0.8	3.4	1.4	3.0	1.3
1985–6	1.9	0.5	2.0	3.4	0.8	3.8
1986–7	6.3	2.3	10.5	2.2	3.5	7.0
1987–8	3.6	3.7	7.3	2.8	2.9	3.3

References

Cousens, R D (1987). 'Theory and reality of weed control thresholds.' *Plant Protection Quarterly* **2**, 13–20.

Cussans, G W & Moss, S R (1982). 'Population dynamics of annual grass weeds.' In: R B Austin, (Ed.) *Decision Making in the Practice of Crop Protection, BCPC Monograph no. 25*, pp. 91–98.

Cussans, G W, Cousens, R D & Wilson, B J (1987). 'Progress towards rational weed control strategies.' In: Brent, K J and Atkin, R K (Eds.) *Rational Pesticide Use*. Cambridge University Press. pp. 301–314.

Eggars, T & Niemann, P (1980). 'Zum Begriff des Unkrauts und uber Schadshwellen bei der Unkrautbekampfung.' *Berichte uber Landwirtschaft* **58**, 264–272.

Froud-Williams, R J, Pollard, F & Richardson, W G (1980). 'Barren brome: a threat to winter cereals?' *Report of the Weed Research Organization 1978–79*, pp. 43–51.

Marshall, E J P (1987). 'Using decision thresholds for the control of grass and broad-leaved weeds at the Boxworth E H F.' *Proceedings of the 1987 British Crop Protection Conference—Weeds* **3**, 1059–1066.

Marshall, E J P (1988). 'Field-scale estimates of grass weed populations in arable land.' *Weed Research* **28**, 191–198.

Moss, S R (1981). 'The agro-ecology and control of black-grass, *Alopecurus myosuroides* Huds., in modern cereal growing systems.' *ADAS Quarterly Review (1980)* **38**, 170–191.

Tottman, D R & Wilson, B J (1989). 'Weed control in small grain cereals.' In: Hance, R J & Holly, K (Eds.) *The Weed Control Handbook: Principles*, 8th. Edn. Blackwell Scientific Publishers, Oxford. pp. 301–328.

Wilson, B J (1986). 'Yield responses of winter cereals to the control of broad-leaved weeds.' *Proceedings of the European Weed Research Society Symposium 1986, Economic Weed Control* pp. 75–82.

Diseases

David J Yarham and **Barry V Symonds** (ADAS, Cambridge)

5

Introduction

The boulder clay plateau of west Cambridgeshire on which Boxworth EHF is situated is near enough to the Fenland basin to be at some risk from the yellow rust epidemics which often start in the Wash basin and move down through the west Fens into the area. However, it cannot be considered as a 'high risk' area for any other of the major diseases of cereals or oilseed rape, though all the common pathogens of these crops are regularly recorded on the farm. On its well-structured, chalky clays, root diseases seldom present serious problems, and its generally open landscape, distant from any coastal influence, lacks the broad river valleys where mists favour the development of foliar pathogens such as the *Septoria* species.

The fungicide programme applied to the Full Insurance area was designed to control (so far as was possible) stem base, leaf and ear diseases throughout the year (see Appendix I). It had been planned that in this area the fungicide programme should be kept the same throughout the Project. In the event, however, the withdrawal of clearance for the use of captafol after ear emergence necessitated a change to the programme after the 1986 season.

Crops in the Supervised and Integrated areas were monitored through the season and sprays were applied only when appropriate disease thresholds were reached or meteorological criteria met. At the beginning of the Project guidelines were laid down for the chemicals to be used at stem extension, flag leaf emergence and ear emergence (see Appendix I). However, the final choice of products and the details of timing were kept very flexible so that it was possible to respond to particular circumstances (such as the late development of mildew in the Integrated area in 1986).

In the Integrated area the policy was to exploit the resistance of crop varieties to pathogens in order to reduce the need for fungicides. This included:

a) the use of varieties resistant to particular diseases;

b) the use of several varieties with different genes for resistance to yellow rust and mildew, which would prevent a pathogen strain adapted to one variety from spreading into the crop of a different variety in an adjacent field;

c) the use of blends in which three crop varieties with different resistance genes are grown as a mixture (Wolfe, Barrett & Jenkins, 1981).

The disease resistance ratings of the National Institute of Agricultural Botany (NIAB) for the varieties grown over the trial period (Anon., 1985–1988) are given in Table 5.1.

In the replicated plot trial in Shackles Aden field, it was envisaged that the fungicide programmes applied to the Full Insurance and Supervised plots should be similar to those in the corresponding fields. However, strict comparability was not always achieved, because weather conditions sometimes prevented the spraying of the small plot trial after the fields had been sprayed. Apart from an organomercury seed treatment, no fungicides were applied to the 'Minimum Input' plots in the trial.

Disease assessment and spray decisions

During the treatment years regular disease assessments of stem base and foliar diseases were carried out at sampling points selected to allow not only field-to-field comparisons but also comparisons between different topographical areas within the same field. Leaf disease assessments were

Table 5.1   NIAB disease resistance ratings of varieties of wheat for the years in which the varieties were used. The higher the rating, the more resistant is the variety. Figures in brackets indicate greater susceptibility to certain strains of the pathogen.

	Mildew	Yellow rust	Brown rust	Septoria tritici	Septoria nodorum	Eyespot
Aquila	4	9	7(4)	–	6	6
Avalon	7	9(5)	5	4	3	5
Brimstone	7	9(6)	9	5	5	6
Brock	8	8	4	6	6	4
Fenman	8	9	6	4	4	5
Galahad	7	5	7	6	5	5
Mercia	8	9	8	5	5	5
Norman	7	9(4)	7	4	4	4
Rendezvous	7	9	9	6	7	8

based on the percentage area of leaf blade affected, and stem disease assessments on the percentage of stem bases bearing lesions. In the Supervised and Integrated areas these assessments were used to determine when spray thresholds had been reached using criteria set out in the ADAS Managed Disease Control System (Anon., 1986a).

The need for fungicide applications was assessed at the following growth stages:

a) 'a leaf sheath erect' to 'first node detectable' (Growth Stages 30 to 31)

b) 'flag leaf beginning to emerge' to 'flag leaf fully emerged' (Gowth Stages 37 to 39)

c) 'early ear emergence' to 'beginning of flowering' (Growth Stage 59)

d) 'end of flowering' to 'grain watery ripe' (Growth Stage 71).

For oilseed rape, disease thresholds were used to determine the need to spray against light leaf spot (*Pyrenopeziza brassicae*) and dark leaf spot (*Alternaria* spp.). For wheat they were used to determine the need for sprays against eyespot (*Pseudocercosporella herpotrichoides*) and mildew

(*Erysiphe graminis*). The thresholds for spraying against yellow and brown rusts (*Puccinia striiformis* and *P. recondita*) were not reached in any of the treatment years.

To determine the need to spray against the leaf spots caused by *Septoria tritici* and *S. nodorum* and against the complex of 'ripening diseases' (including the ear blights caused by *Fusarium* spp. and *Botrytis cinerea*) meteorological criteria were used. For *Septoria*, a spray was applied about the time of flag leaf emergence if four or more days each with at least 1 mm rain, or one day with more than 5 mm rain, had occurred in the previous two weeks (Anon., 1986b). Spraying against ear blights was carried out if the weather was warm and wet during flowering.

If crops in the Supervised and Integrated areas had not been sprayed at flag leaf emergence, the ADAS Managed Disease Control System allowed for the application of an inexpensive prophylactic fungicide at ear emergence.

In 1984 and 1985 isolates of eyespot both from field crops of wheat and from the replicated plot trial were tested *in vitro* for resistance to the methyl benzimidazole carbamate (MBC) fungicides which had been used for eyespot control earlier in the Project. Information from these tests was used to determine which fungicide should be used for eyespot control in subsequent years.

Diagnosis of the pathogens involved in the ripening disease complex can be particularly difficult, especially in wet seasons when the senescence of leaves is accelerated by a number of weakly pathogenic organisms (eg *Fusarium* spp., *Didymella exitialis*). In 1987 leaf lesions from plants in the replicated plot trial were cultured on nutrient agar and the resulting isolates were identified in the laboratory to obtain a clearer picture of the range of fungi present.

Seasonal development of crop diseases

The last 'baseline' year—1982–1983

Observations during the 'baseline' years were aimed at establishing whether there were any fundamental differences in disease development between the prospective treatment areas. In 1981–1982, such observations could not be made as there were no comparable crops in the two areas. In 1982–1983 observations were made on crops of wheat (variety Norman) sown in mid to late October in Backside and Thorofare fields, situated in what were to become the Full Insurance and Supervised areas.

During the winter and early spring, incidence of *Septoria tritici* was higher in Backside than in Thorofare field. Mildew levels tended to be higher in Thorofare throughout the season. At flag leaf emergence the fields were sprayed with different fungicides. This is likely to have been the main cause of the differences subsequently observed. In late June and July *Septoria nodorum*, brown rust and yellow rust all built up rapidly in Backside field but remained at very low levels in Thorofare, where carbendazim was used. Amounts of eyespot and take-all (*Gaeumannomyces graminis*) were comparable in the two fields.

The 'baseline' observations thus showed the spectrum of disease that could be expected to occur on the wheat crops in the treatment years, against which fungicides would need to be applied.

The first treatment year—1983–1984

Since yellow rust had been present on the wheat (variety Norman) in 1983, it was decided to employ varietal diversification to reduce the risks of this disease in the Integrated area in 1984. In addition to a single field of variety Norman, two other varieties possessing different genes for rust resistance (Fenman and Aquila) were grown in the Integrated area.

In the event, 1984 was (for the most part) a dry year, and yellow rust did not develop, while little *Septoria* was seen until *S. nodorum* began to build up after warmer and wetter weather in late June.

The main diseases in the wheat crops in 1984 were eyespot and mildew. On average, levels of both were lower in the Full Insurance area than in the Supervised and Integrated areas (eg Figure 5.1) but there was considerable field-to-field and within-field variation in disease development. In the Supervised area, for example, Top Pavements and Thorofare fields both run down from a fairly exposed plateau into a sheltered valley, and in both fields mildew was most severe in the valley bottom, while there was more eyespot on the plateau.

In the Integrated area, mildew levels in varieties Norman and Fenman remained remarkably low throughout the season. In the two fields of variety Aquila, on the other hand, mildew posed much more of a problem and in late June it was decided to spray these fields against the disease. As it was late in the season (after ear emergence), long-term control of the disease was considered unnecessary so fenpropimorph was applied at a reduced rate of 0.5 litres per hectare, but carbendazim + mancozeb was added to broaden the spectrum of activity and afford protection against the later development of 'ripening diseases'.

Testing of eyespot isolates in the spring (before any sprays had been applied) showed that in the Supervised and Integrated areas about 70% of the isolates were resistant to MBC fungicides. In the Full Insurance area only 58% were MBC-resistant. However, there was considerable field-to-field variation in the structure of the populations. By the end of the season the average proportion of

Figure 5.1   Average percentage of the area of leaf 3 affected by mildew, based on all the sampling points in each treatment area in 1983–1984.

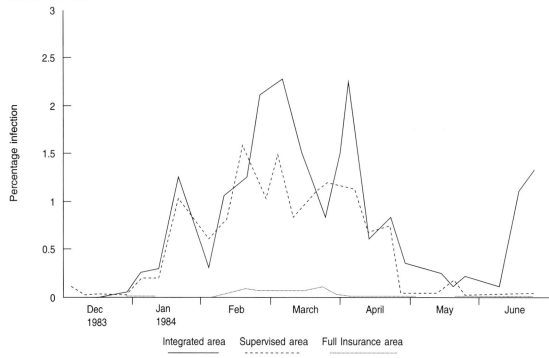

Integrated area   Supervised area   Full Insurance area

isolates resistant to MBC was 92% in the Supervised area, 87% in the Integrated area and 85% in the Full Insurance area.

The second treatment year—1984–1985

In this year a blend of varieties, devised in consultation with the Plant Breeding Institute in Cambridge and comprising Brock, Brimstone and Norman, was grown in one of the two wheat fields in the Integrated area to reduce the risks of infection with yellow rust and mildew.

In all but two of the fields (Knapwell and Extra Close East) there was evidence of a further increase in the proportion of MBC-resistant isolates of the eyespot fungus. From this year onwards prochloraz was used to control the disease in the Supervised and Integrated areas. In the Supervised area a formulated mixture of prochloraz and carbendazim was used in 1984–1985. This was done because carbendazim was thought likely to increase the effectiveness of the fungicide against eyespot strains still susceptible to MBC fungicides. This policy was later abandoned when trial evidence suggested that in a population dominated by MBC resistant strains the application of carbendazim may actually increase eyespot levels (Griffin & King, 1985).

Of the cereal diseases, only *Septoria* was present at higher levels than in 1984. Throughout the winter and early spring, the incidence of diseases was noticeably higher in the Supervised and Integrated areas than in the Full Insurance area where a seed treatment containing fuberidazole + triadimenol had been used. Thereafter, the fungicide programmes in the Supervised and Integrated areas appeared to give good control both of leaf diseases and of eyespot. In late June and July, *S. tritici* infections built up more rapidly in the Full Insurance fields than in the other two areas, although, at least on variety Norman, development of late season *S. nodorum* was delayed in the Full Insurance fields (Figure 5.2). Senescence of leaf 3

Figure 5.2 Average percentage of the area of leaf 2 affected by *Septoria nodorum*, for all fields of variety Norman in 1984–85. Dates of sprays are shown by arrows.

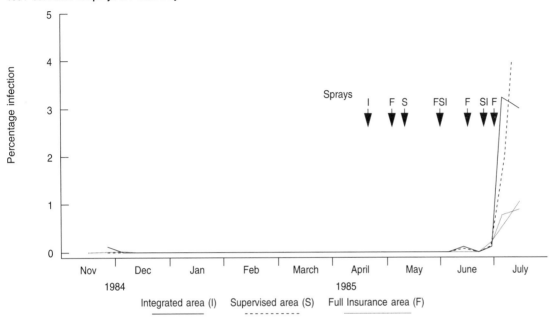

occurred more than a week later in the Full Insurance area than in the Supervised and Integrated areas.

Take-all assessments were carried out in May, July and August. In general, levels were lower in the Full Insurance fields. It was noted that the incidence of take-all in May was higher where straw had been incorporated than where it had been baled and removed from the field.

On oilseed rape, *Phoma* leaf spot, downy mildew and dark leaf spot (*Alternaria*) remained at low levels throughout the season. Light leaf spot affected up to 20% of the surface area of pods in the Integrated area but incidence in the Full Insurance and Supervised areas was much lower. There was no close relationship between the amount of disease and use of fungicides. Three fields suffered badly from pigeon damage in the winter, and foliar disease levels were generally lower in the affected areas than elsewhere.

The third treatment year—1985–1986

In 1985–1986 blends of varieties were grown in all three fields in the Integrated area.

Disease levels were low for most of the season. In the Supervised and Integrated areas it was not necessary to apply a spray for the control of eyespot at the stem extension first node stage (Growth Stages 30 to 31). However, it was decided to use prochloraz at flag leaf emergence because this would not only control *Septoria* (weather conditions conducive to its spread had occurred in late May) but should also guard against any late development of eyespot (ADAS trials, unpublished).

The presence of mildew in the Supervised area prompted the application of propiconazole + tridemorph after ear emergence. In the varietal blend in the Integrated area, the amount of mildew remained very low until the end of June and no ear emergence spray was applied. However, the disease later built up very rapidly and a late application of fenpropimorph on 8 July failed to control the epidemic.

Carbendazim + maneb was applied in the Integrated area with the 8 July fenpropimorph spray as an insurance against late *Septoria* and 'ripening diseases'. At the end of the season, however, there was more *S. tritici* in the Integrated

Figure 5.3   Average percentage of the area of leaf 3 affected by *Septoria tritici*, based on all sampling points in each treatment area in 1985–86.

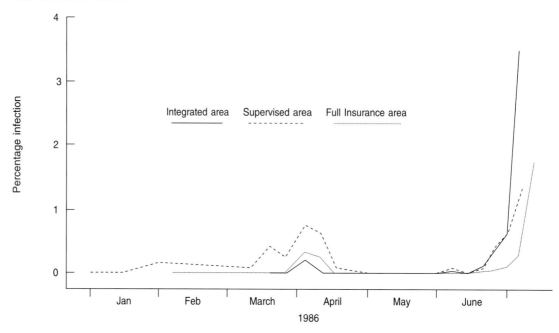

than in the Supervised fields that had received the propiconazole spray at ear emergence (Figure 5.3).

Take-all levels remained low throughout the season (never exceeding 10% of roots infected in any field). Take-all incidence was particularly low in crops that followed oilseed rape.

The fourth treatment year—1986–1987

After the mildew problems encountered in 1986, it was decided to abandon the use of varietal blends. Instead, the eyespot-resistant variety Rendezvous (which is also moderately resistant to mildew) was grown in the Integrated area (varieties Avalon and Galahad being grown in the other two areas). In the Integrated fields this obviated the need to apply a spray for the control of eyespot, though some build up of the disease in these fields was noted later in the season.

Mildew and *Septoria tritici* were active in the early spring but declined as the crops grew away. Wet weather in mid-summer prompted the application of propiconazole sprays at both the flag leaf and ear emergence growth stages in the Super-

vised and Integrated areas. In the Supervised area this appeared to give as good a control of *Septoria tritici* as the more expensive fungicide programme used on the same varieties in the Full Insurance area. During June, however, *S. nodorum* built up in all three areas to levels higher than had been recorded in any of the previous three years. Observations made on the replicated plot trial in 1987 highlighted the difficulties inherent in the measurement of *S. nodorum*. As the ageing leaves are invaded by weak pathogens such as *Didymella exitialis* and *Fusarium* spp. it may be virtually impossible to determine how much of the senescence seen is due to *S. nodorum* and how much to these other fungi.

Take-all levels were rather higher than they had been in 1986, the disease being most severe in the lower-lying and more poorly-drained parts of the fields. No differences between the three areas were observed.

Light leaf spot had been the main oilseed rape disease in previous years, and therefore the resistant variety Mikado was grown in the Integrated area. Unfortunately this variety is very sus-

Figure 5.4   Average percentage of leaf area of oilseed rape plants that was infected with dark leaf spot in each treatment area in 1986–87.

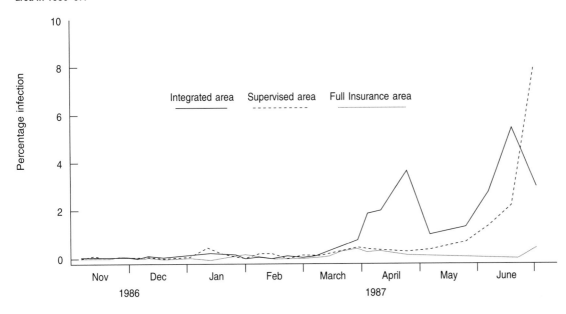

Figure 5.5   Average percentage of tillers (plants up to 22 April) infected with eyespot based on all sampling points in each treatment area in 1987–88.

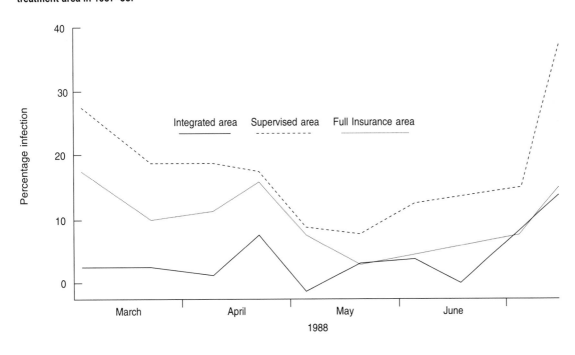

ceptible to dark leaf spot and it was this, rather than light leaf spot, which provided the main disease problem in 1987 (Figure 5.4). In early June it became necessary to apply iprodione to control dark leaf spot in the Integrated area, but no fungicides were needed on variety Bienvenu in the Supervised area.

The fifth treatment year—1987–1988

As in 1986–1987, the use of variety Rendezvous obviated the need to apply a spring spray for the control of eyespot to the wheat in the Integrated area, where eyespot incidence was low (Figure 5.5). However, when wet weather prompted the application of a spray at flag leaf emergence it was decided to use prochloraz which, in addition to controlling *Septoria*, would also give some protection against late development of eyespot. Since at that time mildew was also very active in the Intergrated area (despite the relatively high resistance score of variety Rendezvous), prochloraz was applied in combination with fenpropimorph.

In the Supervised area propiconazole was used as the spray at flag leaf emergence. In Knapwell Field, where mildew was active, it was used in combination with tridemorph. As the wet weather continued, and active mildew pustules could still be found, flutriafol + tridemorph was applied at ear emergence to all the Supervised and Integrated fields with the exception of Knapwell, where mildew was still especially active and where fenpropimorph + chlorothalonil was used instead.

Though present in the crops over winter, *Septoria tritici* and particularly *S. nodorum* built up later in the season. *S. tritici* and particularly *S. nodorum* proved particularly difficult to diagnose in 1988 as the wet summer favoured the development of weak pathogens such as *Didymella exitialis* and *Fusarium* spp. on the ageing leaves after anthesis. These fungi accelerated leaf senescence, often invading and aggravating lesions caused initially by *Septoria*.

Throughout the season take-all levels were lowest in crops of variety Rendezvous in the Inte-grated area. In all three areas, incidence of take-all was lowest in those crops which followed oilseed rape.

Yield losses caused by the main diseases

The percentage yield losses due to diseases which occurred in the three areas, despite spraying, were estimated using methods described in Appendix A.

An indication of how much higher these losses might have been had no sprays been applied can be gained by comparing them with the percentage yield increases obtained from full protectant programmes in fungicide trials carried out at Boxworth during the years of the Project (Table 5.2). The 1984–1986 data given in Table 5.2 are for variety Galahad and the 1987–1988 data for variety Mercia; the two varieties are very similar in their susceptibility to the main diseases and in their responsiveness to fungicides.

It should be noted that responses to fungicide programmes are sometimes greater than would be expected from estimated yield losses due to disease. The fungicides not only affect major diseases but can also suppress the weakly pathogenic organisms which accelerate the senescence of ageing leaves. This could explain the delay in senescence noted in the Full Insurance area in 1985. This factor is difficult to quantify and has not been taken into consideration in calculating the yield losses given in Table 5.2.

Table 5.2 Combined effects of eyespot, mildew and *Septoria* spp. on yield of wheat.

The values show yield losses in the three treatment areas as percentages of losses in unsprayed trial plots at Boxworth, and yield gains attributable to fungicides in NIAB trials

	1984	1985	1986	1987	1988
Full Insurance area	0.11	1.20	0.00	1.00	3.14
Supervised area	0.76	3.39	1.39	0.65	6.28
Integrated area	0.44	3.05	5.99	2.32	5.28
% yield increase from fungicide programmes in NIAB trials	5.42	24.5	4.80	13.10	17.08

Yield losses caused by brown foot rot

Although sprays were not applied specifically to control brown foot rot (*Fusarium* spp.), assessment of this disease in the replicated plot trial suggested that both the seed treatment used in the Full Insurance area (containing triadimenol + fuberidazole) and the fungicides applied for the control of eyespot could give some suppression of brown foot rot. Estimates of the effect of this disease on yield were necessarily very tentative, but suggested that, over the last four years of the study, losses caused by the disease averaged no more than about 1.9% in the Full Insurance area, and 2.4% in the Supervised and Integrated areas.

Summary of disease losses

Over the five treatment years of the Project, yield losses caused by eyespot and the foliar diseases against which fungicides were applied averaged 1.09% in the Full Insurance area, 2.49% in the Supervised area and 3.42% in the Integrated area. If brown foot rot is included in the calculation then the figures become, respectively, 2.99%, 4.89% and 5.82%. Thus the yield benefit from the use of the Full Insurance fungicide treatment, compared to the Supervised treatment, appeared to be no more than 2.0%.

The replicated plot trial

Incidence of the two main diseases in the replicated plot trial (eyespot and *Septoria tritici*) is shown in Figures 5.6 and 5.7. As the replicated plot trial was situated in a field receiving an intensive fungicide programme, the reduced pathogen pressure would probably have affected the level of disease within the plots.

A formula based on that of Scott & Hollins (1974) was used to calculate disease indices for plants with late-season eyespot. Plants were categorised according to the severity of the disease and the number of stems in each category was used in the formula.

Effects of eyespot were recorded before any spray was applied in the spring in the second

Figure 5.6   Incidence of eyespot at Growth Stage 75 in the replicated plot trial, 1984–88.

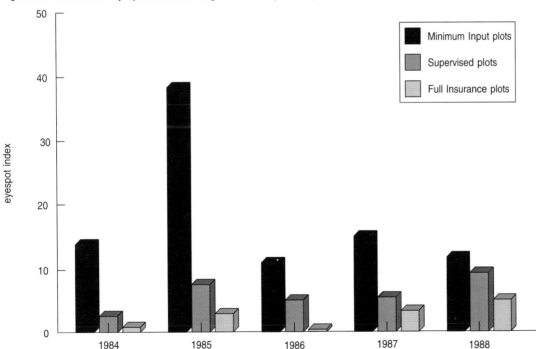

Figure 5.7   Incidence of *Septoria tritici* on leaf 2 (percentage of the leaf area affected) in the replicated plot trial, 1984–88.

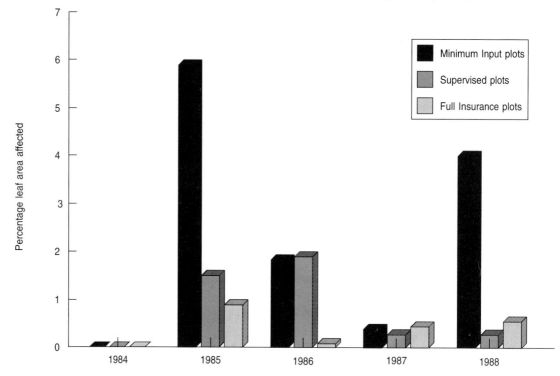

and fifth treatment years of the Project. Eyespot levels were lowest in the Full Insurance plots in all treatment years, which could have been due to the triadimenol + fuberidazole seed treatment. In all years the Supervised plots also had a lower incidence than the Minimum Input plots, suggesting a cumulative effect of treatment on disease. Data from the first year of the trial would have helped to validate this hypothesis, but unfortunately the trial was not assessed in spring 1984.

There was less consistency in the incidence of other diseases. For example, differences between treatments varied greatly from year to year for *Septoria tritici* (Figure 5.7).

Data collected in 1986, 1987 and 1988 suggested that, in addition to controlling the diseases at which they were aimed, the fungicide programmes reduced the incidence of brown foot rot.

Discussion

Table 5.3 summarises the frequency of fungicide applications in each area. Over the five years of the Project, the numbers of fungicidal active ingredients applied annually to wheat fields averaged 9.8 in the Full Insurance area, 4.7 in the Supervised area and 4.6 in the Integrated area (this includes seed treatment; also, an active ingredient is counted twice if it was applied twice in the same year). The Supervised area had very little more disease than the Full Insurance area, and it gave as satisfactory an economic return, indicating the success of the ADAS Managed Disease Control approach to fungicide usage.

This study highlights some of the problems associated with the exploitation of genetic diversity in integrated disease control programmes, but shows how a high level of varietal resistance to a major disease (in this case eyespot on variety Rendezvous) can reduce the need for fungicides.

Table 5.3 Summary of applications of fungicides to fields in the three treatment areas.

Figures indicate the numbers of fields of winter wheat treated each year. In two years (1984–5 and 1986–7), oilseed rape was grown in one field in each area (*), or in one case, two fields (**).

Disease Problem	Full Insurance area (4 fields)				
	1983–4	1984–5*	1985–6	1986–7*	1987–8
Seed-borne diseases (seed treatment)	4	3	4	3	4
Eyespot (at stem extension)	4	3	4	3	4
Septoria (at flag leaf emergence)	4	3	4	3	4
Septoria (at ear emergence)	4	3	4	3	4
Mildew (at flag leaf emergence)	4	3	4	3	4
Mildew (at ear emergence)	0	0	0	0	0
Ear blights and 'ripening diseases' (at ear emergence or after flowering)	4	3	4	3	4

Disease Problem	Supervised area (3 fields)				
	1983–4	1984–5*	1985–6	1986–7*	1987–8
Seed-borne diseases (seed treatment)	3	2	3	2	3
Eyespot (at stem extension)	3	2	0	2	3
Septoria (at flag leaf emergence)	3	2	3	2	3
Septoria (at ear emergence)	0	2	3	2	3
Mildew (at flag leaf emergence)	0	0	1	0	1
Mildew (at ear emergence)	0	0	3	0	3
Ear blights and 'ripening diseases' (at ear emergence or after flowering)	0	2	0	0	0

Disease Problem	Integrated area (4 fields)				
	1983–4	1984–5**	1985–6	1986–7*	1987–8
Seed-borne diseases (seed treatment)	4	2	4	3	4
Eyespot (at stem extension)	4	2	0	0	0
Septoria (at flag leaf emergence)	4	2	4	3	4
Septoria (at ear emergence)	0	2	0	3	4
Mildew (at flag leaf emergence)	0	0	0	0	4
Mildew (at ear emergence)	4[1]	0	4	0	4
Ear blights and 'ripening diseases' (at ear emergence or after flowering)	2	2	4	0	0

NOTE: [1] Two fields of variety Aquila were treated after flowering

References

Anon. (1985–1988). *NIAB Farmers Leaflet No. 8— Recommended Varieties of Cereals*. NIAB, Cambridge.

Anon. (1986). 'Managed disease control'. In: *Use of fungicides and insecticides on cereals* ADAS Booklet 2257. MAFF Publications, Alnwick. pp. 28–38.

Clarkson, J D S (1981). 'Relationship between eyespot severity and yield loss in winter wheat'. *Plant Pathology* **30**, 125–131.

Griffin, M J & King, J E (1985). 'Benzimidazole resistance in *Pseudocercosporella herpotrichoides*: results of ADAS random surveys and fungicide trials in England and Wales 1982–1984'. *EPPO Bulletin* **15**, 485–494.

King, J E, Jenkins, J E E & Morgan, W A (1983). 'The estimation of yield losses in wheat from severity of infection by *Septoria* species'. *Plant Pathology* **32**, 239–249.

Scott, P R & Hollins, T W (1974). 'Effects of eyespot on the yield of winter wheat'. *Annals of Applied Biology* **78**: 269–279.

Wolfe, M S, Barrett, J A & Jenkins, J E E (1981). 'The use of cultivar mixtures for disease control'. In: Jenkyn J F & Plumb R T (Eds). *Strategies for the Control of Cereal Disease*. Blackwell Scientific Publications, Oxford. pp. 73–80.

Calculation of yield losses due to pests, weeds and diseases

Introduction

The performance of crops under different pesticide regimes depends on how well the damaging effects of pests, weeds and diseases are controlled. This is difficult to assess on a field-scale, because there is no opportunity to compare pesticide-treated areas with untreated control areas known to have similar levels of infestations. However, an attempt was made to estimate the scale of yield losses indirectly, using a variety of approaches, which are different for pests, weeds and diseases.

Weeds

Weeds reduce the yield of crops by competing for light, space and nutrient resources, even at low densities (Cousens, 1985). Spray decision thresholds used in the Boxworth Project were based on the predicted competitive effects of weed populations, with built-in margins of safety to avoid economic losses. Accordingly, yield losses can be estimated from the densities of weeds shortly before harvest. This was measured directly for grass weeds in July, but had to be based on March surveys of broad-leaved weeds, assuming little subsequent change in density.

For grass species, yield losses at a mean density of one inflorescence per m² were taken to be 0.6% (wild-oats), 0.18% (brome grasses), 0.1% (couch) and 0.07% (blackgrass). Expected losses due to broad-leaved weeds were estimated from the following formula:

$$\% \text{ yield loss} = \frac{100 \times C}{240 + C}$$

in which C is the total crop equivalent per m² from the weediest quarter of each field, and 240 is assumed to be the density of wheat plants per m².

Losses were calculated in this way for each field separately, and were then averaged to provide a value for each treatment area, ignoring fields

sown to oilseed rape, and Shackles Aden, where the replicated plot trial was sited. There was great variation among fields, for both grass weeds (0.02 to 18.6% yield loss) and broad-leaved weeds (0 to 18.8%).

It must be stressed that these figures are approximate, and depend on a series of assumptions, several of which may not be true. For example, the competitive effects of cleavers (*Galium aparine*) were represented by a value of 1.7 crop equivalents. More recent work (see Chapter 8) has shown that this species is much more competitive, and a value of 7 crop equivalents might be more appropriate.

Pests

Assessment of yield losses caused by pests was simpler than for weeds, because with one exception there was no significant damage to the crop when pest populations were below their treatment thresholds. In 1986 aphid densities in the Supervised and Integrated areas increased in late summer. Their effect was estimated by extrapolation of results from previous ADAS experimental trials which contrasted yields in insecticide-treated and untreated plots. This indicated that field-scale losses in 1986 might have been about 5%, but on all other occasions would have been less than 1%.

Diseases

For wheat diseases, a series of formulae was available from previous ADAS trials, providing the means to calculate percentage yield losses from the extent to which stems or leaves were affected by lesions of eyespot, mildew, *Septoria tritici* and *S. nodorum*, the four main diseases against which fungicide treatments were applied. These formulae are listed in Table A. In addition, data from the replicated plot trial suggested that some of the

---

Table A   Formulae used to calculate yield losses caused by stem and foliar diseases in winter wheat.

Eyespot   (Clarkson, 1981)
Yield loss = 0.1 (% stems with moderate lesions)
            + 0.36 (% stems with severe lesions)

Mildew   (M R Thomas, unpublished)
Yield loss = 3.5 (area of leaf 2 affected at the milky ripe stage)

Septoria tritici   (M R Thomas, unpublished)
Yield loss = 0.42 (area of leaf 2 affected at the milky ripe
                stage)

Septoria nodorum   (King, Jenkins & Morgan, 1983)
Yield loss = 0.55 (area of leaf 2 affected at the milky ripe
                stage)

---

fungicides may have reduced brown foot rot. Estimates of yield losses due to diseases are discussed in Chapter 5.

References

Clarkson, J D S (1981). 'Relationship between eyespot severity and yield loss in winter wheat.' *Plant Pathology* **30**, 125–131.

Cousens, R D (1985). 'A simple model relating yield loss to weed density.' *Annals of Applied Biology* **107**, 239–252.

King, J E, Jenkins, J E E & Morgan, W A (1983). 'The estimation of yield losses in wheat from severity of infection by *Septoria* species.' *Plant Pathology* **32**, 239–249.

Performance of crops and economics of production

6

Robin H. Jarvis (ADAS, Boxworth EHF)

Introduction

All fields involved in the Boxworth Project have an alkaline clay soil of the Hanslope series. Typical analyses of Boxworth soils are given in Table 6.1. Although the amount of soil phosphate is inherently low, it has been increased by 40 years of regular application so now only maintenance dressings once every three years are needed. On the other hand soil potash levels are naturally high and none has been used commercially on the farm since 1949, with no evidence of soil or crop deficiency. During the Project nitrogen was applied to all crops in spring, following standard ADAS recommendations.

Most of the fields in the Project have a long history of predominantly arable cropping. The only exceptions were parts of Eleven Acre Extra and Bushes & Pits which were broken from long-term grass in 1979 and 1980 respectively. Cropping programmes in all the other fields have shown a strong emphasis on the production of winter wheat, with relatively small areas each year of winter barley, field beans and oilseed rape.

Methods

Field-scale

Measurements of the yield of wheat from each field were available for the period 1974–1983 and it was decided to use this information to provide a 'baseline' against which yields obtained during the Project itself could be assessed (Figure 6.1). All crops were harvested using a standard farm combine and the yield of grain from each field was measured using a weighbridge. Yields of oilseed rape were determined in 1985 and 1987, the two years when this crop was grown. From 1984 onwards, samples of wheat grain were taken for the assessment of thousand grain weight, grain bulk density, protein content and Hagberg falling number, which are measures of grain quality and sale value. A high Hagberg falling number, which is inversely related to the activity of a starch-degrading enzyme, is needed for a good bread-making wheat.

Replicated plots

As the replicated plot trial in Shackles Aden contained Supervised and Minimum Input treatments, the overall field yields and grain quality parameters for this field are excluded from the Full Insurance treatment averages (Table 6.2 and Figures 6.2 to 6.6). The plots of wheat were harvested by a plot combine and for each plot the grain from a 2.62 m × 24 m strip was weighed. Samples were taken for the measurement of the grain quality parameters set out above.

Table 6.1 Typical properties of Boxworth soils.

Soil particle type	Composition	Plant nutrient content of soil (ppm)		Organic matter content of soil	Soil pH
Coarse sand	15%	Phosphorus	42	3.8%	7.6
Fine sand	29%	Potassium	316		
Silt	16%	Magnesium	119		
Clay	31%				

Figure 6.1   Average yields of winter wheat (tonnes per hectare) in the Project fields during 1974–83, before the introduction of contrasting pesticide regimes.

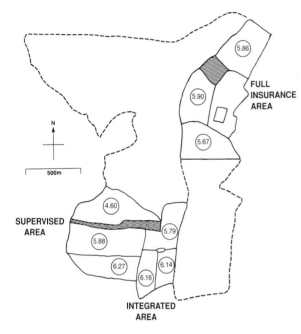

Results

Winter wheat

Yield. There was considerable variation in the yield of grain from year to year, both in the replicated plots and on a field-scale. In both cases the highest mean yields were recorded in 1984, when weather conditions in the period from May to mid August were ideal both for the timely application of fungicides and for the later stages of crop development. Some of the lowest yields were recorded in 1987 (Figure 6.2), when the summer months were cool and very wet, and cereal harvesting was not completed until mid September. On average, in the replicated plots the Full Insurance treatment significantly outyielded the Minimum Input treatment (Table 6.2) (Analysis of Variance, $P < 0.001$). The differences between the Supervised and Minimum Input treatments in 1986, all three treatments in 1987, and the Full Insurance and Supervised treatments in 1988, were smaller and not significant. In comparison with the replicated plots, the field-scale yields showed a similar (although not identical) pattern, with the Full Insurance treatment outyielding the Supervised treatment by greater margins in

Table 6.2   Mean yields and grain quality parameters of wheat in field-scale and replicated plot treatments for the treatment phase of the Project (1984–1988).

Standard errors for all the replicated plot treatments are given in brackets.

	Replicated plots			Field-scale		
	Full Insurance	Supervised	Minimum Input	Full Insurance	Supervised	Integrated
Yield of grain (t/ha) at 85% dry matter	8.33	7.86 (±0.04)	7.25	7.74	6.82	6.39
Thousand grain weight (g)	48.9	47.9 (±0.22)	45.7	48.7	45.1	43.3
Grain bulk density (kg/hl)	72.3	71.6 (±0.12)	70.2	73.8	71.7	70.9
% protein in dry matter	11.0	11.1 (±0.04)	11.1	11.7	12.3	12.3
Hagberg falling number	185	188 (±3.70)	196	144	156	174

Figure 6.2 Yields of winter wheat (tonnes per hectare) in the replicated plots and on a field-scale, in the period 1984–88.

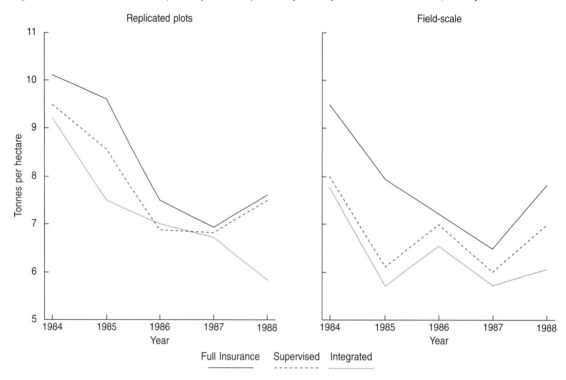

1984, 1985 and 1988 than in 1986 and 1987, whilst the Supervised treatment consistently outyielded the Integrated treatment. The patterns of yield were more consistent in the field-scale Supervised and Integrated treatments than in the plot-scale Supervised and Minimum Input treatments (Figure 6.2). This result can be explained by differences between the Integrated and Minimum Input treatments as the latter treatment received no insecticides or fungicides, irrespective of the incidence of pests or diseases.

Thousand grain weight. Year-to-year variations in thousand grain weight were similar in the replicated plots and on the field-scale (Figure 6.3). In the plots, the Full Insurance treatment gave a higher average thousand grain weight than the Supervised treatment, which in turn gave a higher figure than the Minimum Input treatment (Table 6.2). Both these differences were statistically significant (Analysis of Variance, $P < 0.01$ and $P < 0.001$ respectively).

Looking at individual years, thousand grain weights from the Full Insurance plots significantly exceeded those from the Supervised in 1984, 1985 and 1987 and bettered those from the Minimum Input plots in all years except 1984 (Analysis of Variance, $P < 0.001$ and $P < 0.01$ respectively). Thousand grain weights from the Supervised plots were similar to those from the Minimum Input treatment in 1984 and 1987 but better in 1985, 1986 and 1988 ($P < 0.001$, $P < 0.01$ and $P < 0.001$ respectively).

Grain bulk density. In the replicated plots, average grain bulk density was significantly higher in the Full Insurance treatment than in the Supervised ($P < 0.001$), which in turn significantly exceeded that in the Minimum Input plots ($P < 0.001$) (Table 6.2). This pattern was seen in all years except 1986 when the difference in grain bulk density between the Minimum Input and Supervised plots was not statistically significant, and 1987 when the Full

Figure 6.3   Thousand grain weight (g) at 85% dry matter in the replicated plots and on a field-scale, in the period 1984–88.

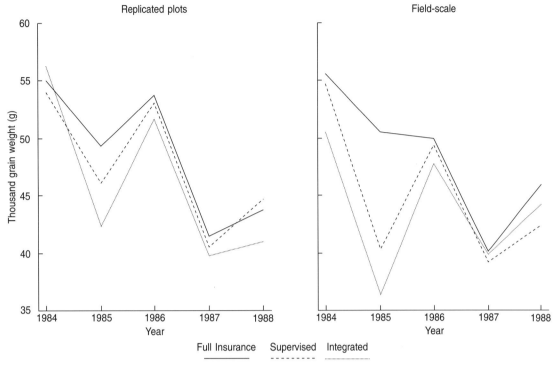

Figure 6.4   Grain bulk density (Kg/hl) in the replicated plots and on a field-scale, in the period 1984–88.

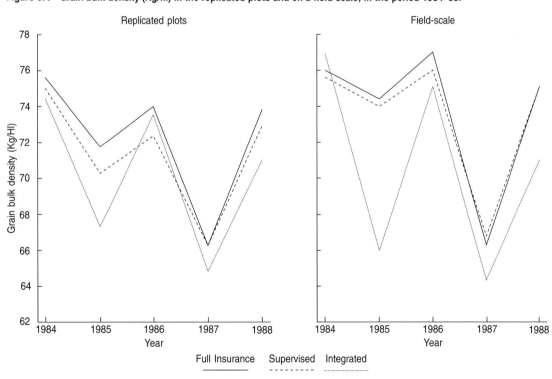

Insurance and Supervised plots gave identical grain bulk densities (Figure 6.4).

Apart from exceptionally high figures from the Full Insurance and Supervised treatments in 1985, field-scale grain bulk densities closely mirrored those obtained in the replicated plots (Figure 6.4).

Protein content. In the replicated plots the treatments had no overall effect on the protein content of the grain (Table 6.2) but there were statistically significant differences between years (P<0.001) (Figure 6.5).

On the field-scale there was an indication of a lower protein content in grain from the Full Insurance area when compared both with Supervised and Integrated fields, in 1984 and 1985. Year-to-year variations in protein content were similar to those observed in the replicated plot trial, except that the grain protein content was lower in the plots in 1986 than in 1987, whereas on the field-scale the result was the reverse (Figure 6.5).

Hagberg falling number. There were no overall statistically significant effects of the treatments on Hagberg falling number, the differences between the treatments being fully explained by chance variation, although there were considerable year-to-year differences (Figure 6.6). In the replicated plots the highest figures were recorded in 1988 and the lowest in 1987. On the field-scale the best Hagberg falling numbers were obtained in 1984 and 1988, the worst in 1987.

Economic appraisal. In order to compare the economic performances of the field-scale Full Insurance, Supervised and Integrated treatments, gross margins were prepared by ADAS Farm Management Advisers at Cambridge. The gross margin of an enterprise is defined as its enterprise output (ie the market value of the produce) less its variable costs. The main variable costs of crop production in the Boxworth Project were seed, fertilizer and pesticides. It is important to remember

Figure 6.5   Protein content of the grain (%) in the replicated plots and on a field-scale, in the period 1984–88.

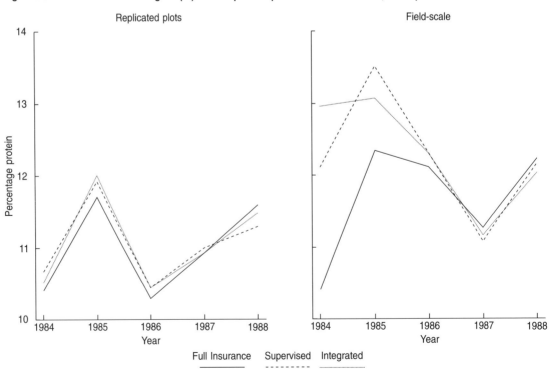

Figure 6.6 Hagberg falling numbers of grain in the replicated plots and on a field-scale, in the period 1984–88.

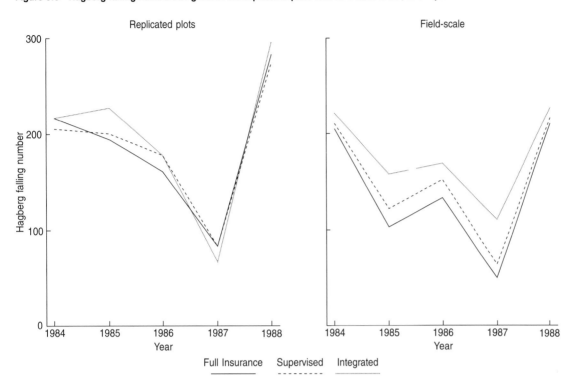

Full Insurance   Supervised   Integrated

that the gross margin is not a measure of profit since it takes no account of fixed costs such as rent, labour and machinery. Nix (1989) provides a full description of the calculation of gross margins.

Over the five-year treatment phase of the Project the average annual outputs per hectare were £818, £717 and £665 from the Full Insurance, Supervised and Integrated treatments respectively. The corresponding figures for variable costs were £364, £240 and £242, giving gross margins of £454 (Full Insurance), £477 (Supervised) and £423 (Integrated). Components of the gross margins calculations for wheat are given in Table 6.3.

Oilseed rape

Oilseed rape crops were grown only on the field-scale, in 1985 and 1987. The varieties were Jet Neuf in all three treatments in 1985 and, in 1987, Bienvenu in the Full Insurance and Supervised treatments and Mikado in the Integrated. In 1985

the Supervised and Integrated treatments out-yielded the Full Insurance by 0.5 t/ha while in 1987 the Full Insurance treatment gave the highest yield and the Integrated the lowest. On average, there were only small differences in yield between treatments (Table 6.4). However, although the Supervised and Integrated treatments gave similar gross margins (£613 and £625 respectively), both exceeded that from the Full Insurance treatment (£442) by a considerable sum.

Discussion

Despite the fact that Integrated and Minimum Input treatments were not identical, similar patterns in yields and grain quality measurements were obtained in the replicated plots and on the field-scale. This gives confidence that differences recorded as statistically significant in the replicated plot experiment are indicative of real field-scale differences due to the pesticide treatments.

Table 6.3 Economic appraisal of winter wheat production, 1984–88. Differences between the treatment areas are discussed in the text.

	Full Insurance	Supervised	Integrated
Yield (t/ha)	7.74	6.82	6.39
Value (£/t)	105.49	104.94	103.87
Total output (£/ha)	818	717	665
Variable costs (£/ha)			
Seed	48	39	55
Fertilizer: nitrogen	63	67	65
phosphate	18	18	18
manganese	1	–	–
Herbicide	121	66	54
Fungicide	68	40	37
Insecticide	41	6	9
Growth regulator	4	4	4
Total variable costs (£/ha)	364	240	242
Gross margin (£/ha)	454	477	423

Table 6.4 Economic appraisal of winter oilseed rape production in 1985 & 1987.

	Full Insurance	Supervised	Integrated
Yield (t/ha)	2.95	3.11	2.97
Value (£/t)	265	265	265
Total output (£/ha)	765	826	795
Variable costs (£/ha)			
Seed	17	20	21
Fertilizer: nitrogen	79	69	67
phosphate	17	19	19
Herbicide	103	99	49
Fungicide	39	1	9
Insecticide	23	–	–
Other*	45	5	5
Total variable costs (£/ha)	323	213	170
Gross margin (£/ha)	442	613	625

*includes a pigeon repellent in one year on the Full Insurance treatment

The mean field yields in the period 1974–1983 provide a base-line against which yields from the same fields in the period 1984–88 can be assessed (Table 6.5). Appreciably higher yields were seen in the latter period. This is to be expected owing to improvements in plant breeding, disease control and crop management in general which have come about in the last 15 years. There is no reason to believe that the yield differences in 1974–83 were related to differences in pesticide usage because at that time inputs to the prospective treatment areas were broadly similar. Data available from the 1974–83 period show that a difference of 0.57 t/ha between yields obtained in different treatment areas represents a statistically significant effect. In view of this, the main difference in 1984–1988, which was between the Full Insurance and Integrated treatments, is likely to be real.

Table 6.5   Mean field yields of winter wheat for the periods 1974–83 and 1984–88

Treatment in 1984–88	1974–83 (t/ha)	1984–88 (t/ha)	1984–88 as % of 1974–83
Full Insurance	5.82	7.74	133
Supervised	5.40	6.82	126
Integrated	6.00	6.39	107

In the replicated plot experiment, yields were higher than on the field-scale. This is often the case when comparisons are made between plots and whole field data, although the reasons for this are not clear.

The disappointing performance of the Integrated treatment needs to be discussed. With this treatment the objective was to obtain the best possible yields with minimum pesticide inputs. The cropping programme was the same as in the Full Insurance and Supervised areas but was combined with attempts to rely on cultivation rather than herbicides to control weeds, particularly grasses. This led to appreciable delays in sowing winter wheat in four of the five years, which undoubtedly reduced yields. So too did damage to soil structure by late cultivations in wet autumns, which resulted in poor plant establishment.

Disease control in the Integrated area relied mainly on varietal diversity in the early years and subsequently on the high levels of resistance (particularly to eyespot) of the variety Rendezvous. However, most of the varieties used produced relatively low yields and their grain rarely commanded a quality premium. It was primarily for this reason that the yields and gross margins in the Integrated area were appreciably below those in the other treatment areas, outweighing the potential for high performance which the prospective Integrated fields demonstrated in the years 1974–83.

Estimates of yield losses caused by the failure to fully control pests, diseases and weeds have been made by entomologists, plant pathologists and weed biologists (Appendix A). To these may be added yield losses due to the lower-yielding ability of most varieties grown in the Integrated area. When these estimates are totalled and added to the mean treatment yields recorded on the field-scale, the 'potential' yields are increased to 8.17 t/ha (140% of the average 1984–83 base-line yield) 7.25 t/ha (134%) and 6.94 t/ha (116%) for the Full Insurance, Supervised and Integrated treatments respectively.

Despite the overall yield increase that would be obtained by excluding these losses, differences between the treatment areas are still very similar to those given in Table 6.5. However, this does not mean that the yield differences in 1984–88 were entirely due to the considerable differences in pesticide inputs. The difference in yield in 1974–83 between the prospective Full Insurance and Supervised areas was 7.8%, and was probably due to an inherently lower-yielding ability of the Supervised fields. This difference was 13.5% in 1984–88, and the balance of 5.7% is probably attributable to the difference in pesticide regimes.

Table 6.3 lists the input and output costs of the three pesticide regimes. The differences in outputs were due almost entirely to yield, with very little variation in grain quality. Differences in seed costs were due to the use of a broader-spectrum fungicide dressing in the Full Insurance treatment and the use of disease-resistant varieties in the Integrated treatment area, both of which were more expensive than the seed used in the Supervised treatment. Similar amounts of nitrogen, phosphate and manganese fertiliser were used in each treatment.

Although output from the Full Insurance treatment exceeded that from the Supervised by an average of £101 per ha, the variable costs of the former were £124 higher. As a result the Supervised treatment had a gross margin £23 per ha higher than that from the Full Insurance treatment. The variable costs of the Integrated treatment were similar to those of the Supervised. Although there were some savings in pesticide costs, seed costs were appreciably higher because some of the varieties grown for their disease resistance were new to the market.

The costs of the 'crop walking' which was essential for decision-making in the Supervised and Integrated treatment areas must also be taken into account. Direct costing of the time actually spent walking crops does not provide the appropriate figure because ADAS monitoring staff also collected information not directly related to the Project. Costs of field walking vary with the frequency of visits as well as with time spent in the field on each visit but in East Anglia the current (1990) average figure is £10 per ha (source: ADAS Business Management Advisory Service, Cambridge).

The three margins adjusted for crop walking charges therefore become £454, £467 and £413 per ha for the Full Insurance, Supervised and Integrated treatments respectively. The results show clearly that there was no reduction in margin with a Supervised rather than a Full Insurance pesticide regime; indeed, a small increase was obtained. However, the margin from the Integrated treatment was £48 per ha (10%) lower than the mean result of the other two treatments.

The results obtained from oilseed rape in the two years in which it was grown can be dealt with more briefly, being based on only six field-years compared with thirty-nine for winter wheat. These results indicate that low input treatments are likely to give better gross margins, due chiefly to reduced pesticide costs. However, the results need to be treated cautiously because the differences were not consistent. In 1985 the Supervised and Integrated treatment areas gave almost identical yields and both outyielded the Full Insurance treatment by 20% (0.53 t/ha). In 1987 the Supervised treatment yielded 10% (0.27 t/ha) more than the Integrated, and the Full Insurance treatment yielded 7% (0.22 t/ha) more than the Supervised. However, in both years the Integrated treatment gave a slightly higher gross margin than the Supervised and the latter gave a much higher gross margin than the Full Insurance treatment.

Summary and conclusions

Over a five-year period on the clay soil at Boxworth EHF, crop performance demonstrated that although very high inputs of pesticides to winter wheat (Full Insurance treatment) usually led to the highest yields, they were not necessary to obtain the highest financial benefits. Results from a Supervised treatment, in which pesticides were applied only when a clear need for them had been identified, showed that although yields were lower, gross margins were improved when compared to those in the Full Insurance treatment. After allowing for the additional costs of crop monitoring, there was still a small advantage in margin for this treatment.

In the Integrated treatment area attempts to reduce pesticide usage still further led to problems with grass weeds. Both yields and gross margins were appreciably lower than in the other two treatment areas and it seems unlikely that this treatment would produce acceptable results in an intensive cereal rotation on heavy land.

The general conclusion is that high inputs of insecticides or herbicides are unlikely to be required in a well-managed crop. However, a farmer should be aware of the disease susceptibility of the varieties he is growing and be prepared to take action as necessary.

Acknowledgements
I am grateful to Dr. D. J. Pike, of the Department of Applied Statistics, University of Reading, for advice on the statistical treatment of crop yield and quality data; to Mr. W. J. Ridgman, formerly of the Department of Applied Biology, University of Cambridge, for the calculation of mean wheat yields and standard errors 1974–83; to Fengrain Ltd., Wimblington, Cambs. for the determination of grain quality parameters; to Mr. P. Skeels, Boxworth EHF for help in computation as well as day-to-day work on the Project; to Mr. K. Butterworth, ADAS, Cambridge for the computation and interpretation of gross margins; and to Mr. L. H. Webb, formerly Farm Manager at Boxworth EHF.

Reference
Nix, J. (1989). *Farm Management Pocketbook*, 19th Edition. Department of Agricultural Economics, Wye College.

Distribution and persistence of sprayed pesticides, and residues in drainage water

Peter W Greig-Smith * *, **David Eagle** #, **Charles T. Williams**† and **Peter M. Brown** *(MAFF Central Science Laboratory, * *Worplesdon and *Tolworth, #National Pesticide Residues Unit, Cambridge, and †Institute of Terrestrial Ecology, Monks Wood)

Introduction

The environmental risks of pesticide use depend on the type of chemical, how much is applied, and where it goes. In theory, correct application methods should result in most of the active ingredient reaching its target. In practice, however, there are likely to be losses at several stages, owing to imperfections in application machinery, weather conditions that promote drift of droplets or vapour, accidental spillage, and so on. Even if these losses are minimized or eliminated, it is still important to know how much of the pesticide is present on various parts of the vegetation or the ground, and for how long after application. This provides an essential background for understanding the exposure of wild animals and plants to a potentially hazardous chemical.

In the Boxworth Project it was not feasible to monitor the fate of all the pesticides applied. Two insecticides that were considered likely to affect some animals were chosen for detailed study; these were triazophos (sprayed against yellow cereal fly in early spring) and demeton-S-methyl (sprayed against aphids in mid-summer). Spray applications were examined because of the potential for heavy losses through drift.

In addition, it was decided to examine samples of ground water from fields at the end of the treatment phase of the Project. This was undertaken to indicate whether or not there was any accumulation of residues as a consequence of the sustained high inputs of the Full Insurance regime.

Spring applications of triazophos

Triazophos was applied routinely in the Full Insurance area to control yellow cereal fly, usually in late February or early March. Studies were undertaken in 1984 to investigate spray drift at the time of application.

Sampling was carried out by the Operator Protection Group (OPG) at the Central Science Laboratory, Harpenden. A variety of 'targets' was placed in and around the hedgerow on Backside field on the day of triazophos application (1 March 1984). These included pieces of stainless steel wire or mesh, and polythene tubing, from which it was known that residues of triazophos could be efficiently recovered. After the field was sprayed, the targets were collected, and analysed for the presence of residues.

The results require complex calculations to determine the deposition of spray per unit area of the target surfaces. The outcome is summarised in Table 7.1, which demonstrates that there was consistently less than 0.35 μg of triazophos per cm^2. These figures can be interpreted as estimates of the volume of spray drift flowing at different parts of the hedge profile (Figure 7.1). It appears that there was very little difference between windward and leeward sides of the hedge, suggesting that the

Table 7.1 Distribution of spray drift on and around a hedge at the boundary of Backside field when the field was treated with triazophos in March 1984.

	Average spray deposit (μg triazophos/cm^2)		
	Windward side	Centre	Lee side
Outside the hedge	0.25	–	0.16
Within the hedge	0.16	0.27	0.33
On top of the hedge	0.06	0.04	0.01
On the ground	0.04	0.02	<0.01

Figure 7.1   Distribution of spray drift of triazophos in a hedgerow adjacent to Backside field when triazophos was applied in March 1984. Figures show ml of spray per m² of hedge cross-section. Arrows indicate the direction of drift.

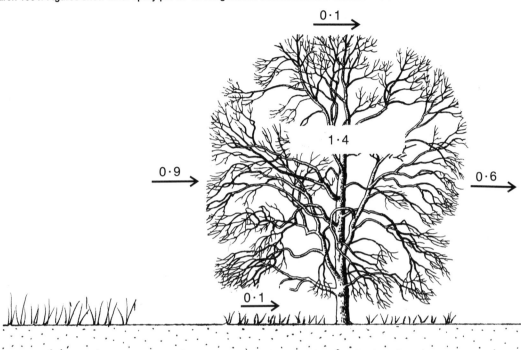

plants provided little resistance to drift when free of leaves at the end of the winter. However, drift was greater at hedge height (approximately 0.3–1.5 m) than at ground level or above the top of the hedge. This indicates that the spray cloud was travelling consistently at the height of its release from the sprayer.

Previous studies of drift by the OPG have established drift profiles for a range of application equipment operated under different meteorological conditions. From these data it is possible to calculate an expected volume of drift, and the amount deposited on objects such as hedges. It is necessary to make particular simplifying assumptions, but in this case the agreement between the predicted level of contamination of the hedge (1.015 ml per m²) and the measured value (0.9 ml per m²) was impressive.

This study indicates that a substantial amount of the triazophos applied to the field in March was lost in drift beyond the crop margin. In view of the fact that drift was hardly diminished by passage through the hedge at that time of year, it seems likely that any adverse effects of this contamination on hedgerow flora and fauna might affect the entire field margin. Because of the rapid loss of triazophos from the crop plants, it was not considered worthwhile to measure the decay of residues in field margin samples. However, one possible effect was examined, by analysing earthworms collected from the hedgerow after the field was sprayed with triazophos. This revealed a detectable residue of triazophos (0.14 mg per kg) in only one of seven samples analysed, indicating a potential, but not substantial risk.

In 1985, a study was carried out to determine the persistence of triazophos on the treated crop after spraying. This study was in Pamplins North field, where spraying of the experimental area occurred at midday on 9 March. Samples of 20 wheat plants from each of three rows, 10 m apart, were analysed to detect the presence of residues, on several occasions up to 24 days after spraying. In addition, strips of absorbent paper tape were placed 10 m apart between the rows of plants before spraying, and were collected later the same day for analysis. The method of analysis, using gas chromatography, was shown to be able to detect at least 96% of residues, on both wheat and paper.

The concentration of triazophos in the sprayed formulation would give an expected distribution of 63 mg per m². The measured residues on paper strips at the time of spraying were lower than this, averaging 38 mg per m². Residues on the wheat plants were substantial on the day of spraying, but declined rapidly, to less than 1 ppm at 17 days after spraying (Figure 7.2). This rapid fall is unlikely to have been caused by heavy rain; average daily rainfall for the first ten days after spraying was only 0.7 mm per day. These results demonstrate clearly that triazophos was lost rapidly after it was deposited on the fields.

Applications of summer aphicides

In most years there were two applications of demeton-S-methyl to all the fields in the Full Insur-

ance area, and occasionally to some in the Supervised and Integrated areas, to control populations of aphids. Residues on the crop were assessed in 1984, using strips of absorbent paper as described above, and in 1985 by direct analysis of wheat plants. The mean residue on paper strips was equivalent to 4.2 g per hectare (which is about 3.5% of the predicted deposition) and it is likely that demeton-S-methyl evaporated from the strips before they were collected from the field. Residues on plants in 1985 declined to 5% of their initial value within 24 hours after spraying, and after this were barely detectable.

No detailed direct measurements of demeton-S-methyl spray drift were attempted, but a study was carried out to examine possible effects of drift on hedgerow invertebrates. This work was undertaken in 1986, by staff from the Institute of Terrestrial Ecology at Monks Wood Experimental Station. Their results will be published in detail elsewhere, and only a summary is presented here. The study was part of a broader investigation of spray drift problems, and was also used to examine

Figure 7.2 Average residues of triazophos detected on winter wheat plants in 1985 up to 24 days after spraying.

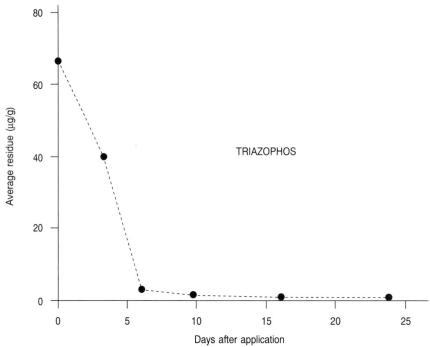

a field bioassay technique (developed by Sinha *et al.*, 1990), involving larvae of the large white butterfly (*Pieris brassicae*).

For the first demeton-S-methyl spray, applied to the Full Insurance area on 27 June 1986, the hedge on the west side of Grange Piece was monitored and for the second spray, on 14 July, the hedge on the east side of Backside was used. Both were matched with 'control' hedges of similar type, located next to unsprayed wheat fields, outside the Full Insurance area.

For the first spray, D-vac samples were taken from the hedge-bottom vegetation on the sprayed field side, shortly before, and on three occasions after spraying (at 5 hours, 3 days and 10 days); for the second spray, samples were taken at 1½ hours before, 22 hours after and 10 days after application. Pitfall traps were also used to monitor the second spray event.

The *P. brassicae* bioassay was conducted at the second application, and involved placing potted kale plants in the crop and on both sides of the downwind hedge. On the near side, pots were 25–50 cm from the end of the sprayer boom, and were 20, 65 and 110 cm above ground level; on the far side, pots were c.2.5 m from the sprayer and 65 cm above ground.

Each kale plant had 20 larvae on one leaf. There were four replicates of each treatment, spaced at 50-pace intervals along the hedge. Eight control replicates were kept outdoors, well away from the sprayed field. Mortality was assessed seven days after spraying.

Residual toxicity was tested by moving groups of unexposed larvae onto plants five hours and seven days after the plants had been exposed to the spray at ground level on the nearside of the hedge.

Table 7.2 summarises the results of D-vac sampling, based on the numbers of arthropods caught in the treated fields relative to those caught in the untreated control fields, both corrected for changes from the pre-spray levels.

Despite light winds at the time, total numbers of invertebrates collected after five hours in

Table 7.2 Percentage changes in post-spray D-vac samples of arthropods from treated (T) and control (C) hedges, relative to their pre-spray levels.

(a) First demeton-S-methyl spray (27 June 1986)

	After 5 hours		After 10 days	
	C	T	C	T
Total Insects	111%	41%	195%	75%
Total Arachnids	139%	66%	1547%	61%
Total Arthropods	113%	44%	282%	74%

(b) Second demeton-S-methyl spray (14 July 1986)

	After 22 hours		After 10 days	
	C	T	C	T
Total Insects	111%	63%	74%	82%
Total Arachnids	66%	44%	64%	125%
Total Arthropods	104%	58%	73%	91%

hedges immediately downwind of the sprayed fields were reduced to 44% of the pre-application level after the first spray and to 58% after the second. Total catches from control hedges showed a small increase at the same time. All groups of invertebrates showed much variation in numbers within hedges, so that many of the apparent changes must be interpreted with caution. However, insects appeared to suffer more than arachnids. The group most affected was the Hymenoptera, predominantly small Parasitica, which were reduced by 89% on the first and second sprays.

Samples taken three days after the first spray showed that numbers of most taxa were already increasing in the treated hedge. The rapid initial recovery suggests that reinvasion from unsprayed areas immediately adjacent (especially the roadside verge) was mainly responsible, rather than recruitment from within the sampled area. Ten days after the first spray, the total insect catch had partially recovered, whereas 10 days after the second spray, recovery of numbers was apparently complete, although the taxonomic composition of the samples was different.

For both Diptera and Hymenoptera, small species (<2 mm) were more severely reduced initially, and recovered more slowly than large ones, possibly because of their greater surface area: volume ratios and slower rates of immigration into fields. Parasitic Hymenoptera were the most severely affected, probably owing to their high physiological susceptibility to organophosphorus compounds, and their habits of actively searching for hosts on the foliage, where they would be exposed to the spray deposit. Despite demeton-S-methyl's systemic properties, populations of sap-sucking Hemiptera were not markedly reduced. This may be because initial concentrations deposited in the hedge-bottom soil were too low for effective systemic action. Translocation within the plant may then have diluted the insecticide to sub-lethal doses.

Table 7.3 presents the results of pitfall trapping in two periods after the second spray. The data appear to be less reliable than the D-vac samples, because there were large changes in the numbers of most of the main invertebrate groups in the control hedge.

The pitfall-trap catch for the treated hedge showed a greater decline relative to the control in the four days after spraying than did the D-vac sample one day after the spray. The most severely affected of the main ground fauna were Isopoda (woodlice), Carabidae (ground beetles) and Araneae (spiders), catches of which were reduced by over 50% compared to the control hedge.

Table 7.3 Percentage changes in post-spray pitfall-trap catches of arthropods (based on the mean number per trap per day), in the treated (T) and control (C) hedges, relative to their pre-spray levels, for the second demeton-S-methyl spray.

	1–4 days after spray		5–10 days after spray	
	C	T	C	T
Total Insects	124%	104%	41%	46%
Total Arachnids	183%	84%	89%	59%
Total Arthropods	137%	91%	49%	57%

Figure 7.3 Survival of larvae of the large white butterfly on kale plants placed in and around a hedgerow during spraying of a wheat crop with demeton-S-methyl. Values show percentages on the treated site. In an untreated control field survival was 88%; statistically significant differences are indicated by asterisks.

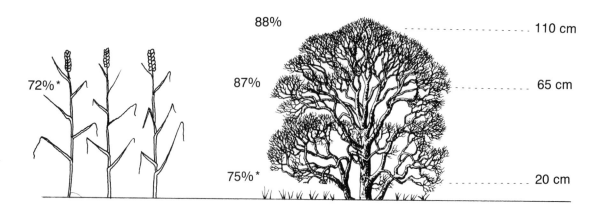

Mortality in the *P. brassicae* bioassay is shown in Figure 7.3. The larvae were relatively resistant to demeton-S-methyl; even those exposed at crop height within the treated field showed over 70% survival. Larvae exposed low down in the hedge had slightly higher survival than those within the crop, while those placed higher in the hedge, or on its far side, had comparable mortality to the control.

The large differences between treatments at different heights suggest that invertebrates on herbaceous vegetation in the hedge bottom are more vulnerable to short-range drift than those at and above crop height; those on low vegetation receive nearly the full field dose. This may be because some large droplets from the outermost nozzle miss the crop and continue downwards into the hedge bottom. Therefore, an increase of as little as a metre from the outer nozzle to the crop edge should greatly reduce mortality of invertebrates in the hedgerow.

When unexposed larvae were transferred, five hours after spraying, to plants from the hedge bottom, mortality was similar to that of the control larvae, demonstrating that residual toxicity to *P. brassicae* was negligible after a few hours.

Overall, this experiment demonstrates that short-range spray drift can adversely affect important invertebrate groups in downwind hedge-bottom vegetation, but in this case recovery was rapid, by immigration from nearby unsprayed areas. The results are not completely consistent with those obtained for triazophos, which showed very little drift reaching the hedge bottom (Figure 7.1). This demonstrates the caution needed in interpreting data from different stages of crop growth and different weather conditions.

Pesticide residues in drainage water

In order to detect any major differences between the high-input and lower-input areas that might have built up in the treatment years of the Project, samples of water were taken from drain outlets in Extra Close East (Integrated area) and Grange Piece (Full Insurance area). Sampling started when the drains began to run in autumn 1987 and continued at approximately two-week intervals until flow ceased in the spring. Sampling was resumed in autumn 1988 and continued until late December 1988.

These water samples were delivered to the ADAS National Pesticides Residues Unit at Cambridge, where they were analysed for residues of the pesticides applied during the 1987–1988 autumn and winter, and during autumn 1988. Analyses were by gas-liquid chromatography except for isoproturon and metoxuron, which were determined by high performance liquid chromatography (HPLC). Tri-allate was confirmed by mass spectrometry at the Central Science Laboratory, Harpenden. Metoxuron was analysed by both normal and reverse-phase HPLC.

The results of the analyses are summarised in Table 7.4. No residues of the molluscicide methiocarb, nor of the insecticide cypermethrin were detectable. The only other insecticide involved was chlorpyrifos, of which a trace (less than 0.1 μg/litre) was detected in only one sample, 12 days after application.

Residues of several herbicides were detected (Table 7.4, Figure 7.4). Tri-allate was detected in four of nine samples, but at very low levels, the highest being 0.25 μg/litre.

Metoxuron was detected in all samples obtained from Grange Piece during the 1987–88 winter. Amounts ranged from 3.3 μg/litre in the first sample taken about six weeks after application down to 0.06 μg/litre in the last sample taken in early March.

Isoproturon was detected in the two samples from Grange Piece taken following its application in March 1988. The amount was 2.16 μg/litre nine days after application and this had dropped to 0.32 μg/litre two weeks later.

Simazine was detected in the three samples from Extra Close East following application on 31 October to the 1988–89 winter bean crop. The level was 35 μg/litre in the first sample

Table 7.4   Detection of pesticide residues in drainage water from Extra Close East (Integrated area) and Grange Piece (Full Insurance area) in the winters of 1987–88 and 1988–89.

	Date of application	Residues detected
Extra Close East		
methiocarb	5 Nov 87	Not detected (limit of detection 0.01 μg/litre)
simazine	31 Oct 88	0.39–35.0 μg/litre
Grange Piece		
methiocarb	14 Oct 87	Not detected (limit of detection 0.01 μg/litre)
tri-allate	26 Oct 87	0.05–0.25 μg/litre
metoxuron	7 Nov 87	0.06–3.30 μg/litre
cypermethrin	7 Nov 87	Not detected (limit of detection 0.1 μg/litre)
chlorpyrifos	5 Dec 87	0.06 μg/litre
triazophos	18 Feb 88	Not detected (limit of detection 0.1 μg/litre)
isoproturon	14 Mar 88	0.32–2.16 μg/litre
methabenzthiazuron	1 Nov 88	0.07–0.40 μg/litre

Figure 7.4   Drain water residues of pesticides recorded in Grange Piece in the Full Insurance area, and Extra Close East in the Integrated area.

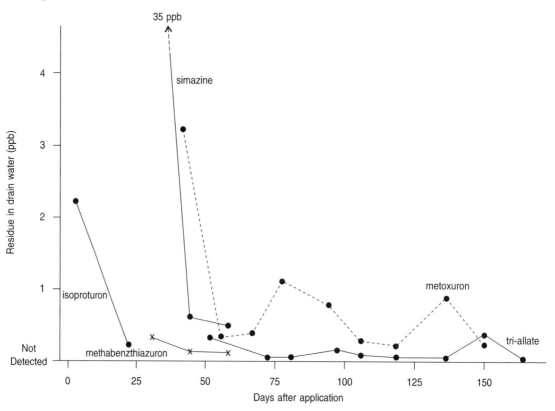

taken one month after application but was very much lower at around 0.4 μg/litre in the subsequent two samples.

Methabenzthiazuron was detected in the three samples from Grange Piece following application on 1 November 1988. The level was 0.4 μg/litre in the first sample taken four weeks after application but was less than 0.1 μg/litre in subsequent samples.

These results demonstrate great variation in the appearance of pesticides in drainage water, both in comparisons of different chemicals, and in the patterns of decline over the sampling period. Generally, residues of insecticides were almost absent, whereas herbicides were detectable at levels significantly above the EC drinking water limit in the first sample taken after application, declining slowly thereafter. The only pesticide to reach a level considerably above this limit was simazine, for which preliminary results of comparable studies elsewhere indicate short-term peaks.

The number of pesticides monitored in the Integrated area field was too few to allow any comparison between the high- and reduced-input regimes.

Although initially the residues of simazine in the drainage water were high, the available evidence suggests this to be a transient effect. The possibility that such residues might contaminate surface and ground water bodies exists, but the implications of this are currently unknown.

ment by immigration from unsprayed areas. A bioassay using larvae of the large white butterfly showed that mortality was greatest in the crop and in the hedge bottom.

Residues of pesticides were measured in drainage water during the final year of the Project. Residues of insecticides applied to the fields were almost absent, whereas herbicides were generally detectable only in the first sample taken after application. The only substantial levels recorded were for simazine, which showed a transient peak of 35 μg/litre, 30 days after application.

Reference

Sinha, S N, Lakhani, K H & Davis, B N K (1990). 'Studies on the toxicity of insecticidal drift to the first instar larvae of the Large White butterfly *Pieris brassicae* (Lepidoptera: Pieridae).' *Annals of Applied Biology* **116**, 27–41.

Summary

Measurements were made of the distribution and persistence of pesticide residues, for selected spray applications in the Full Insurance area at Boxworth. Residues of triazophos deposited on the crop in early spring, and of demeton-S-methyl in mid-summer, were lower than expected from the application rate, and were lost rapidly over the succeeding few days. Spray drift of demeton-S-methyl reduced the catches of arthropods in an adjacent hedgerow, although there was replace-

Patterns of distribution of plant species in the fields and their margins

8

Jon Marshall (Long Ashton Research Station)

Introduction

The flora of cereal fields and their margins is important for many kinds of animals. Little was known at the start of the Boxworth Project about the effects of agricultural practices on the flora, or on the dispersal of wild plants into the crop area.

The field margins at Boxworth, as defined by Greaves & Marshall (1987), are typical of those found on intensively-farmed arable land throughout eastern England, comprising low thorn hedges, woodland edges and the grassy verges of tracks or roads.

Surveys of the flora at Boxworth were made in the cereal fields and along 10 km of hedgerows to investigate the influence of boundary flora on weed populations and the consequences of herbicide use. For example, plants in a hedgerow might be adversely affected by drift of herbicide sprays from the adjacent crop. Headlands (crop margins) were also studied to examine the spread of hedgerow plants into fields. Dispersal from boundaries is thought to be a major route of ingress into crops for wild plants, some of which may be serious weeds. This might be influenced by herbicides, either directly, or through changes in the hedgerow flora.

In addition to studies of seedlings and mature plants, the 'seed bank' (viable seeds in the soil) was examined. This can give an indication of species' reproductive potential and dispersal ability, and should reflect the effects of herbicides through the amount of viable seed produced each year.

Broad-leaved plants and grasses, and annuals and perennials, were considered separately. These groups, which show broad ecological differences, and received different treatments (Chapter 4), might exhibit different responses to herbicides.

Floras of the fields, their headlands and boundaries were compared in the three treatment areas using statistical techniques to identify any patterns which might be due to differences in the herbicide programmes.

Seed banks in the soil

A series of soil samples was taken from 50×100 m areas within fields in order to assess changes in weed seed populations. Areas were selected in the weediest patches of fields, identified in quadrat assessments made in 1982 and 1983.

Soil samples were taken annually after harvest for five years, starting in September 1984, using a 2.5 cm diameter corer to a depth of 20 cm. Sampling followed the method of Roberts & Chancellor (1986). Six transects and eight sampling locations along each transect were selected at random. Each batch of eight cores was bulked and placed in a shallow earthenware pan and kept moist in an unheated glasshouse. Germinating seedlings were removed and identified and the soil was stirred periodically, following the method of Roberts (1981). The pans were kept for two years to allow two cold periods to break seed dormancy.

The results of seed bank assessments are summarised in Figure 8.1, which shows seed densities, excluding seeds of oilseed rape and wheat. In fields following oilseed rape, seed density after harvest increased markedly in all but one field. The number of species in soil samples also increased in some fields after rape, but not in all. The large increase in seed density in Pamplins North in 1987 was entirely due to chickweed (*Stellaria media*). Analysis of seed density in each field over the five-year treatment period showed significantly less viable seed in the Full Insurance area (Analysis of Variance with logarithmic transformation, $P < 0.05$) though there was marked field-to-field variation within each area.

Numbers of species in each field differed significantly between years, but not between areas.

Figure 8.1 Densities (per m²) of viable seed from 0.5-ha areas within each field at Boxworth, between 1984 and 1988.

Integrated area

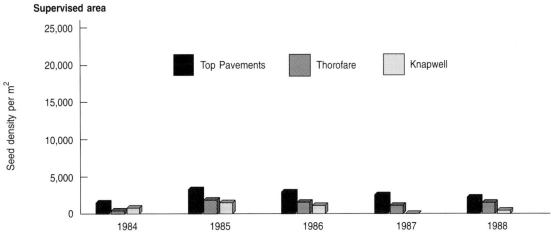

Supervised area

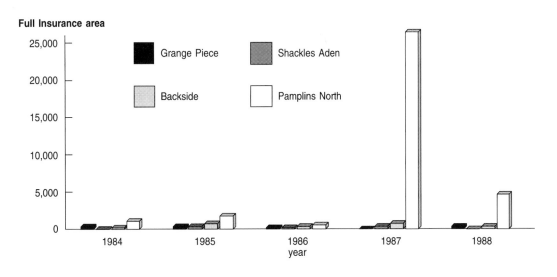

Full Insurance area

Table 8.1 Average numbers of plant species represented in the seed banks of Boxworth fields.

The standard error of the difference between means is 1.72 (8 degrees of freedom).

Treatment	1984	1985	1986	1987	1988	Mean
Integrated	5.5	10.0	8.0	8.3	6.8	7.7
Supervised	6.3	10.0	11.7	7.7	7.7	8.7
Full Insurance	3.5	6.8	5.8	4.3	5.8	5.2

Seed banks in the Supervised and Integrated areas tended to contain more species than the Full Insurance area (Table 8.1). Although no data were collected before 1984, the first treatment year, it is likely that the trend existed at the outset of the experiment.

In order to examine the composition of the plant seed communities in more detail the data were analysed by two-way indicator species analysis (TWINSPAN) and de-trended correspondence analysis (DECORANA) (Malloch, 1988). These are multivariate statistical techniques which divide the sampling sites (fields) into groups according to similarities and differences in their plant communities. Detailed results from these analyses will be published elsewhere.

The analyses failed to show any clear evidence of differences related to the pesticide regimes. However, most samples with large numbers of grass weed seeds came from the Integrated and Supervised areas, reflecting the efficiency of grass weed control in the Full Insurance fields.

Densities of viable seed at Boxworth were generally low in comparison to many published figures (Roberts, 1981). This perhaps reflects the heavy clay soil conditions, since wetter soils encourage fungal diseases and the death of seeds (Harper, 1977).

Plant distributions at the field boundaries

Wild plants at the edge of arable fields are likely to include arable weed species, together with species characteristic of the relatively undisturbed hedge bottom and field boundary. Initial studies at Boxworth and elsewhere have identified several patterns of distribution of individual species (Marshall, 1985). Further investigations at Boxworth were made to map the occurrence of seed and seedlings of boundary plants, and to study their dispersal into the crop.

Five sites were chosen for intensive study of plant distributions. These were in Extra Close East, Knapwell and Grange Piece (coinciding with intensive insect study sites) and also in Thorofare and Top Pavements fields (studies of the Top Pavements site were started after the hedgerow was accidentally burnt on 19 August 1983). Plant surveys were carried out in November, March and July each year. Counts were made on a sampling grid based on a particular 50 m length of hedgerow. Grid points were parallel to the hedge at distances of 0 m, 5 m, 10 m and then at 10 m intervals, up to 100 m into the crop. In autumn and spring, plants were counted in quadrats at each grid point. In summer, grass inflorescences were counted at each point and broad-leaved species were given an abundance score on a scale of 1 to 3.

The distribution of plants close to the hedgerow was examined in greater detail by counting plants in contiguous 0.01 m squares in autumn and spring. In summer, the presence of any vegetative part was recorded in each square. Fixed markers allowed a measure of how the width of uncultivated ground close to the hedge varied from year to year.

If plants spread into crops from boundaries such as hedgerows they might be expected to be more abundant in the headlands (field margins close to the boundaries) than in open fields. However, the distribution of weeds in a crop is likely to be influenced by the herbicides applied there. Populations of grasses and broad-leaved weeds were examined to ascertain whether there was a correlation between the floras of the fields and headlands and their boundaries, which might give an insight into the effects of herbicides on the distribution of weeds.

Figure 8.2   Plant densities (per m²) in Top Pavements field from within the boundary to 100 m into the crop in December 1987 and March 1988.

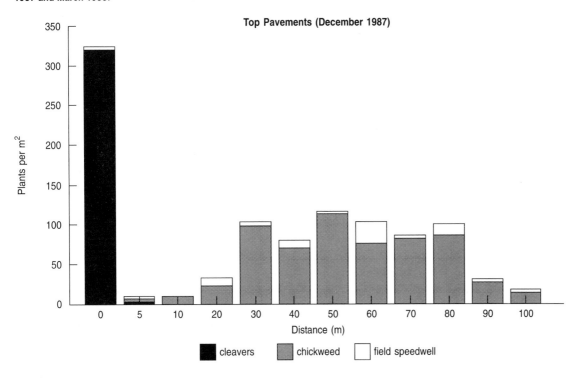

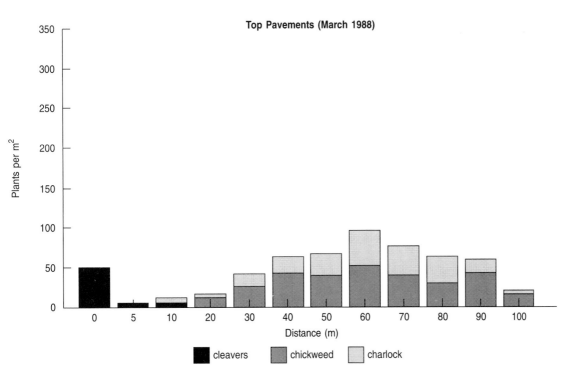

The densities of plants of several species at different distances into the crop in Top Pavements field in one autumn and the following spring are compared in Figure 8.2. Plant densities declined over winter, particularly in the headland, though charlock was only present in spring counts. The density of cleavers was high only in or close to the hedge. The annual weed flora of the field, represented by chickweed, common field speedwell and charlock, was similar to that of the headland but was usually not recorded in quadrats in the hedgerow.

A comparison of the densities of grass inflorescences in July at different distances into the crop was made for each field in each year. In general, densities of barren brome (*Bromus sterilis*) in the headland declined between 1984 and 1988, while those of meadow brome (*Bromus commutatus*) increased. Changes in boundary populations of grasses appeared to be poorly correlated with changes in grass weed populations within the main field area. Nevertheless, populations of brome grasses within the 3 m of crop closest to the field edge were occasionally high and strips of crop 2 m wide were mown or rotovated in June 1986, 1987 and 1988 to reduce the risks of seed being spread into the fields.

Within five metres of the field boundary, there were marked differences in the distributions of a number of species, as reported by Marshall (1989). In autumn and spring, total weed density (broad-leaved species and grasses) was highest at the field boundaries and negligible at 5 m (Figure 8.3). Numbers of plant species varied between fields, the Knapwell field edge being the most diverse. This site bordered a ditch and felled woodland, while sites in other fields bordered thorn or mixed hedges. Plant diversity was often greatest in the first metre of cultivated ground, where both hedge and field species occurred.

Most plant species typical of the boundary were absent from the crop beyond 5 m and only a quarter occurred in the crop as weeds. Data for 1985 from the five intensive study sites are shown in Table 8.2. In all sites, there were fewer species common to hedge and crop than the number limited to the hedge. Two important weeds found in the hedgerow that were also common in the crop are barren brome and cleavers.

Occurrence of buried seed banks relative to hedgerows

The soil seed bank was examined in four fields, using soil cores within the hedge bottom and at distances up to 100 m into the crop, in January 1984 and in each September from 1985 to 1988. Cores were 2.5 cm in diameter and 20 cm deep. At each distance, 40 cores were taken and bulked into five samples, which were then placed in shallow clay pans in an unheated glasshouse and watered regularly over a two-year period. Germinating seeds were identified and removed, and the soil was stirred periodically.

The density of viable seed and the number of species recorded in the seed bank at different distances from the hedgerow are shown in Figure 8.4. Seed densities in the hedge bottom were much higher than in the adjacent crop area, probably reflecting the effects of regular tillage and herbicide applications within the field. Numbers of species represented in soil cores were also greater in the hedge than the field, generally declining with distance from the field boundary. Patterns in Thorofare were not as clear as in the other fields, showing larger seed banks at 25 m and 100 m into the crop, which may reflect high production of weed

Table 8.2 **Numbers of wild plant species found only in the hedge, in both hedge and crop, and only in the crop, in five fields at Boxworth in 1985.**

Field	Hedge only	Crop & hedge	Crop only (>5 m)	Totals
Extra Close East	14	10	4	28
Knapwell	38	6	4	48
Grange Piece	18	3	4	25
Thorofare	12	4	14	30
Top Pavements	12	7	13	32

Figure 8.3 Total densities of wild plants (per m²) and numbers of species from within the boundary to 5 m into the crop in five fields in November 1984.

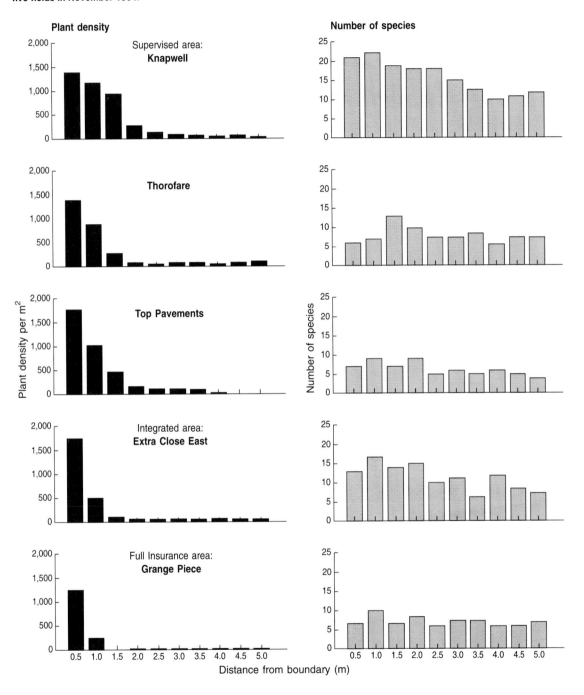

Figure 8.4   Density of viable seeds in the soil (per m²) and numbers of species in the seed bank in the hedge bottom at three distances into the crop in four fields, averaged over four years (1984 to 1988).

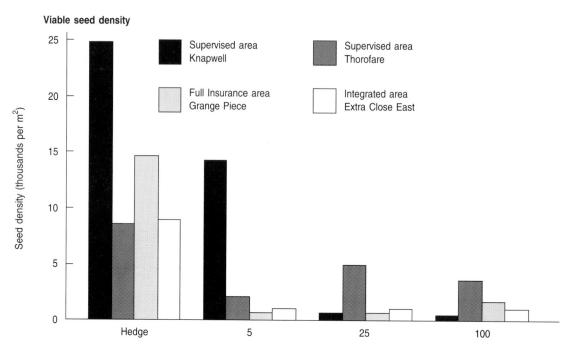

Viable seed density

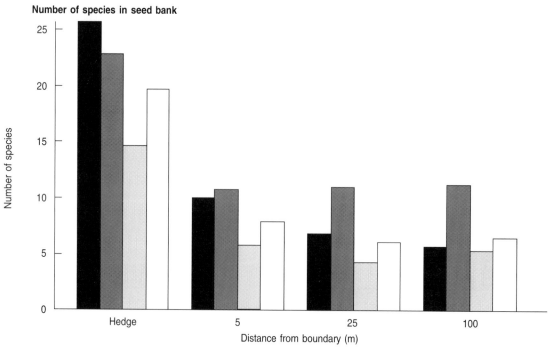

Number of species in seed bank

Figure 8.5 Examples of plant distribution patterns from the hedgerow to 5 m into the crop in Extra Close East field in 1984: a) limited to the boundary, b) usually limited to the crop, c) spreading from the boundary and d) at greatest density in the crop edge.

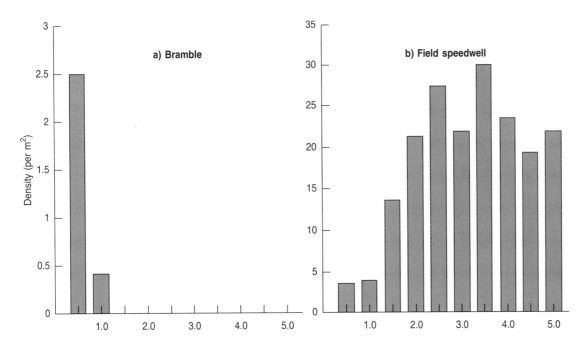

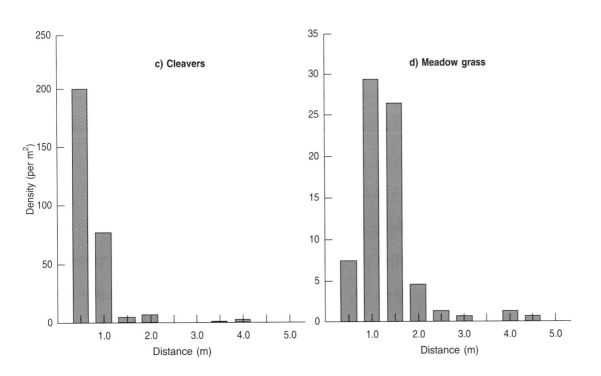

Figure 8.6   Seed and seedling densities (grasses and broad-leaved species) in the hedgerow and at three distances into the crop in four fields, averaged for 1985, 1986 and 1987.

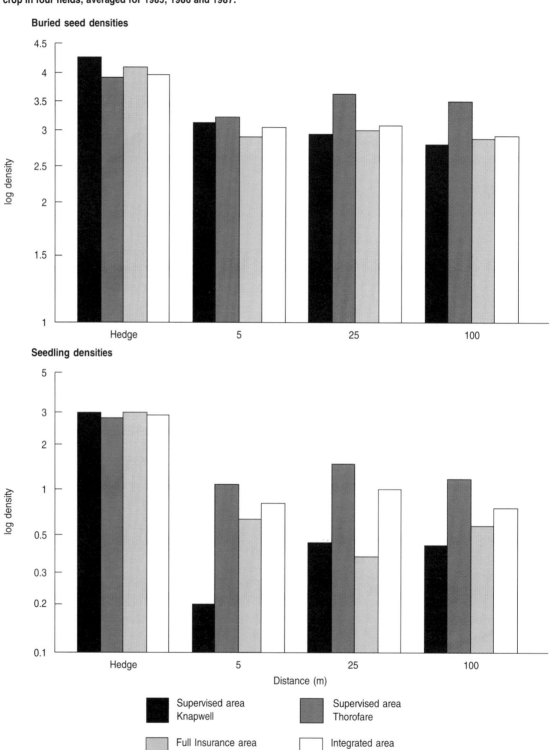

Buried seed densities

Seedling densities

seed after oilseed rape was grown during the 1985 harvest year, whereas the other three fields had continuous winter wheat crops.

The species composition of the seed bank varied with distance into the crop. Perennials were particularly poorly represented in the crop, the principal exception being couch grass, which regenerates from rhizome fragments and was notably abundant in the samples taken at 5 m from the boundary. Annuals dominated the seed bank, and few biennials were recorded. Many annual weed species, such as chickweed, were abundant in the hedge bottom samples.

Plant distribution patterns

Patterns of distribution of individual species fall into four categories (Marshall, 1989): (a) limited to the hedge; (b) limited to the crop; (c) occurring at greatest density in the hedge and declining with distance into the crop; and (d) occurring with greatest density within the crop headland. These distribution patterns are illustrated in Figure 8.5, using data collected in Extra Close East field. While some boundary species shed seed into the crop edge where their seedlings can grow, other species such as cocksfoot (*Dactylis glomerata*) will not grow in the crop area, and most do not have seedlings beyond 2.5 m from the hedge. A comparison of seed densities (in September) and seedlings (the following March) is shown in Figure 8.6.

The densities of seeds and seedlings were lower in the crop than in the hedge, though there was no marked decline with increasing distance into the crop. The major implication of these results is that most boundary plants do not constitute a threat as field weeds. Management should therefore aim to keep in check those few plant species (such as cleavers and couch) capable of dispersing and becoming a problem, while encouraging the majority. Most boundary species do not harm the adjacent crop and may be important for a variety of other wildlife, including birds and insects, some of which may benefit the crop by helping to control pest populations (Wratten, 1988).

Field boundary floras

Surveys of the flora of field boundaries were made in order to analyse any trends associated with the three systems of management applied to the adjacent crop areas. The presence or absence of species was recorded in 2 m-wide strips located every 50 m around field perimeters. Each year from 1983 to 1988, the field margins were examined on two or three occasions (March, May and July). The combined lists of species at each sample point were assessed to determine their similarity, using the statistical technique of cluster analysis (Wishart, 1970).

The results for 1983, in the baseline period, indicated twelve different types of field boundary (Table 8.3). For example, type 1 comprised sites which were typically low thorn hedges, type 2 were either thin hedges or areas adjacent to tracks or banks, and type 3 were woodland edges with ash or elm trees.

Table 8.3   Types of Boxworth field boundaries identified using cluster analysis.

Type	Number of sites	Boundary structure
1	89	Low hawthorn hedges
2	43	Track and bank sides, thin hedges
3	46	Woodland edges, elm or ash trees
4	15	Wood or tall hedge with ditch
5	6	Bank and ditch
6	5	Species-poor thorn hedge
7	10	Tall hedge, many shrub species
8	3	Thin hedge
9	10	Ditch sides
10	13	Grass tracks and roadsides
11	3	Road side
12	3	Disturbed grassland

These data were used as the basis for later assessments of the abundance of plant species, using a scoring from 1 (rare) to 6 (covering more than 50% of the ground).

During the study more than 170 higher plant

Figure 8.7 Frequencies of selected species in extensive surveys of field margins in the three treatment areas at Boxworth, 1983–88.

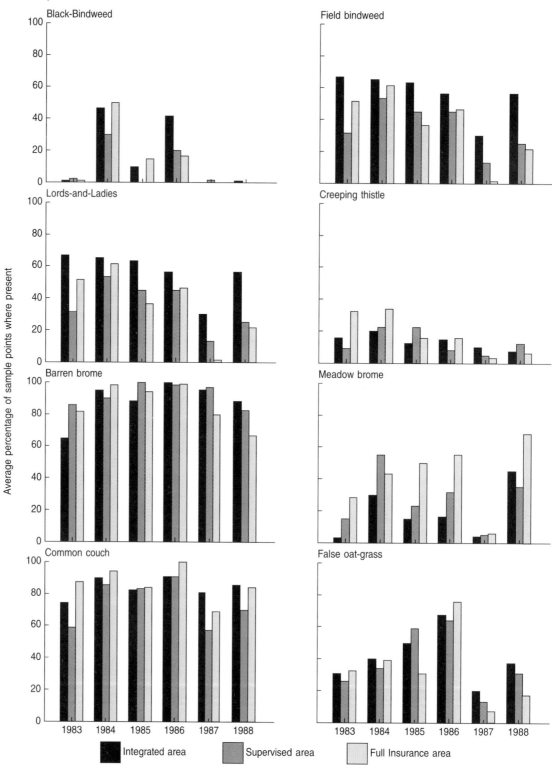

species were recorded in the fields and their margins at Boxworth (Appendix IV). This reflects the variety of habitats present, including arable land, grassland, woodland, ditches and streams. Many species were found only rarely, though some of the ground flora were sufficiently abundant to examine in more detail. The commonest species were barren brome, cleavers, cow parsley, (*Anthriscus sylvestris*), hogweed (*Heracleum sphondylium*) and nettle (*Urtica dioica*), which dominated the ground flora. All of these are robust species, and are indicative of various forms of disturbance, reflecting soil conditions, fertilizer enrichment or the effects of other agricultural operations.

Changes through the Project in the frequencies of selected field margin species are shown in Figure 8.7. Certain species, such as black-bindweed (*Fallopia convolvulus*), fluctuated widely from year to year. Others, such as barren brome, showed little change. Usually there were significant differences between years. However, there were no significant differences between treatment areas, which may be a result of the often high field-to-field variation. Differences between treatment areas for lords-and-ladies (*Arum maculatum*), perennial ryegrass (*Lolium perenne*), nipplewort (*Lapsana communis*), and cleavers were already present at the outset of the experiment. Overall, there were few consistent trends, the exceptions being a decline in creeping thistle (*Cirsium arvense*) in Full Insurance fields (not statistically significant) and meadow brome which tended to increase, apart from a sharp decline in 1987, in all areas. An initial increase in amounts of false oat-grass (*Arrhenatherum elatius*) from 1983 to 1986 was followed by a dramatic decline in all fields. The reasons for this are unclear, because there was no single treatment applied to all the fields.

The average numbers of species found at the extensive survey points in each field, which include tree, shrub and climbing species, show that the number of ground flora species at any point was modest. Analysis indicated no significant effect of the field treatments on average species number, which ranged from 13.6 in the Full Insurance area

to 14.7 in the Supervised area.

Species presence/absence data and cover scores of hedgerow plants indicated that boundary structure was a major factor in the species composition of field margins. No trends associated with the levels of herbicide use in the adjacent field areas were found. This may indicate that, at Boxworth, herbicide drift from field applications was not an important influence on botanical composition. The number of wild plant species, and their individual frequencies, varied from year to year. There was also marked field-to-field variation, which would probably have limited the likelihood of detecting all but major differences associated with the pesticide treatments.

Drift from field applications does occur (Chapter 7) but is unlikely to be significant beyond 10 m under normal spray conditions (Elliott & Wilson, 1983). Lagerfelt (1988) examined a range of hydraulic spray nozzles in different wind speeds and found that much less than 1% of applied material reached a distance of 10 m. Nevertheless, field margins are often within 1 m of the sprayed crop and glasshouse studies have demonstrated direct effects of small quantities of herbicides on boundary plant species (Marshall, 1987). Therefore, these results are consistent with the hypothesis that boundary floras are generally resilient to herbicide drift. However, it is equally possible that botanical composition may have become adapted to periodic spray drift over many years of intensive cropping before the Project began. Indeed, the boundary flora at the start of the Project, including trees and shrubs, averaged only fourteen species at each survey location.

Data from the intensive study fields identified a factor affecting the species-richness of the boundary flora. The number of species present was related to the width of the uncultivated hedge bottom (Figure 8.8). With few changes in the flora associated with pesticide applications, this indicated that close cultivations, which could greatly reduce the width of the hedgerow base, had more effect than herbicide drift on boundary floras at Boxworth.

Figure 8.8   The effect of the width of unculviated field boundary on the numbers of plant species present in field-edge transects at Boxworth.

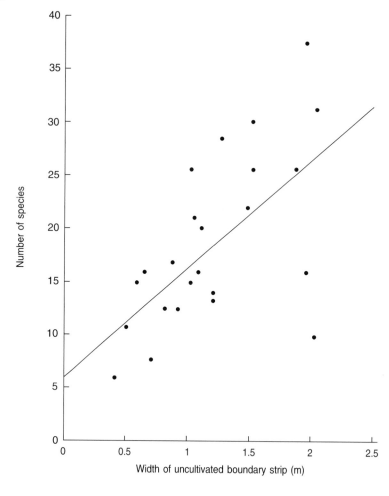

Discussion

The number of herbicides used in each of the three treatment areas during the course of the Project, and the effects of the treatment regimes on the most important species of weeds at Boxworth, are considered in Chapter 4.

Studies of the distribution of plants in the field edge and in the crop area have demonstrated a number of important points. Firstly, four distribution patterns were found, with some species limited either to the crop or to the boundary, some spreading from the boundary and a small number of species showing peak densities in the crop head-

land. Secondly, most boundary species do not occur as weeds, or are found only within 2 to 3 m of the crop edge. Thus, most plant species in the boundary can be regarded as innocuous. Thirdly, those species which are capable of spreading into crops tend to be annual species which produce large amounts of seed. Therefore, control measures should aim to limit these few species, particularly by reducing disturbance and the opportunity for them to regenerate from seed. Examination of the soil seed bank indicated that the crop areas supported low seed densities. There were large increases in the number of dicotyledonous weed seeds following oilseed rape, which was grown as a break crop. Hedge soils contained more species and viable seeds, including more annual weed species which typically have persistent seed banks. The significance of these areas for the reintroduction of annual weeds, and for the conservation of nationally-rare arable weeds which may survive in the soil, has yet to be determined.

Assessments of the boundary flora at Boxworth did not show any clear trends associated with the Full Insurance, Supervised or Integrated treatment regimes. This may reflect the previous history of the fields, and suggests that herbicide drift was of little significance. Variation in the boundary flora, which was not initially diverse, was such that few conclusions can be drawn. If, as suggested above, the flora was adapted to periodic herbicide drift, a relaxation in herbicide pressure might not result in major botanical change, or it might take many years to become apparent. Other factors, particularly the habitat structure and how close to the hedge cultivations are made, appear to be more important in determining the botanical composition of the boundaries.

Acknowledgements

My thanks to the following sandwich course students who contributed to the study of the flora at Boxworth: Gavin Bird, Helen Oakes, Christina Steveni, Yvonne Jones, Elaine Spearing, Claire Munkley and Meg Dowlen.

References

Elliott, J G & Wilson, B J (Eds) (1983). '*The influence of weather on the efficiency and safety of pesticide application.*' Occasional Publication No. 3. British Crop Protection Council, BCPC Publications, Croydon, 135 pp.

Greaves, M P & Marshall, E J P (1987). 'Field margins: definitions and statistics'. In: Way, J M and Greig-Smith, P W (Eds) *Field Margins*. BCPC Monograph No. 35. BCPC Publications, Croydon. pp. 3–10.

Harper, J L, (1977). *Population Biology of Plants.* Academic Press, London.

Lagerfelt, P (1988). 'Spray drift deposits during application with field crop sprayer.' *Proceedings of an International Symposium on Pesticide Application*, ANPP, Paris. pp. 30–39.

Malloch, A J C (1988). *VESPAN II. A computer package to handle and analyse multivariate species data and handle and display species distribution data.* University of Lancaster. 154 pp.

Marshall, E J P (1985). 'Weed distributions associated with cereal field edges—some preliminary observations.' *Aspects of Applied Biology* **9**, 49–58.

Marshall, E J P (1987). 'Herbicide effects on the flora of arable field boundaries.' *Proceedings of the British Crop Protection Conference—Weeds 1987*, 291–298.

Marshall, E J P (1989). 'Distribution patterns of plants associated with arable field edges.' *Journal of Applied Ecology* **26**, 247–257.

Roberts, H A (1981). 'Seed banks in soils.' *Advances in Applied Biology* **6**, 1–55.

Roberts, H A & Chancellor, R J (1986). 'Seed banks of some arable soils in the English midlands.' *Weed Research* **26**, 251–257.

Wishart, D (1970). *CLUSTAN 1A User Manual*. University of St. Andrews.

Wratten, S D (1988). 'The role of field boundaries as reservoirs of beneficial insects.' In: Park, J R (Ed.) *Environmental Management in Agriculture. European Perspectives.* Belhaven Press, London. pp. 144–150.

The effects of different pesticide regimes on the invertebrate fauna of winter wheat

G. Paul Vickerman (University of Southampton)

Introduction

From an environmental point of view one of the most important changes that has taken place in cereal growing over the last two decades has been the dramatic increase in the use of pesticides (Chapman *et al.*, 1977; Steed *et al.*, 1979; Sly, 1986). Pesticides are now used on a large scale and there is concern about possible adverse side-effects, particularly of insecticides, on wildlife, both in the short- and in the long-term.

For example, evidence has accumulated that beneficial species of invertebrates inhabiting cereals may play an important role in reducing the numbers of aphids. These species include the so-called 'aphid-specific' predators (eg ladybirds), polyphagous predators (eg species of ground beetles and spiders that feed on a wide range of prey items) and parasitoids (Potts & Vickerman, 1974; Sunderland & Vickerman, 1980; Chambers, Sunderland, Wyatt & Vickerman, 1983). Therefore, there is a possibility that the use of broad-spectrum compounds which kill natural enemies could exacerbate problems with cereal aphids and other cereal pests.

Many birds inhabiting farmland utilise the invertebrates found in crops as a source of food. For example, chicks of the grey partridge (*Perdix perdix*) feed almost exclusively on invertebrates for the first few days of life (Vickerman & O'Bryan, 1979) and tree sparrows (*Passer montanus*) fed on a variety of invertebrates in cereal fields during the Boxworth Project (Chapter 15). Clearly, pesticides that reduce the availability of invertebrates may have indirect side-effects on other wildlife species.

Monitoring studies carried out by the Game Conservancy suggest that populations of some groups of invertebrates inhabiting cereals have declined over the last two decades (Vickerman, 1977, 1980; Potts, 1986). The declines may be due to the increasing use of pesticides (Potts, 1986), but the evidence is circumstantial. Some could be due to changes in other agricultural practices, the weather or other factors. Several studies have been carried out to determine the effects of individual pesticides on invertebrates in cereals (eg Vickerman & Sunderland, 1977; Vickerman *et al.*, 1987a, 1987b) but there have been no long-term studies, either of individual compounds or of complete pesticide 'packages'. Thus, there was a need to investigate the side-effects of pesticides on invertebrates, without other variations in agricultural practices.

One aim of the Boxworth Project, therefore, was to investigate long-term effects of different pesticide regimes on the invertebrates inhabiting winter wheat fields. This chapter provides a synthesis of the main effects of the treatment regimes on the large assemblage of invertebrates found in wheat fields.

As the pesticide inputs in the Integrated and the Supervised areas were similar in most years (Chapter 2 and Appendix II) the data on invertebrates in these two areas were combined for comparison with those from the Full Insurance area. In 1984 and 1988, winter wheat fields in the Supervised area were treated with demeton-S-methyl, which had marked within-season effects on some invertebrate groups such as predatory flies (Empididae) and predatory ground beetles (Carabidae) (see later). For present purposes the data obtained from these fields were excluded from the analyses.

Apart from pesticides, the invertebrates inhabiting cereal fields may be affected by a range of other factors such as the weather, the nature of the preceding crop, sowing date and crop density, some of which are discussed here.

Methods

Sampling

Estimates of the density of the invertebrate taxa inhabiting the crops were obtained using a Dietrick vacuum insect sampler (D-vac) (Dietrick, 1961). D-vac samples were taken from all the winter wheat fields in the three treatment areas, chiefly between mid-April and the date of harvest. Depending on the weather, the samples were usually taken at intervals of seven to ten days throughout the entire sampling period. All samples were collected between 1100h and 1500h, when the vegetation was usually dry.

The samples were taken along the tramlines (tractor wheelings) in the middle of each field, with the first sample at a minimum distance of 25 m from the edge of the field. On each occasion five samples, each consisting of five 0.092 m² sub-samples taken 10 m apart, were collected along a transect across the middle of each of the three intensive study fields (Extra Close East, Knapwell and Grange Piece). Three samples were also taken from each of the remaining fields in the three treatment areas.

All samples were taken to the laboratory and frozen for storage prior to sorting and identification of their contents.

Sorting and identification

Before sorting, samples were transferred from the freezer to tubes containing 80% alcohol. Some samples contained large amounts of soil, particularly when conditions were very dry. The invertebrates present in such samples were first extracted from the soil by flotation in 10 cm-diameter plastic beakers containing a saturated salt solution.

All samples were subsequently examined in alcohol under a binocular microscope and the invertebrates were removed and counted. Apart from the lucerne-flea (*Sminthurus viridis*), springtails (Collembola) and mites (Acari) were not counted routinely. For most invertebrate groups identification was made at the species level, although for present purposes some taxonomic groups were combined to summarise the complete data set.

Feeding habits of invertebrates

The invertebrates found in the D-vac samples, with the exception of springtails and mites, were assigned to three groups, according to their feeding habits; herbivores, detritivores and carnivores (Table 9.1). The carnivores were further subdivided into predators and parasitoids (parasitic insects whose larvae develop by feeding on or within a host animal and almost always kill it). Although numbers of the lucerne-flea (a plant-eating springtail) were counted, this species was excluded from the overall analyses because other springtails (which are detritivores) were not counted routinely.

In many of the invertebrate taxa, feeding habits were similar in the adult and immature stages. However, in others, particularly most of the Hymenoptera (eg parasitic wasps and sawflies), feeding habits differed between life-stages. In these cases taxa were described as herbivores, detritivores or carnivores according to the feeding habits of their longer-lived, immature stages.

Species were allocated to the categories in Table 9.1 on the basis of dietary studies carried out by the author, or reported in the scientific literature. A large majority of the species feed on only one type of food, and could be assigned with confidence to a particular category. However, a few species feed on more than one kind of food and should be regarded as omnivores. For the following analyses, the numbers of these species were divided equally between the appropriate groups.

Pest species and their control

The target pest species comprised three groups of invertebrates: (1) slugs, mainly the grey field slug (*Deroceras reticulatum*); (2) cereal aphids, includ-

Table 9.1 Allocation of invertebrates to different trophic groups.

HERBIVORES
Lepidoptera
Thysanoptera — Thripidae
Hemiptera — Aphididae, Cicadellidae, Delphacidae, Psyllidae, Aleyrodidae, Cercopidae, Miridae, Tingidae
Diptera — Tipulidae, some Cecidomyiidae, Anthomyidae, Agromyzidae, most Chloropidae, Tephritidae, Opomyzidae
Coleoptera — Elateridae, Nitidulidae, Chrysomelidae, Curculionidae
Hymenoptera — Cephidae, Tenthredinidae

DETRITIVORES
Isopoda
Diplopoda
Symphyla
Psocoptera
Diptera — Trichoceridae, Psychodidae, Chironomidae, Bibionidae, Mycetophilidae, some Cecidomyiidae, Scatopsidae, Lonchopteridae, Phoridae, Lauxaniidae, Drosophilidae, Ephydridae, Muscidae, Sepsidae, Sphaeroceridae, some Chloropidae
Coleoptera — Hydrophilidae, Ptiliidae, some Staphylinidae, Cryptophagidae, Lathridiidae, Phalacridae, some Coccinellidae.

CARNIVORES—PREDATORS
Chilopoda
Araneae
Chelonethi
Opiliones
Neuroptera — Chrysopidae, Hemerobiidae
Coleoptera — Carabidae, some Staphylinidae, Cantharidae, most Coccinellidae
Diptera — Empididae, Dolichopodidae, Rhagionidae, Syrphidae, Culicidae, Ceratopogonidae
Hempitera — Nabidae, Anthocoridae

CARNIVORES—PARASITOIDS
Hymenoptera — all Hymenoptera 'Parasitica'
Diptera — Tachinidae

ing grain aphid (*Sitobion avenae*) and rose-grain aphid (*Metopolophium dirhodum*) in the summer, and these species together with bird cherry-oat aphid (*Rhopalosiphum padi*) in the autumn and (3) the larvae of stem-boring flies, the yellow cereal fly (*Opomyza florum*) and frit fly (*Oscinella frit*). Further details of these pests and their control are given in Chapter 3.

Within the Supervised and Integrated areas routine asessments were made by ADAS staff of the need to apply molluscicides or insecticides to control pests, whereas in the Full Insurance area pesticides were applied prophylactically. Details of the pesticides applied are given in Appendix II, and their usage is summarised in Table 9.2.

During the first pre-treatment year (1982) it was necessary to apply a molluscicide to two fields,

but no insecticides were required. In 1983, however, molluscicides were applied to six fields and a pyrethroid insecticide (permethrin) was applied to two fields in the Integrated area to control aphids transmitting barley yellow dwarf virus. Numbers of the grain aphid also reached threshold levels during the summer and pirimicarb was applied to half the fields in both treatment areas in late June or early July.

During the treatment phase of the Project, pesticides were used in the Supervised and Integrated areas to control slugs and grain aphids in 1984 and 1988 (Table 9.2). Thresholds for the other pest species were not exceeded and therefore no other insecticides were required. Within the Full Insurance area there were applications of methiocarb, triazophos (or dimethoate, in 1986) and

Table 9.2 Percentage of the area of winter wheat grown in the Full Insurance (FI) and Supervised + Integrated (S + I) areas that received a molluscicide to control slugs and insecticides to control cereal aphids and stem-boring flies, 1982–1988.

		Pre-treatment period		Treatment period				
		1982	1983	1984	1985	1986	1987	1988
methiocarb	FI	52.3	74.1	100	100	100	100	100
(slugs)	S + I	17.0	57.2	100	0	0	0	75.2
pyrethroid	FI	0	0	100	100	24.4	0	75.6
(cereal aphids)	S + I	0	17.7	0	0	0	0	0
chlorpyrifos	FI	0	0	100	0	0	100	75.6
(frit fly)	S + I	0	0	0	0	0	0	0
triazophos or dimethoate	FI	0	0	100	100	100	100	100
(yellow cereal fly)	S + I	0	0	0	0	0	0	0
demeton-S-methyl or pirimicarb	FI	0	52.1	100	100	100	100	100
(grain aphid)	S + I	0	49.8	100	0	0	0	91.4
demeton-S-methyl	FI	0	0	100	100	100	100	100
(rose-grain aphid)	S + I	0	0	0	0	0	0	0

NOTE: In 1983 pirimicarb was used in all treated fields. Over the period 1984–1988 demeton-S-methyl was used in the Full Insurance and Supervised areas and pirimicarb in the Integrated area.

demeton-S-methyl in all wheat fields every year. In some years, however, it was not possible to apply the pesticides at the correct time, because of adverse weather conditions. Also, applications of a pyrethroid and chlorpyrifos were not made in every field each year (Table 9.2 & Chapter 3).

Consequently, although the overall inputs of molluscicides and insecticides in the Full Insurance area were much higher than in the Supervised + Integrated area (Table 9.2) there was some variability which was reflected in the effects of the crop protection regimes on invertebrates.

Influence of sowing date on invertebrates

During the first pre-treatment year (1982) there were differences between fields in the density of some invertebrates. It seemed that this was due mainly to differences in sowing dates, which varied considerably in 1981, from 23 September to 13 November.

The influence of sowing date on the numbers of some herbivores found in D-vac samples in June 1982 is illustrated in Figure 9.1. Numbers of both grain aphids and bird cherry-oat aphids were higher in the earlier sown crops whilst rose-grain aphids were most numerous in fields that were sown late (Figure 9.1).

The numbers of two other sap-sucking herbivores, thrips (*Limothrips* spp.) and the cereal leafhopper *Javesella pellucida* (Delphacidae) were also higher in crops that had been sown relatively late. In contrast, adults of the yellow cereal fly were much more numerous in crops sown in late September or early October than in those drilled later in the autumn (Figure 9.1).

Sowing dates also appeared to influence the numbers of some of the carnivores found in the samples (Figure 9.2). For example, the common predatory flies *Tachydromyia arrogans* and *Platypalpus* spp. (Empididae), were significantly more abundant in crops sown in late September or early October than in later sown crops. Similarly, the

Figure 9.1   Mean numbers per m² of some herbivores in D-vac samples taken in June 1982 from winter wheat fields sown in late September/early October, late October, or early November.

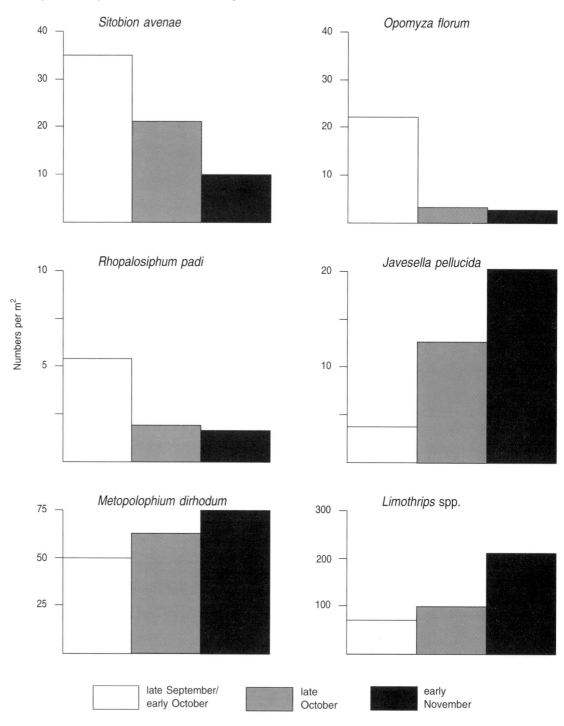

Figure 9.2 Mean numbers per m^2 of some carnivores in D-vac samples taken in June 1982 from winter wheat fields sown in late September/early October, late October, or early November.

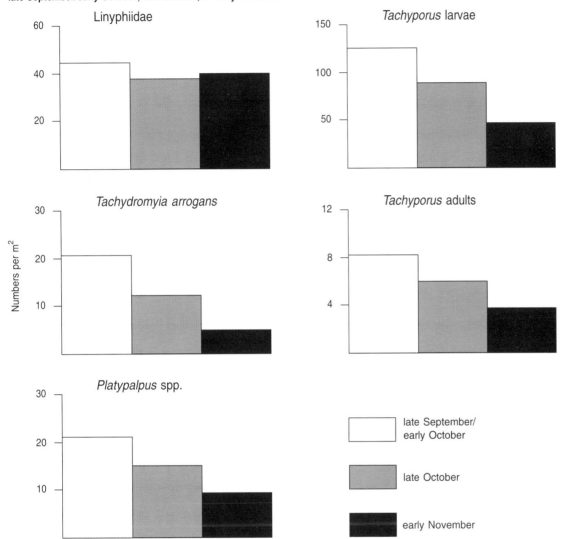

numbers of the common rove beetles *Tachyporus* spp. (Staphylinidae), were least in November-sown wheat, an effect that was most pronounced for the larvae (Figure 9.2). In contrast, the numbers of carnivorous money spiders (Linyphiidae) did not vary markedly with respect to sowing date (Figure 9.2).

The influence of sowing dates on certain herbivorous pests is well known (eg Vickerman, 1982; Dewar & Carter, 1984). Crops sown earlier in the autumn are more likely to be colonised by pests such as grain aphids and yellow cereal flies. This is because adverse weather conditions later in the autumn reduce the opportunities for flight and hence the colonisation of late-sown crops. Conversely, many of the pests that do not colonise crops until the spring, such as rose-grain aphids and cereal leafhoppers, tend to prefer crops at an early growth stage and hence are more likely to affect late-sown crops.

Effects of sowing dates on carnivores are not well documented. However, rove beetles such as *Tachyporus* spp. and predatory flies prefer crops that provide conditions of high humidity and it is possible that fields at a more advanced growth stage (earlier sown crops) are colonised preferentially, as they would provide more equable environmental conditions.

These results emphasised the need to match the timing of husbandry operations in the different treatment areas in order to minimise variability between fields. The adoption of the triplet arrangement (see Chapter 2) worked well in all years apart from 1988. In that year some crops in the Supervised + Integrated area were sown later than crops in the Full Insurance area, with possible effects on invertebrates.

Populations of herbivores

During both the pre-treatment years (1982 and 1983) and the five treatment years (1984–88), over 99% of the herbivores found in the D-vac samples were of species that feed on wheat plants rather than on the weeds in the crop. Therefore, weed-feeding invertebrates, which consisted mainly of blossom beetles (Nitidulidae), weevils (Curculionidae), leaf beetles (Chrysomelidae) and aphids (Aphididae) are not considered in detail in this chapter.

Of the crop-feeding herbivores in the samples, the numbers of Hymenoptera (sawfly larvae), Lepidoptera (moth caterpillars) and Coleoptera (mainly leaf beetles) were always very low and their contribution to the total herbivores never exceeded 0.4%, 0.6% and 2.6% respectively in any year. Therefore, the herbivore community consisted almost exclusively of Thysanoptera (thrips), Hemiptera (aphids and plant bugs) and Diptera (flies) (Figure 9.3).

The most numerous species were thrips (*Limothrips cerealium, L. denticornis* and

Figure 9.3 Average percentage composition of the herbivores found in D-vac samples taken during the pre-treatment (1982–1983) and treatment (1984–1988) periods.

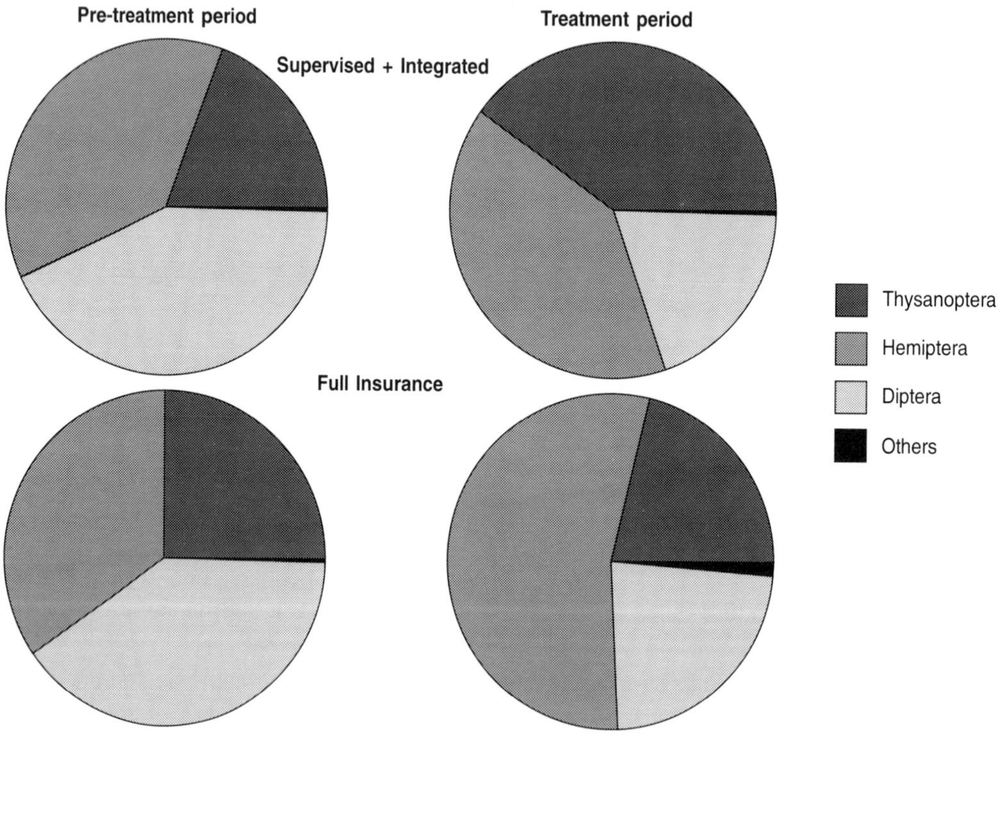

Aptinothrips rufus), cereal aphids, orange wheat blossom midges (*Sitodiplosis mosellana*), yellow cereal flies, cereal leaf miners (*Agromyza* spp.) and cereal leafhoppers. The lucerne-flea, a plant-eating springtail, was also very common in the samples but, as explained previously, this species was excluded from the analyses of community structure.

A number of other recognised pests, such as stem-boring flies (eg frit fly, gout fly (*Chlorops taeniopus*) and wheat bulb fly (*Delia coarctata*)) yellow wheat blossom midges (*Contarinia tritici*), wheat stem sawflies (*Cephus pygmaeus*), cereal leaf beetles (*Oulema melanopa*) and spotted crane flies (*Nephrotoma* spp.) were also found in the samples but in very low numbers; none of these species ever constituted more than 1% of the total herbivores sampled in any year.

During the pre-treatment years, the average numbers of herbivores found in the samples were very similar in the Supervised + Integrated and Full Insurance areas. Diptera (mainly orange wheat blossom midges) and Hemiptera (mainly grain aphids) were the most abundant herbivores. Together these two groups comprised about 80% and 75% of the herbivores found in the Supervised + Integrated and Full Insurance areas respectively (Figure 9.3).

During the treatment phase of the Project, herbivore populations differed markedly between the two treatment areas. Details of individual taxa are considered later but the overall situation during the treatment years can be summarised as follows.

During the first treatment year (1984) thrips predominated in the Supervised + Integrated area and grain aphids in the Full Insurance area. In 1985, grain aphids were the most numerous herbivores in the Supervised + Integrated area and orange wheat blossom midges in the Full Insurance area. In 1986, 1987 and 1988 rose-grain aphids, thrips and grain aphids respectively were the commonest herbivores in both treatment areas.

On average, total numbers of herbivores were about 50% lower in the Full Insurance area, although there was considerable variation between

years. At one extreme, total numbers were 94% lower in 1987, but in 1986 numbers were about 170% *higher* in the Full Insurance area than in the Supervised + Integrated area. In other years, numbers ranged from 68% to 77% lower.

There was no overall difference between areas in the numbers of Coleoptera (beetles) found in the samples. The beetles consisted mainly of blossom beetles (*Meligethes* spp.) from oilseed rape crops, and in two years numbers were higher in the Full Insurance area than in the Supervised + Integrated area. Otherwise, total numbers of beetles were 24% to 69% lower in the Full Insurance area.

The numbers of Hemiptera (aphids and plant bugs) were only about 25% lower on average, but this was due to the fact that in 1986 numbers were 146% *higher* in the Full Insurance area than in the Supervised + Integrated area; in other years numbers were 50% to 85% lower.

Although Lepidoptera (caterpillars, butterflies and moths) were relatively scarce in the samples, they declined in the Full Insurance area relative to the Supervised + Integrated area, so that by the end of the treatment period numbers were only 5% of those found in the latter area.

On average, the numbers of Diptera (flies) and Thysanoptera (thrips) were respectively 70% and 75% lower in the Full Insurance area than in the Supervised + Integrated area. Apart from 1986 (thrips) and 1988 (flies) the numbers of both these groups of herbivores were at least 63% lower in the Full Insurance area.

Hemiptera

Grain aphids and rose-grain aphids

Grain aphids *Sitobion avenae* and rose-grain aphids *Metopolophium dirhodum* were the main targets for insecticides, and a large difference between the treatments might be expected. However, average numbers of these two species were only 34% and 11% lower in the Full Insurance area than the Supervised + Integrated area (Figure 9.4). This was due to two factors; firstly, the numbers of

Figure 9.4   Numbers of some target and non-target herbivores in the Full Insurance area as a percentage of those in the Supervised + Integrated area during the treatment period (1984–1988).

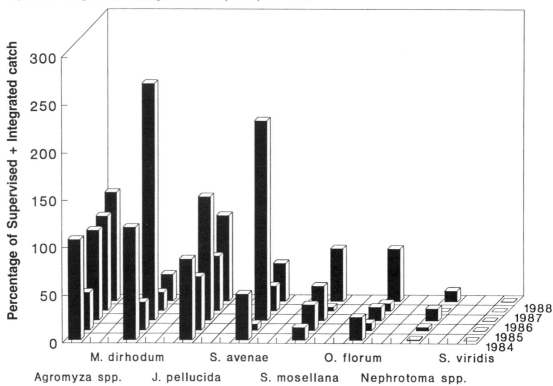

Insurance area in 1986 and secondly, insecticides were also applied to the Supervised + Integrated area in 1984 and 1988 when treatment thresholds for the grain aphid were exceeded (see Chapter 3).

The grain aphid was the commoner of the two species in six out of the seven years of the Project. In 1983 numbers increased rapidly in both areas throughout June (Figure 9.5) and although the threshold at flowering (Growth Stage 60) was not quite exceeded, insecticides were applied to half the fields between 28 June and 2 July. Populations in the untreated fields declined naturally after 4 July.

In the first treatment year (1984), grain aphid numbers in the Full Insurance area were initially very low, owing to the application of insecticides during autumn 1983 and the 1983–84 winter, but increased rapidly from late May until the third week of June when insecticides were applied

prophylactically. Populations also increased in the Supervised + Integrated area and, because the thresholds for treatment were exceeded in all fields, insecticides were applied. However, crops in both areas were recolonised quickly by winged grain aphids. Consequently, peak densities were recorded after the dates of insecticide application in both areas, and differences in population size were small (Figure 9.5).

In 1985, grain aphid population development was later and although the threshold for treatment in the Supervised + Integrated area was not exceeded, peak populations at the end of July were exceptionally high (Figure 9.5). Numbers were lower in the Full Insurance area throughout the sampling period.

In 1986 grain aphid population development was even later. Few aphids were present in either area when demeton-S-methyl was applied to

Figure 9.5 Mean numbers of grain aphids *Sitobion avenae* in D-vac samples taken from the Supervised + Integrated area (solid line) and Full Insurance area (dotted line) from 1983 to 1988. Filled arrows indicate dates on which pirimicarb was applied to the Integrated area; open arrows indicate dates of demeton-S-methyl applications in the Full Insurance area.

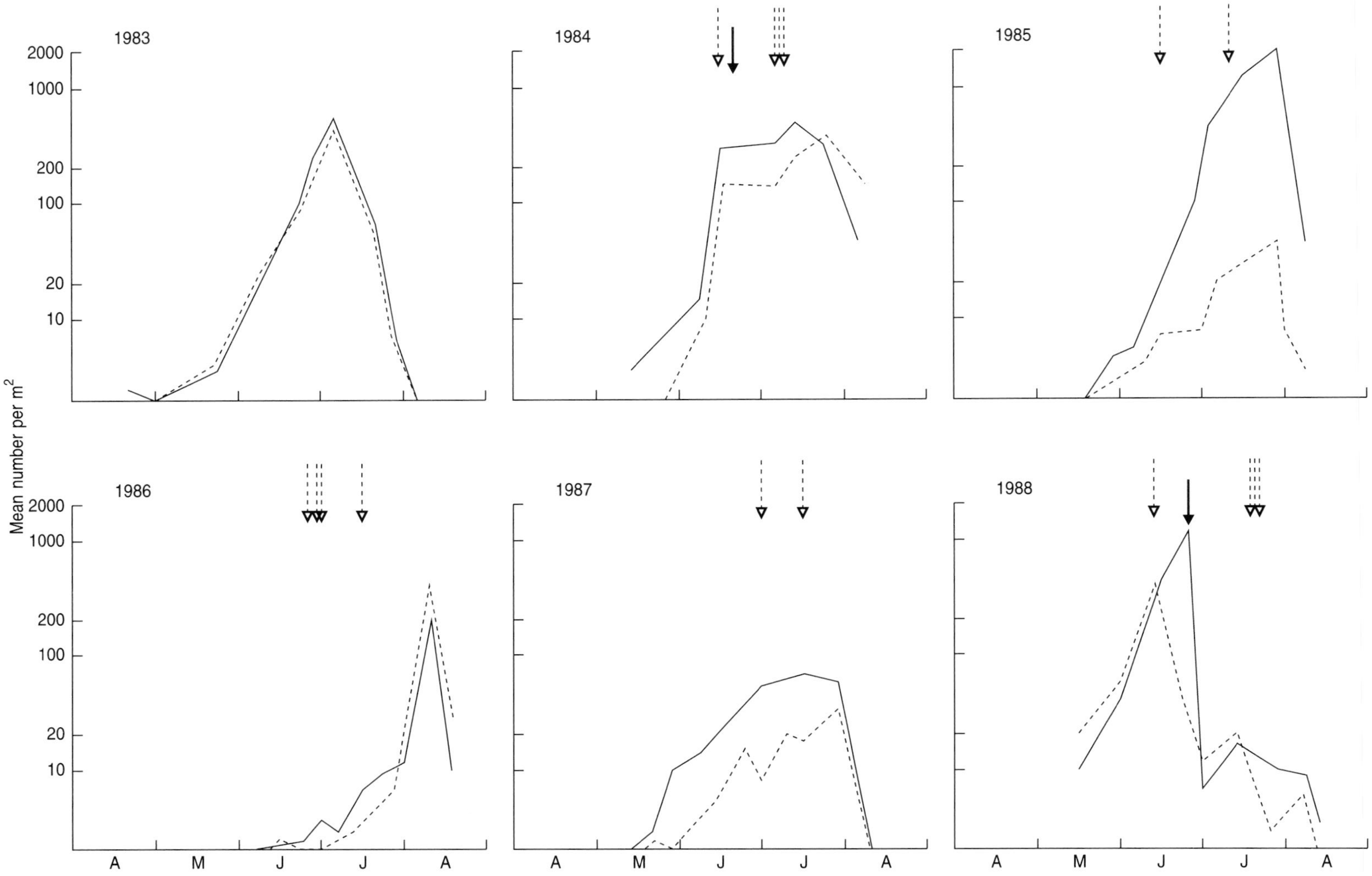

the Full Insurance fields, but numbers continued to increase and were subsequently *highest* in the Full Insurance area (Figure 9.5).

In 1987 there were few grain aphids and although crops within the Full Insurance area were recolonised quickly following insecticide applications, populations remained lower than in the Supervised + Integrated area (Figure 9.5).

During the 1987–1988 winter the survival of grain aphid populations was good. In both areas numbers increased rapidly in 1988 until demeton-S-methyl was applied in the Full Insurance area and thresholds were exceeded in all fields apart from one in the Supervised + Integrated area.

As indicated previously, the rose-grain aphid was usually present in relatively low numbers but in 1986 it was by far the commonest cereal aphid species found in the crops. As with the grain aphid in that year, the population build-up was late and few aphids were present when demeton-S-methyl was first applied. However, numbers rose rapidly in both areas in July and the second application of demeton-S-methyl checked the increase in the Full Insurance area for only one week. Crops were recolonised rapidly by large numbers of winged rose-grain aphids and in late July and August the numbers were significantly *higher* in the Full Insurance area. Over the whole sampling period in 1986, numbers were 150% *higher* in the Full Insurance area (Figure 9.4).

Cereal leafhoppers

Although the cereal leafhopper *Javesella pellucida* was common in the samples it was not a target for insecticides. As with the cereal aphids, average numbers in the Full Insurance area were only about 16% lower than in the Supervised + Integrated area. The difference was least in 1984 and 1988, while in 1986 populations were highest in the Full Insurance area (Figure 9.4).

Unlike cereal aphids, *Javesella* spends the winter in hedgerows, from which cereals are re-colonised by adults in spring. Eggs are laid and nymphs are present in the crops from late June until August, when they leave the fields. Therefore,

this species would have little exposure to insecticides applied in the autumn and winter, but would be exposed to summer applications of demeton-S-methyl.

In the second pre-treatment year *Javesella* populations were similar in both areas, before and after pirimicarb applications (Figure 9.6). In the first treatment year populations were also similar up to the date of the first application of demeton-S-methyl in the Full Insurance area. However, in subsequent years pre-spray populations were higher in the Full Insurance area, and between 1984 and 1987 this difference between the two areas increased (Figure 9.6).

Comparisons of the numbers of *Javesella* immediately before and after applications of demeton-S-methyl suggested that density was reduced on average by about 50% by the first application in June and by about 62% by the second in July. Over the whole period following the first insecticide treatment, numbers in the Full Insurance area were between 16% and 57% lower than in the Supervised + Integrated area, apart from 1986, when they were 20% higher (Figure 9.6). This pattern suggests the existence of strong compensatory mechanisms in *Javesella* populations, producing higher pre-spray densities in the Full Insurance area despite reductions due to pesticide use.

Diptera

Apart from the Thysanoptera and Hemiptera the only other insect order that contained both target and non-target pest species of numerical significance was the Diptera (flies).

Cereal leaf miners

Cereal miners (*Agromyza* spp.) were not regarded as important pests at Boxworth. Their numbers were relatively low during the first three years of the treatment period, but in both 1987 and 1988 numbers were substantially higher and leaf mines were common on the wheat. In all but one year, the

Figure 9.6 Mean numbers of the cereal leafhopper *Javesella pellucida* in the Full Insurance area as a percentage of the mean numbers in the Supervised + Integrated area.

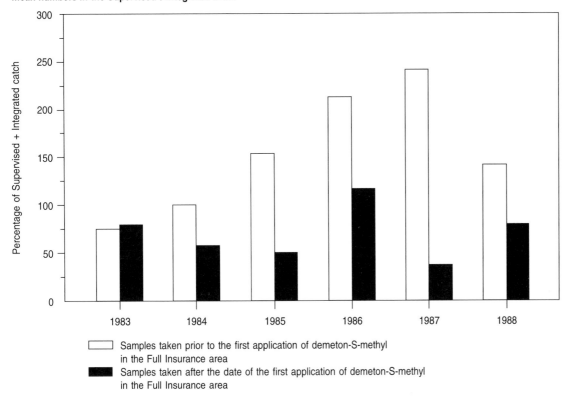

☐ Samples taken prior to the first application of demeton-S-methyl in the Full Insurance area

■ Samples taken after the date of the first application of demeton-S-methyl in the Full Insurance area

numbers of these flies in the Full Insurance area were similar to or higher than in the Supervised + Integrated area (Figure 9.4).

Yellow cereal flies

The yellow cereal fly *Opomyza florum* was the main dipteran pest. This species lays eggs in the autumn, particularly in early sown crops (Vickerman, 1982), but the larvae do not hatch until late winter or early spring. Larvae invade the shoots of cereals and the first adults of the new generation are found in the crop in late June. Applications of triazophos were intended to reduce the numbers of larvae of this species, whereas the adults would have been exposed both to pyrethroid sprays applied in the autumn and to the second of the demeton-S-methyl treatments applied in the summer to control cereal aphids.

Although there was considerable field-to-field variation in the numbers of this species, popu-

lations were, on average, about 78% lower in the Full Insurance area than in the Supervised + Integrated area (Figure 9.4). In 1988 some fields in the Supervised + Integrated area were sown late and the difference in population levels was not as apparent as in the previous years (Figure 9.4).

Orange wheat blossom midges

The orange wheat blossom midge *Sitodiplosis mosellana* was relatively common, sometimes exceeding 1000 insects per m^2. This species also spends its entire life cycle in fields. Adults were sometimes present in fields in late May but numbers usually reached a peak in June and July, when the crops were at anthesis (Growth Stages 60 to 69). Eggs are laid on the ears and the larvae feed on the developing grains. Once fully fed the larvae leave the ears and drop to the ground, and the winter is spent in a cocoon below the soil surface.

Some larvae pupate the following spring and give rise to adults. In contrast to the yellow cereal fly, both the adults and the larvae of this species were exposed directly only to the two summer applications of demeton-S-methyl in the Full Insurance area.

Populations were always lower in the Full Insurance area than in the Supervised + Integrated area, the difference varying from 96% to 45% (Figure 9.4). Adults were exposed to both demeton-s-methyl applications, but in some years, such as 1988, a high percentage of the population had emerged before the date of the first spray. The reduction in numbers in the Full Insurance area was smaller when the percentage of flies that had emerged before the date of the first spray was greater.

Spotted crane flies

The numbers of adult spotted crane flies (*Nephrotoma* spp.) in the Full Insurance area were 95% lower than in the Supervised + Integrated area (Figure 9.4). Adults were exposed directly to the demeton-S-methyl applications in June and July whereas the larvae (leatherjackets) would have been exposed, directly or indirectly, to the insecticides used during the autumn/winter period.

Springtails

The lucerne-flea *Sminthurus viridis* was very common in the samples in spring and early summer, when populations of up to 1000 per m² were recorded. These insects were active on vegetation throughout the whole year and therefore were exposed directly to all the pesticide applications.

Following the introduction of the treatment regimes in autumn 1983 numbers of this species decreased rapidly in the Full Insurance area, where it was virtually eliminated (Figure 9.4).

Summary

The overall effect of the Full Insurance regime was to reduce the total numbers of herbivores by 50%, but the effects were variable. Numbers of caterpil-

lars, butterflies and moths, thrips, yellow cereal fly, orange wheat blossom midge, spotted crane fly and the lucerne-flea were reduced substantially. In contrast, the overall effects on beetles, aphids and plant bugs and cereal leaf miners were smaller.

The Full Insurance area differed from the lower input areas in the composition of the herbivore fauna during the treatment period of the Project. The percentage contribution of thrips (Thysanoptera) was lower in the Full Insurance area in four out of the five years, whereas the contribution of aphids and plant bugs (Hemiptera) was higher. Over the period as a whole, the contributions of Thysanoptera and Hemiptera to the total herbivores were similar in the Supervised + Integrated area whereas Hemiptera predominated in the Full Insurance area (Figure 9.3).

Populations of carnivores

The carnivores found in samples consisted of predators and parasitoids, many of which are natural enemies of pests. Effects of the treatment regimes on these two groups of carnivores will be considered separately.

Predators

The predators included species of Coleoptera (beetles), Diptera (flies), Arachnida (spiders, harvestmen, false scorpions), Hemiptera (bugs), Chilopoda (centipedes), Neuroptera (lacewings) and Dermaptera (earwigs). However, in all years three groups (Coleoptera, Diptera and Arachnida) constituted at least 97% of the predators in D-vac samples. The other groups each contributed less than 1.6% to the total each year.

Individual taxa are discussed later, but the overall effects of the treatment regimes on the main predatory groups can be summarised as follows.

During the two pre-treatment years the average total numbers of predators were similar in the two treatment areas. The average numbers of Coleoptera and Arachnida were also similar but

Diptera were less abundant in the Full Insurance area in 1982, due to the influence of late sowing dates (see above). Consequently, the percentage contribution of Diptera to the total predators was slightly lower in the Full Insurance area than in the Supervised + Integrated area (Figure 9.7).

During the treatment period the total numbers of predators found within the Full Insurance area were 53% lower on average than in the Supervised + Integrated area, ranging from 39% to 70% lower in individual years (Figure 9.8). Of the three main groups of predators, Coleoptera appeared to be least affected by the Full Insurance regime. The numbers of Diptera and Arachnida were, on average, 73% and 70% lower in the Full Insurance area. In some years the numbers of these groups were up to 85% lower (Figure 9.8).

During the treatment phase of the Project, Arachnida predominated in the Supervised + Integrated area while Coleoptera were the most abundant predators in the Full Insurance area

(Figure 9.7). The average percentage contribution of Diptera was also consistently higher in the Supervised + Integrated area.

Arachnida

In both treatment areas spiders (Araneae) constituted 99% of the Arachnida found in the samples, the remaining individuals being harvestmen (Opiliones) and false scorpions (Chelonethi). Money spiders (Linyphiidae) were the dominant family within the Araneae, constituting 98% and 99% respectively of the spiders found in the Full Insurance and Supervised + Integrated areas.

The adult money spiders were mainly *Lepthyphantes tenuis*, *Bathyphantes gracilis*, *Meioneta rurestris*, *Erigone atra* and *Porrhomma convexum*, but immature individuals, which could not be identified with certainty, predominated in the D-vac samples. Both adults and immatures inhabit the crop throughout the year, although there are two main dispersal periods, in spring and in late

Figure 9.7 Average percentage composition of the predators found in D-vac samples during the pre-treatment (1982 & 1983) and treatment (1984–1988) periods.

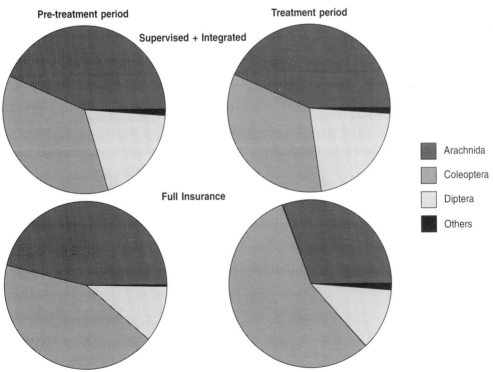

Figure 9.8 Numbers of predatory Coleoptera, Arachnida and Diptera and of total predators and total carnivores, in D-vac samples taken from the Full Insurance area as a percentage of those in the Supervised + Integrated area during the pre-treatment (1982 & 1983) and treatment (1984–1988) periods.

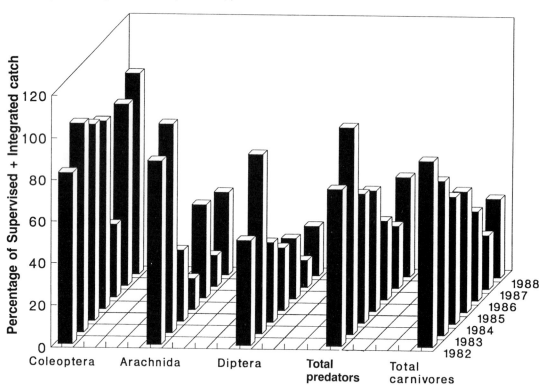

summer/autumn. They inhabit the ground zone and field layer and therefore were exposed directly to all pesticide applications.

During the pre-treatment period, populations of money spiders were very similar in the two areas. In all the treatment years, however, populations were lower in the Full Insurance area than in the Supervised + Integrated area (Figure 9.9). Total numbers varied from 54% to 86% lower, and the average reduction in numbers over the entire period was 70%. The trends in money spider populations caught in pitfall traps (in which adults predominate) (Chapter 10) were very similar to those obtained by suction sampling (Figure 9.10).

In most years, populations increased from spring until the crops were harvested in late summer (Figure 9.9). However, population development was markedly different in 1986 and 1988, when the difference between the two areas was

least (Figure 9.9). Nevertheless, in every year the pattern in the Full Insurance area matched that in the reduced-input area.

In early spring 1984, shortly after the introduction of the treatment regimes, the numbers of money spiders in the Full Insurance area were already substantially lower than in the Supervised + Integrated area, and apart from 1986 differences between the two areas were greatest during the spring (Figure 9.9). This suggests that pesticides applied in autumn and winter had a greater impact than those used in the summer, a conclusion supported by results obtained using pitfall traps (see Chapter 10).

Diptera

The predatory flies were represented by species of hoverflies (Syrphidae), dung flies (Scatophagidae), mosquitoes (Culicidae), biting midges (Cerat-

Figure 9.9 Mean numbers of money spiders (Linyphiidae) in D-vac samples taken from the Supervised + Integrated area (solid lines) and Full Insurance area (dotted lines), 1983–1988.

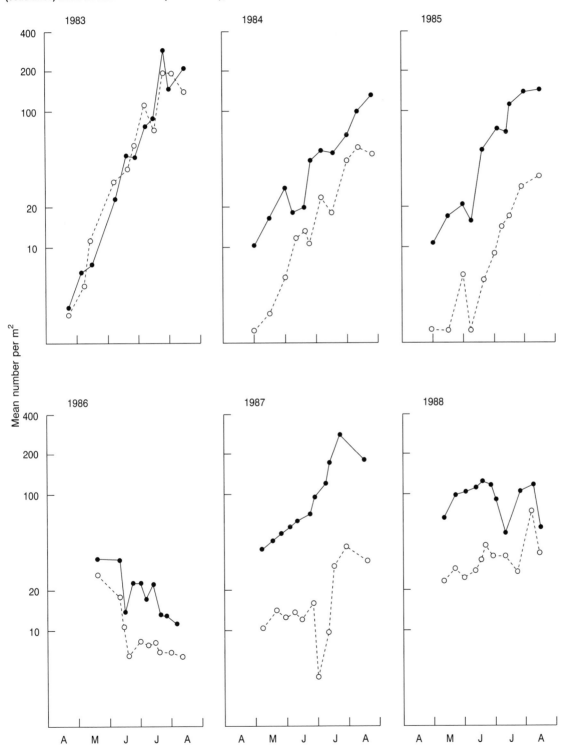

Figure 9.10   Numbers of Linyphiidae, *Bembidion obtusum, Trechus quadristriatus* and *Notiophilus biguttatus* found in D-vac samples (unshaded) and pitfall trap samples (shaded) taken from the Full Insurance area as a percentage of numbers in the Supervised + Integrated area during the pre-treatment (1982 & 1983) and treatment (1984–1988) periods. Asterisks indicate where insufficient data were available for analysis.

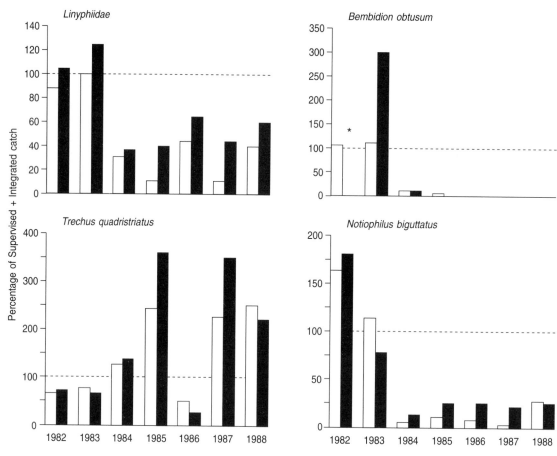

opogonidae), Rhagionidae, Empididae and Dolichopodidae. In both treatment areas Empididae constituted at least 92% of the predatory flies.

The Empididae consisted mainly of *Tachydromyia arrogans, Platypalpus pallidicornis* and *P. pallidiventris*. All these species spend the winter months in the soil as larvae and therefore were not exposed directly to the pesticides applied in autumn and winter. The adults, however, were present in crops throughout the summer and consequently were exposed to demeton-S-methyl applications.

Over the whole treatment period the numbers of Empididae found in samples taken from the Full Insurance area were, on average, 75% lower than in samples from the Supervised + Integrated area. The numbers of *Platypalpus* spp and particularly *T. arrogans*, were reduced after each of the demeton-S-methyl applications to wheat crops in the Full Insurance area.

Coleoptera

The predatory beetles were represented by ground beetles (Carabidae), rove beetles (Staphylinidae), soldier beetles (Cantharidae) and ladybirds (Coccinellidae) but the ground and rove beetles always constituted at least 82% of the total individuals.

During the pre-treatment period, the average numbers of ground beetles, rove beetles and ladybirds were very similar in the two areas, although the numbers of soldier beetles were rather lower in the Full Insurance area. The overall percentage composition of the predatory beetles was very similar in the two areas (Figure 9.11).

During the treatment period the numbers of rove beetles and ladybirds were respectively 26% and 91% lower in the Full Insurance area while the numbers of ground beetles and soldier beetles were, on average, 4% and 24% *higher*. In terms of percentage composition, therefore, the contribution of both rove beetles and ladybirds was less while that of ground beetles was greater in the Full Insurance area (Figure 9.11).

Apart from ladybirds, there was considerable variation from year to year in the response of these groups to the treatment regimes, due in part to very marked differences in the effects of the Full Insurance regime on individual species.

Unlike the predatory Diptera and Arachnida, there was great variation among the Coleoptera species in their life cycle strategies and vertical distribution in the crop. To a great extent these accounted for variations in the effects of the Full Insurance regime.

Species inhabiting the soil surface in fields in winter (when there is little plant cover to give protection from spray applications), having only one generation per year and with relatively poor dispersal powers might be most vulnerable to the Full Insurance regime. This can be illustrated by the effects of the treatment regimes on some of the predatory ground beetles and rove beetles.

Ground beetles
Bembidion obtusum overwinters in the field as an adult, breeds in the spring and the larvae live in the soil during the summer. A new generation of adults emerges in late summer (after the time of demeton-

Figure 9.11 Average percentage composition of the predatory Coleoptera found in D-vac samples during the pre-treatment (1982 & 1983) and treatment (1984–1988) periods.

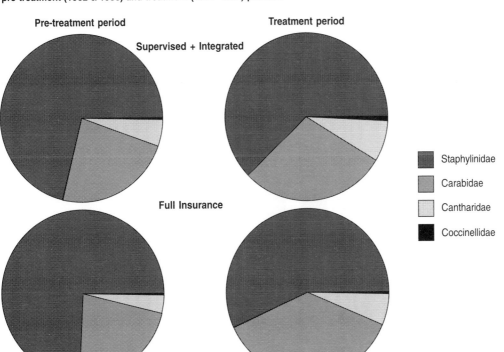

S-methyl applications) and remains in the crop. Although numbers of this species were similar in the two areas before treatment, they decreased markedly in the Full Insurance area following the introduction of the prophylactic pesticide regime in autumn 1983. By 1986 this species had been almost eliminated from the Full Insurance area (Figure 9.12).

Notiophilus biguttatus also spent the winter on the soil surface in the field, but occurred both as adults and larvae. The numbers of *Notiophilus* in the Full Insurance area were 88% lower on average than in the Supervised + Integrated area (Figure 9.12). The slightly better persistence of this species than *B. obtusum* was probably due to the fact that it completes at least two generations per year and consequently has a greater opportunity to recover from pesticide effects.

Both *B. obtusum* and *N. biguttatus* were also caught in pitfall traps (Chapter 10) which showed similar severe effects (Figure 9.10).

Trechus quadristriatus was the commonest ground beetle species in the samples. A small proportion of the population overwintered as adults on the soil surface, and were active in the fields in spring, but the bulk of the adults (derived from overwintering larvae) did not emerge until June and July, and bred in the autumn.

Before the introduction of the treatment regimes, numbers of this species were slightly lower in the Full Insurance area than in the Supervised + Integrated area (Figure 9.12). In the treatment years, adult *Trechus* were rarely found in the Full Insurance area in early spring (in contrast to the Supervised + Integrated area), but were abundant during the summer months. Indeed, total

Figure 9.12 Numbers of some predatory ground beetles (Carabidae) in D-vac samples taken from the Full Insurance area as a percentage of those in the Supervised + Integrated area during the pre-treatment (1982 & 1983) and treatment (1984–1988) periods.

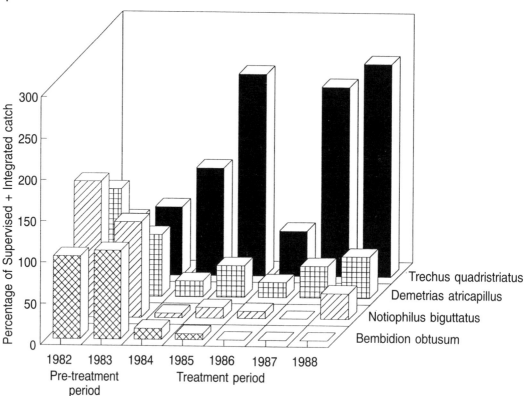

numbers of *Trechus* adults were 83% higher on average in the Full Insurance area. This difference was seen every year except 1986, and numbers were up to 160% greater in the Full Insurance area in 1988 (Figure 9.12).

These data suggest the overwintering *Trechus* adults were adversely affected by the pesticides applied in autumn and winter (similar to *Bembidion obtusum* adults). However, overwintering larvae were not only unaffected, but their survival appeared to be enhanced within the Full Insurance area.

Trechus adults were present in the crops at the time of demeton-S-methyl applications. However, they are not usually found on vegetation and there were adverse effects of these sprays only in 1986, when some crops were exceptionally thin; this probably allowed greater penetration of the insecticide to the soil surface.

Catches of *Trechus* adults in pitfall traps (see Chapter 10) were also much higher in the Full Insurance area than in the Supervised + Integrated area in all the treatment years apart from 1986 (Figure 9.10).

In contrast to these species, which remain in the field throughout the year, some leave the crops in late summer to overwinter in field boundaries. Adults recolonise the crops in spring and breed, and larvae are found during the summer months. Examples are *Agonum dorsale* (see Chapter 10) and *Demetrias atricapillus*, which was more common in the D-vac samples.

Demetrias adults colonised the crops in late May and early June. They are active plant climbers (Vickerman & Sunderland, 1975) and were exposed to the demeton-S-methyl sprays in summer. Their larvae also climb on vegetation but were not found in the samples until the end of June, and thus were exposed only to the second of the demeton-S-methyl applications. The adults were not exposed directly to pesticides applied in autumn and winter, and the effect of the Full Insurance regime on this species was smaller than on either *Bembidion* or *Notiophilus* (Figure 9.12). During the treatment period, the numbers of *Demetrias* in the Full Insurance area were, on average, 67% lower than in the Supervised + Integrated area.

Trechus always constituted at least 85% of all ground beetles found within the Full Insurance area, whereas within the Supervised + Integrated area it never exceeded 67%. Apart from *Trechus*, the total numbers of ground beetles in the Full Insurance area were, on average, 88% lower than in the Supervised + Integrated area.

Rove beetles

There was also variability in the life-cycle strategies of the predatory Staphylinidae and hence in the effects of the Full Insurance regime. For example, the numbers of *Stenus* spp. were greatly reduced within the Full Insurance area whereas the numbers of adults of the highly dispersive *Tachyporus* spp. were, on average, only 35% lower than in the lower input area.

Like *Trechus quadristriatus*, some rove beetles, such as *Xantholinus* spp. and *Lathrobium* spp., were more abundant in the Full Insurance area than in the Supervised + Integrated area during the treatment years of the Project.

Parasitoids

The parasitoids at Boxworth included primary parasitoids that attack a wide range of invertebrate hosts, and secondary parasitoids (or hyperparasitoids) that attack the primary parasitoids. Like the predators, primary parasitoids are regarded as beneficial insects because they may reduce the numbers of pests such as cereal aphids, wheat blossom midges and cereal leaf miners (e.g. Vorley & Wratten, 1985). However, primary parasitoids may also attack other beneficial invertebrates such as spiders.

The parasitoids consisted almost exlusively of Hymenoptera; only a very few dipteran parasitoids, such as Tachinidae, were found in the samples. While it was feasible to identify a few parasitoids to species or genus, most were identified at a broader level, because of the considerable taxonomic problems involved.

In all years, braconids (Braconidae), which are primary parasitoids, and chalcids (Chalcidoidea), which include both primary parasitoids and hyper-parasitoids, constituted at least 88% of the Hymenoptera found in the Full Insurance area and at least 73% of those in the Supervised + Integrated area. None of the other groups contributed more than 22% to the total parasitoids found in the samples each year.

During the pre-treatment years, the numbers of all the main groups of parasitoids were similar in the two areas (Figure 9.13) and their percentage contributions to the total parasitoids were also similar.

During the treatment period, total numbers of parasitoids were 39% to 79% lower in the Full Insurance area than in the Supervised + Integrated area (Figure 9.13). Effects of the Full Insurance regime were more severe on the chalcids (excluding Mymaridae) than on the braconids. On average, the numbers of these chalcids were 62% lower in the Full Insurance area. The Mymaridae (fairy flies), which are parasitoids of insect eggs, was the only group of parasitoids that was more numerous in the Full Insurance area. This was the case in all years apart from 1985 (Figure 9.13).

The braconids were little affected by the Full Insurance regime. This group included *Aphidius* spp., which are primary parasitoids of cereal aphids. Although the numbers of braconids were on average only 39% lower in the Full Insurance area the difference between the two treatment areas was greatest in the last two treatment years (Figure 9.13).

One marked change was in the ichneumons (Ichneumonidae). This family includes species that

Figure 9.13 Numbers of some parasitoids in D-vac samples taken from the Full Insurance area as a percentage of those in the Supervised + Integrated area during the pre-treatment (1982 & 1983) and treatment (1984–1988) periods.

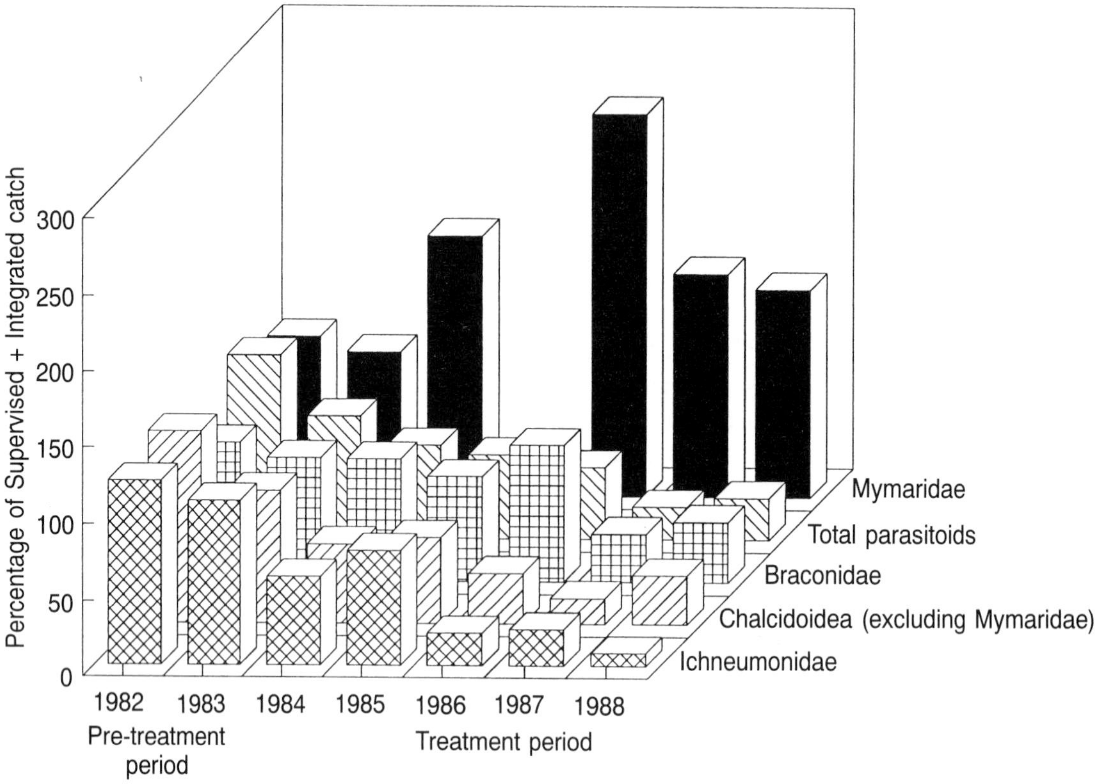

attack caterpillars (Lepidoptera), hoverflies (Syrphidae) and the egg cocoons of money spiders (Linyphiidae). By the end of the treatment period numbers were 92% lower in the Full Insurance area than in the Supervised + Integrated area (Figure 9.13).

Numbers of the other parasitoid groups (Ceraphronoidea, Scelionoidea, Proctotrupoidea and Cynipoidea) averaged 71%, 75%, 83% and 84% lower in the Full Insurance area.

As a result of these changes, the two areas differed in the percentage composition of their parasitoid faunas. In the Full Insurance area, braconids formed a much higher proportion and chalcids a lower proportion of the total than in the Supervised + Integrated area, due to the greater effect of high pesticide inputs on the chalcids.

Populations of Detritivores

In most years about 98% of the detritivores found in the D-vac samples were Coleoptera and Diptera. Other groups, such as book lice (Psocoptera), millipedes (Diplopoda), woodlice (Isopoda) and symphylids (Symphyla) were scarce. During the baseline years and the treatment period, the total numbers of detritivores found in the two areas were very similar, apart from 1985 (Figure 9.14).

In marked contrast to the herbivores and carnivores there was little evidence that this trophic group was adversely affected by the Full Insurance regime.

Coleoptera
During the pre-treatment period, Coleoptera constituted 59% of the detritivores in the Super-

Figure 9.14 Numbers of some detritivores in D-vac samples taken from the Full Insurance area as a percentage of those in the Supervised + Integrated area during the pre-treatment (1982 & 1983) and treatment (1984–1988) periods.

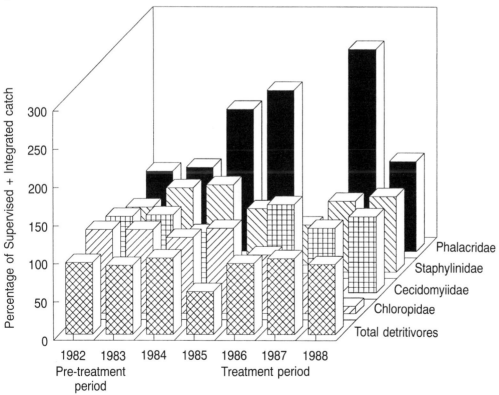

vised + Integrated area and 52% of those in the Full Insurance area. Total numbers of beetles were, on average, 18% lower in the Full Insurance area than in the Supervised + Integrated area.

These beetles were mainly rove beetles (Staphylinidae) and fungus feeding species of plaster beetles (Lathridiidae), Phalacridae and mould beetles (Cryptophagidae), and occurred as both adults and larvae. Although some species overwintered in the fields the adults of many, such as *Stilbus testaceus* (Phalacridae) and *Lathridius* spp. (Lathridiidae), overwintered in field boundaries, woodland or other non-crop habitats.

During the treatment period, the total numbers of beetles were only 2% lower in the Full Insurance area than in the Supervised + Integrated area. There was little evidence that any of the main groups were adversely affected. Some, such as the Phalacridae, appeared to benefit in some way from the Full Insurance regime (Figure 9.14). Apart from 1986, the numbers of Phalacridae were substantially higher in the Full Insurance area than in the Supervised + Integrated area. Thus, like some predatory beetles, certain detritivores were apparently favoured by the Full Insurance regime.

Diptera

The dipteran detritus-feeders belonged to sixteen families and were represented by a wide range of species. They constituted 38% and 45% respectively of the detritivores found in the Supervised + Integrated and Full Insurance areas in the pre-treatment period. The average difference in numbers between the areas was only 4%.

Nematoceran flies were numerically far more important than cyclorrhaphan flies. The former were represented mainly by gall midges (Cecidomyiidae) and fungus flies (Mycetophilidae) while the latter consisted primarily of lesser dung flies (Sphaeroceridae), Lonchopteridae, small fruit flies (Drosophilidae) and Chloropidae.

Unlike the beetles, most of these species remained in fields throughout the year, either as adults or larvae, or both. Perhaps because of this difference in life cycle strategy, the effects of the Full Insurance regime on the Diptera were greater than on the Coleoptera. Their numbers were about 20% lower in the Full Insurance area, but by comparison with other trophic groups the effects were small.

The numbers of most families of nematoceran Diptera, such as the Cecidomyiidae (Figure 9.14), were similar in the two areas in most years, but the situation with the cyclorrhaphan Diptera was more varied. Some groups, such as Drosophilidae, were more numerous in the fields within the Full Insurance area whereas others, such as Lauxaniidae and Chloropidae were lower (eg Figure 9.14).

Discussion

Studies of the side-effects of single pesticide applications on invertebrates can provide detailed information on short-term (within-season) risks. This has been the focus of most previous investigations. They can also give clues about possible longer-term consequences and the effects of repeated applications, which are often ignored. Most fields are treated year after year, and more than once within a season, often with mixtures of chemicals. It could be argued, therefore, that most small plot experiments have provided information only at a scale relevant to what farmers regard as 'spot treatments', used to control patches of grass weeds within a stubble, for example.

The Boxworth Project was able to reduce these limitations by monitoring invertebrate populations for several years on large experimental areas. Because of the size of these areas, it was not possible to adopt a replicated statistical design, which could have measured differences between areas in the density of invertebrate populations. However, the scale of the experiment increased the likelihood that the full range of side-effects would be detected, at least for the less mobile invertebrates. Many of the changes reported in this chapter would not have been identified if small plots had been used. Perhaps more effects would have been discovered if the scale of the Project had been even larger.

Baseline populations

The two-year 'baseline' period allowed an assessment of the similarity of invertebrate populations in the two treatment areas prior to the introduction of the pesticide regimes. In 1982, the first pretreatment year, there was some variability between fields, for example in the numbers of cereal leafhoppers, which were attributable chiefly to differences in sowing date. This variation was largely eliminated in the second year, so that over the whole pre-treatment period, the Supervised + Integrated area and the Full Insurance area were very similar, both in the numbers of the main invertebrate groups and in the relative numbers of herbivores, carnivores and detritivores.

Interpretation of treatment effects

During the treatment phase of the Project, differences between areas reflected both the effects of the Full Insurance regime and variations in the invertebrate populations in the Supervised + Integrated area. These were influenced by applications of insecticides to the Supervised and Integrated fields in 1984 and 1988. Also, heavy rainfall in October 1987 delayed sowing in these areas so that populations of some species were lower than normal in 1988. In all years, population levels were influenced by the weather and factors such as predation, parasitism and disease. Indeed, for some species, such as money spiders, differences between years were as great or greater than within-year differences between treatment areas. In the Full Insurance area, it was not possible to carry out all the scheduled pesticide applications as planned. This too would have contributed to variability in invertebrate populations. Despite these influences, some conclusions about the effects of the Full Insurance pesticide programme on the invertebrate community can be drawn.

The overall effects can be considered as a function of two factors: short-term risks resulting from each pesticide application, which reflect both the extent of exposure and the inherent susceptibility of each species, and the capacity of populations to recover from reductions, which depends on their ecological characteristics (dispersal ability, reproductive potential, diet, and number of generations per year).

Risk to populations

There was little evidence that herbicide or fungicide use had any substantial effects on invertebrates, compared to insecticides and molluscicides. The latter can be divided into summer treatments (demeton-S-methyl) and those applied in autumn and winter (methiocarb, a pyrethroid (permethrin, cypermethrin or deltamethrin), chlorpyrifos and triazophos).

Crop invertebrates can be grouped by aspects of behaviour and ecology that determine their exposure to summer and winter pesticides (Figure 9.15). Species include (1) those that leave the crop in late summer to overwinter in non-crop habitats and recolonise fields in spring, and (2) those that reside in the field throughout the year, or else leave the crops around harvest but recolonise newly-sown crops in the autumn. The first type are normally exposed only to summer insecticides, whereas the second are potentially exposed, directly or indirectly, to all applications. The latter were much more severely affected by the Full Insurance regime at Boxworth.

Vertical distribution was also important in determining pesticide effects. In summer, species that are active in the upper layers of the crop canopy are potentially more exposed to aphicide sprays than those which are active in the soil, on its surface, or in the lower part of the canopy, where they are afforded some degree of protection. For example, many herbivores and their associated parasitoids and predators (see Figure 9.15) were heavily exposed, and were severely reduced in numbers. Among the ground beetles, surface-active species such as adults of *Trechus quadristriatus* were little affected in most years, whereas plant-climbing species such as *Demetrias atricapillus* were adversely affected. In 1986, however, when the crops were exceptionally thin, allowing

Figure 9.15 Examples of invertebrates that differ with respect to their life cycle strategies and vertical zonation in the crop and hence vary in their exposure to pesticide applications.

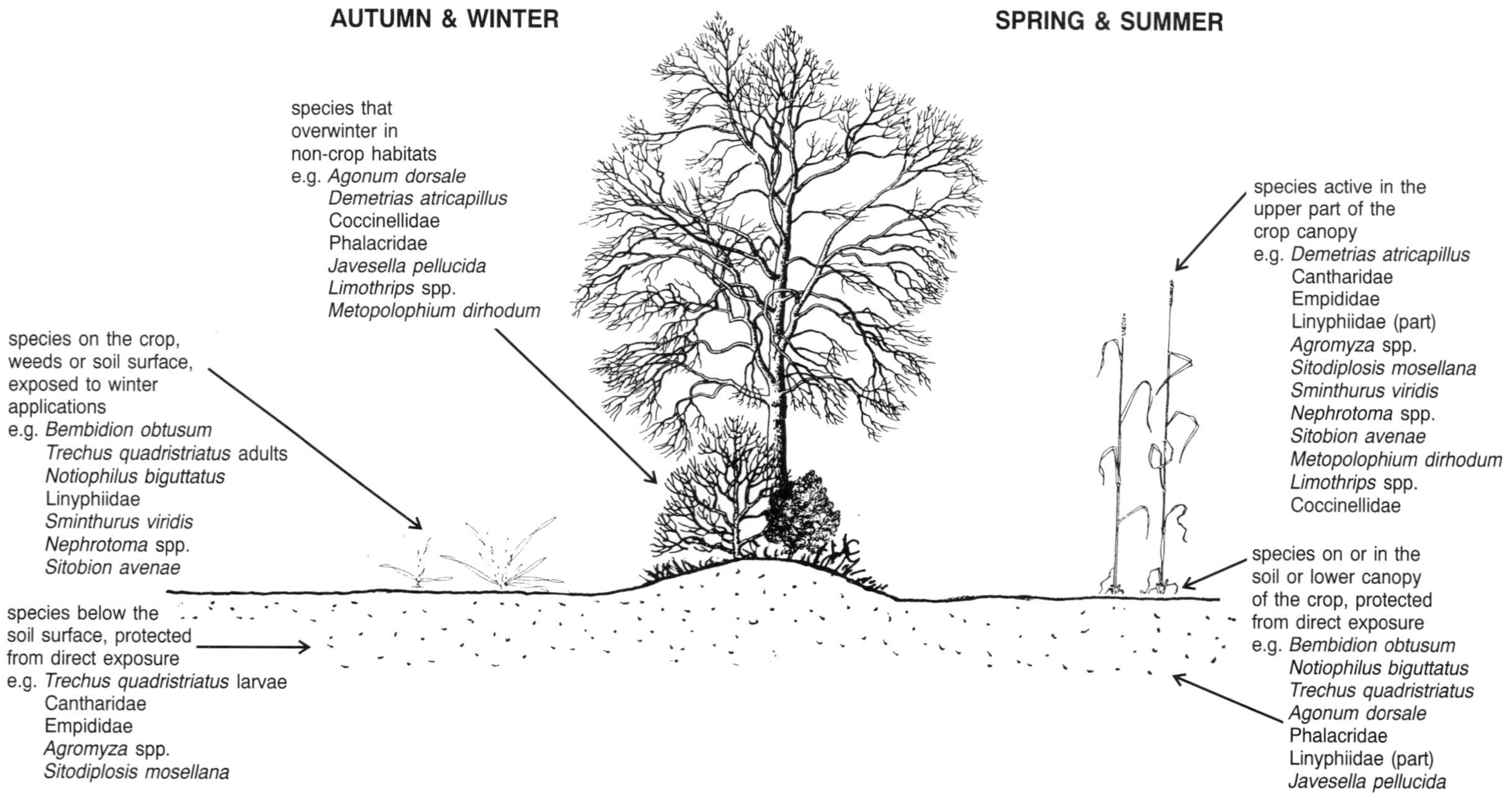

AUTUMN & WINTER

species that
overwinter in
non-crop habitats
e.g. *Agonum dorsale*
Demetrias atricapillus
Coccinellidae
Phalacridae
Javesella pellucida
Limothrips spp.
Metopolophium dirhodum

species on the crop,
weeds or soil surface,
exposed to winter
applications
e.g. *Bembidion obtusum*
Trechus quadristriatus adults
Notiophilus biguttatus
Linyphiidae
Sminthurus viridis
Nephrotoma spp.
Sitobion avenae

species below the
soil surface, protected
from direct exposure
e.g. *Trechus quadristriatus* larvae
Cantharidae
Empididae
Agromyza spp.
Sitodiplosis mosellana

SPRING & SUMMER

species active in the
upper part of the
crop canopy
e.g. *Demetrias atricapillus*
Cantharidae
Empididae
Linyphiidae (part)
Agromyza spp.
Sitodiplosis mosellana
Sminthurus viridis
Nephrotoma spp.
Sitobion avenae
Metopolophium dirhodum
Limothrips spp.
Coccinellidae

species on or in the
soil or lower canopy
of the crop, protected
from direct exposure
e.g. *Bembidion obtusum*
Notiophilus biguttatus
Trechus quadristriatus
Agonum dorsale
Phalacridae
Linyphiidae (part)
Javesella pellucida

much greater penetration of aphicide through the crop canopy than usual, many invertebrate groups, particularly carnivores (including adult *Trechus*) were severely affected.

There were also differences in effects because summer aphicide applications did not always occur at the same time relative to species' occurrence in the crop. For example, the effects of demeton-S-methyl were less marked than usual in 1988, when adverse weather conditions delayed the second application. However, even when sprays were applied when planned, exposure differed between species, depending on the proportion of the population that had emerged from the soil, or had colonised the crop.

A similar range of factors influenced the effects of winter applications. Invertebrates that feed on the crop or weeds, or are active on the soil surface (see Figure 9.15), were exposed directly to these pesticides. Following the introduction of the treatment regimes in autumn 1983, the numbers of many such species (eg *Bembidion obtusum, Notiophilus biguttatus*, money spiders and lucerne-fleas) had declined, even by the spring of 1984. Similar adverse effects on ground beetles and money spiders have been noted in previous short-term studies of the effects of pyrethroid insecticides and of molluscicides in autumn (eg Purvis, Carter & Powell, 1988; Kennedy, 1988).

In contrast, species that spend the winter below the soil surface, such as the larvae of predatory flies, wheat blossom midge cocoons and leaf-mining *Agromyza* pupae (see Figure 9.15), were not directly exposed to winter applications, and showed little evidence of effects by spring 1984. In *Trechus quadristriatus*, the surface-active adults were severely affected, whereas the soil-inhabiting larvae were not.

Invertebrates that spent the winter in non-crop habitats were not normally exposed to winter applications but there was evidence that some were severely affected by triazophos or dimethoate applied after colonisation of fields had commenced in spring 1986. In a prophylactic regime some species are likely to be exposed to such hazards from time to time.

As well as their ecological characteristics, the intrinsic susceptibility of species to pesticide poisoning is likely to vary, affecting the severity of adverse effects. There are few data on the susceptibility of most of the invertebrates occurring at Boxworth. However, if different pesticides had been used, the relative effects of summer and winter applications might have been different. For example, dimethoate can be used to control cereal aphids in summer and is more toxic than demeton-S-methyl, causing major effects even on soil invertebrates (Vickerman & Sunderland, 1977; Vickerman *et al.*, 1987a, 1987b).

Nevertheless, invertebrates that inhabited the soil surface in autumn and winter and the upper crop canopy in summer (eg lucerne-fleas, spotted crane flies, grain aphids and money spiders) were most at risk from the Full Insurance regime. Those that overwintered in non-crop habitats or below soil level in fields and subsequently avoided the summer aphicides by their phenology (eg cereal leaf miners) or by virtue of the fact that they occupied the lower strata of the crop in summer (eg many fungus-feeding beetles and most of the *Trechus* population) were least at risk.

Recovery of populations

Dispersal ability was the most important factor determining the long-term susceptibility of invertebrates. For example, the lucerne-flea and grain aphids were resident in the crop canopy all year and were both heavily exposed to insecticides. However, the lucerne-flea (which can disperse only by walking or hopping short distances) was virtually eradicated from the Full Insurance area, whereas grain aphids (which disperse readily by flight) were able to recolonise the area every year, even following the summer insecticide applications.

Similarly, carnivores such as the ground beetle *Bembidion obtusum* and money spiders were particularly vulnerable to winter pesticides, but whereas *B. obtusum* (which disperses by walking) was quickly eliminated from the Full Insurance fields, money spiders were able to recolonise the crops by 'ballooning' on the wind.

Ladybirds (Coccinellidae) and predatory flies (Empididae) were not exposed to pesticides applied in autumn and winter, but were adversely affected by summer aphicide applications. There was no evidence that the highly mobile ladybirds were depleted, but many predatory flies were unable to recolonise the crops and their populations were adversely affected in the long-term.

Most of the species involved (apart from cereal aphids) complete only one or two generations each year. Consequently, the number of generations each year was much less important than dispersal for the recovery of populations.

There was also evidence that diet influenced both the susceptibility of some carnivores and their ability to recover from the Full Insurance regime. Species that fed on prey items severely depleted by the Full Insurance regime were affected more than those eating food that was renewed annually. Thus the declines of beetles which feed almost exclusively on springtails (eg *Stenus* spp., *Notiophilus biguttutas, Loricera pilicornis*) may have been exacerbated by reduced densities of these insects (see Chapter 10). Similarly, the declines of some parasitoids were probably due in part to a lack of potential hosts after spraying. In contrast, aphid-specific natural enemies (ladybirds, hoverflies and *Aphidius* spp.) whose food was renewed annually were less severely affected.

Interactions Between Species

The effects of the Full Insurance programme on the dynamics of pests and their natural enemies are complex. Cereal aphids and the other pests of wheat at Boxworth are examples of species which are of economic importance from time to time, but are generally held below the carrying capacity of the habitat by the action of natural enemies. Such highly mobile pests are able to colonise crops before specialist predators and parasitoids find them. In these circumstances, resident non-specialist predators can be particularly important in helping to reduce the numbers of pests such as cereal aphids (Edwards *et al.*, 1979; Chiverton, 1986). Since natural enemies were reduced by 50% on average in the Full Insurance fields, and some field residents were more or less eliminated, the likelihood that pest outbreaks would occur was increased. Indeed, the numbers of certain pests, notably the grain aphid and rose-grain aphid, were highest in the Full Insurance area in some years.

The adoption of a prophylactic approach to pest control therefore tends to favour highly dispersive species and shifts the natural control of pests from resident predators and parasitoids with relatively poor powers of dispersal, that were selectively and severely depleted, to more mobile natural enemies. However, the ability of these species to recolonise the Full Insurance area was limited or delayed by adverse weather conditions in some years. Consequently, the likelihood of pest outbreaks might increase not only because there were fewer natural enemies, but also because of a greater delay between the arrival of pests and their natural enemies. The system of natural control would thus become less predictable.

This study addressed the effects of a Full Insurance programme that included considerably higher pesticide inputs than many cereal growers were using. However, the results, and those of other short-term studies, suggest that many of the population changes observed would have occurred even if there were only one or two broad-spectrum compounds applied in autumn/winter, and one aphicide in summer.

Acknowledgements
I would like to thank all the staff at Boxworth EHF for their help throughout the course of this work and the Ministry of Agriculture, Fisheries and Food who financed the studies.

References
Chambers, R J, Sunderland, K D, Wyatt, I J & Vickerman, G P (1983). 'The effects of predator exclusion and caging on cereal aphids in winter wheat.' *Journal of Applied Ecology* **20**, 209–224.

Chapman, P J, Sly, J M A & Cutler, J R (1977). *Pesticide Usage Survey Report 11. Arable farm crops 1974.* MAFF, London.

Chiverton, P A (1986). 'Predator density manipulation and its effect on populations of *Rhopalosiphum padi* (Hom.: Aphididae) in spring barley.' *Annals of Applied Biology* **109**, 49–60.

Dewar, A M & Carter, N (1984). 'Decision trees to assess the risk of cereal aphid (Hemiptera: Aphididae) outbreaks in summer in England.' *Bulletin of Entomological Research* **74**, 387–398.

Dietrick, E J (1961). 'An improved back pack motor fan for suction sampling of insect populations.' *Journal of Economic Entomology* **54**, 394–395.

Edwards, C A, Sunderland, K D & George, K S (1979). 'Studies on polyphagous predators of cereal aphids.' *Journal of Applied Ecology* **16**, 811–823.

Kennedy, P J (1988). 'The use of polythene barriers to study the long term effects of pesticides on ground beetles (Carabidae, Coleoptera) in small-scale field experiments.' In: Greaves M P, Smith B D & Greig-Smith P W (Eds). *Field Methods for the Study of Environmental Effects of Pesticides*. BCPC Monograph No. 40. pp. 335–340.

Potts, G R (1986). *The Partridge. Pesticides, Predation and Conservation.* Collins, London.

Potts, G R & Vickerman, G P (1974). 'Studies on the cereal ecosystem.' *Advances in Ecological Research* **8**, 107–197.

Purvis, G, Carter, N & Powell, W (1988). 'Observations on the effects of an autumn application of a pyrethroid insecticide on non-target predatory species in winter cereals.' In: Cavalloro R & Sunderland K D (Eds) *Integrated Crop Protection in Cereals*, A A Balkema, Rotterdam, pp. 153–166.

Sly, J M A (1986). *Pesticide Usage Survey Report 35. Arable farm crops and grass 1982.* MAFF, London.

Steed, J M, Sly, J M A, Tucker, G G & Cutler, J R, (1979). *Pesticide Usage Survey Report 18. Arable farm crops 1977.* MAFF, London.

Sunderland, K D & Vickerman, G P (1980). 'Aphid feeding by some polyphagous predators in relation to aphid density in cereal fields.' *Journal of Applied Ecology* **17**, 389–396.

Sunderland, K D, Crook, N E, Stacey, D L & Fuller, B J (1987). 'A study of feeding by polyphagous predators on cereal aphids using ELISA and gut dissection.' *Journal of Applied Ecology* **24**, 907–933.

Vickerman, G P (1977). 'Monitoring the insect fauna of cereals and grasses.' *Game Conservancy Annual Review* **8**, 43–49.

Vickerman, G P (1980). 'Important changes in the numbers of insects in cereal fields.' *Game Conservancy Annual Review* **11**, 67–72.

Vickerman, G P (1982). 'Distribution and abundance of adult *Opomyza florum* (Diptera: Opomyzidae) in cereal crops and grassland.' *Annals of Applied Biology* **101**, 441–447.

Vickerman, G P & O'Bryan, M (1979). 'Partridges and insects.' *Game Conservancy Annual Review* **10**, 35–43.

Vickerman, G P & Sunderland, K D (1975). 'Arthropods in cereal crops: nocturnal activity, vertical distribution and aphid predation.' *Journal of Applied Ecology* **12**, 755–766.

Vickerman, G P & Sunderland, K D (1977). 'Some effects of dimethoate on arthropods in winter wheat.' *Journal of Applied Ecology.* **14**, 767–777.

Vickerman, G P, Coombes, D S, Turner, G, Mead-Briggs, M A & Edwards, J (1987a). 'The effects of pirimicarb, dimethoate and deltamethrin on Carabidae and Staphylinidae in winter wheat.' *Mededelingen van de Faculteit Landbouwwetenschappen Rijksuniversiteit Gent* **52**, 213–223.

Vickerman, G P, Coombes, D S, Turner, G, Mead-Briggs, M A & Edwards, J (1987b). 'The effects of pirimicarb, dimethoate and deltamethrin on non-target arthropods in winter wheat.' *Proceedings of the International Conference on Pests in Agriculture* **1**, 67–74.

Vorley, W T & Wratten, S D (1985). 'A simulation model of the role of parasitoids in the population development of *Sitobion avenae* (Hemiptera: Aphididae) on cereals.' *Journal of Applied Ecology* **22**, 813–823.

Interactions between cereal pests and their predators and parasites

10

Alastair J. Burn (Cambridge University)

Introduction

For several species of cereal pests, natural enemies (predators and parasites) play an important role, reducing the need for pesticide applications and other control measures. There is considerable evidence for the importance of natural enemies in controlling both autumn and summer aphid populations (Burn, 1987; Kendall *et al*., 1986). There is concern that adverse effects of pesticides on these predators may increase the need to carry out treatments against pests. For example, a long-term survey of insect populations in cereal fields carried out by the Game Conservancy (Vickerman, 1980) showed a general decline in the numbers of some beneficial insects accompanying an increasing use of pesticides over a 14-year period. However, the interpretation of such patterns is difficult, because it is not possible to separate the influence of other variables operating at the same time.

The direct, short-term effects of pesticides on invertebrate predators have been comparatively well-studied (Burn, 1989). However, few field experiments have examined effects of pesticides persisting over more than one season or on a scale sufficient to minimise the effects of immigration. Nor have such studies examined the possible indirect effects of pesticides on natural enemies (eg changes in food availability) which may also cause long-term population reductions. Many 'polyphagous' predators (those which take a variety of foods) also eat aphids, and are known to be sensitive to the insecticides used in the Boxworth Project (Vickerman & Sunderland, 1977; Powell, Dean & Bardner, 1985; Shires, 1985), while certain classes of their prey, such as springtails, are sensitive to fungicides (Frampton, 1988). Therefore, the present study aimed to examine both long-term and indirect effects of pesticides on predators.

Long-term effects on invertebrate populations may arise either through a single pesticide application at a critical phase of the population cycle (particularly if poor immigration prevents population recovery), or by repeated applications of chemicals having a cumulative effect on the population. It was not the intention of this study to identify the particular chemicals which have the most significant effects, but rather to examine the impact of a complete pesticide programme. Nevertheless, some information was obtained on short-term depressions in animals' activity following particular pesticide applications.

Although many predators and parasites of dipteran pests such as wheat bulb fly and frit fly are known, there is little evidence that they can significantly reduce pest populations in cereals. For this reason, together with the low frequency of outbreaks of stem-boring flies at Boxworth, this study concentrated on examining the effects of the three treatment regimes on natural enemies of cereal aphids. In addition, an attempt was made to measure the effects of the three treatment programmes on the impact of predators on slugs, which occurred regularly at Boxworth. These were principally the grey field slug *Deroceras reticulatum*, which polyphagous predators are known to attack (Tod, 1970).

The principal aim of studying pest-predator interactions was to examine the effects of high pesticide inputs on the ability of predators and parasites to reduce aphid outbreaks. The work fell into two parts: (1) a monitoring exercise was carried out to demonstrate any changes in the populations or activity of the major cereal pests (aphids and slugs) and their natural enemies, and (2) a more detailed study was made of the interactions between predators and aphids in the three treatment areas.

Because the unreplicated design of the Project did not allow statistical evaluation of between-treatment differences within any given year (Chapter 2), this work was based on trends in populations or predation rates over the course of the Project, relative to levels measured during the baseline years.

This Chapter is divided into two parts. The first deals with the effects of pesticide treatments on the natural enemies themselves (principally polyphagous predators; see also Chapter 9). It examines changes in species composition, abundance, and activity of predators, and indicators of indirect effects such as changes in their diet. The second part considers the consequences of these changes for the ability of predators to prey on pests, and describes differences in predation rates on aphids in the different treatment areas, as well as changes in predation measured on artificial prey and on slugs.

Effects of the treatment regimes on predators

Introduction

Both polyphagous predators such as ground beetles, rove beetles and money spiders, and aphid-specific natural enemies such as ladybirds, parasitoids and parasitic fungi, are able to reduce the growth of cereal aphid populations in some years (Chambers et al., 1983; Burn, 1987). However, aphid-specific predators usually do not appear in cereal fields until aphid populations have started to build up so they are less likely than generalists to provide effective control. Moreover, most species are highly mobile and, because of immigration, long-term population effects are unlikely to be seen even in a study on the scale of the Boxworth Project. For this reason, although aphid-specific groups were recorded, detailed monitoring of natural enemies concentrated on those groups of polyphagous predators thought to have a greater influence on predation of aphids. Such predators are generally confined to cereal fields or adjoining habitats, and long-term effects of pesticides could therefore be followed in relatively isolated populations within each of the treatment areas.

Two approaches were used to measure indirect effects of pesticides on selected species of polyphagous predators. For those predators which colonise cereal fields in spring, having overwintered in field boundaries, rates of colonisation were measured in order to reveal any differences between the treatment areas that might be attributable to any changes in plant cover caused by herbicide use, or to the abundance of prey. The effect of variation in the abundance of alternative prey (ie prey other than aphids) on diet and on population recruitment were investigated for the ground beetles *Agonum dorsale* (an abundant species which is an important cereal aphid predator; Griffiths, 1985), and *Loricera pilicornis*, which has a similar ecology to *A. dorsale*.

Sampling methods

Samples of invertebrates were collected using pitfall traps, a Dietrick vacuum insect sampler (D-vac) and searches of the soil surface in fields. Pitfall-trap catches depend on animals' movement and therefore they provide a combined measure of numbers and activity, whereas D-vac samples, which are less affected by movements, can provide an estimate approximating the actual density.

Pitfall traps consisted of cylindrical white plastic beakers 9 cm in diameter and 12 cm deep, each filled to a depth of 5 cm with a weak solution of formaldehyde and detergent. These were sunk in the ground so that the rims were flush with the surface. At least three fields were sampled in each of the treatment areas in all years. Traps were placed near the field centre, with a minimum of 25 m between adjacent traps. The number of traps varied but there was a minimum of eight traps per field, with more in the intensive study fields. Cereal fields were sampled in this way weekly between May and August in each year, the period of aphid population development and decline. In addition, in each of the intensive study fields, rows of five or eight pitfall traps were placed parallel to a hedgerow at distances of 0 m, 5 m, 75 m and 150 m, in order to monitor colonisation of the fields by predators between March and May. Most of the results presented are from the mid-field samples (75 m and 150 m).

Changes in species diversity

In the baseline years there were some differences between the treatment areas in the species present and their abundance in traps. For example, of the ground beetles *Pterostichus melanarius* and *P. macer*, the former species was more abundant in pitfall traps in the Supervised and Integrated areas whereas the latter was more abundant in the Full Insurance area. However, in most cases these differences were slight for the more abundant predators, and variation in catches between fields within each area was usually as great as that between areas. Although predator populations in the Supervised and Integrated areas were slightly different, no further differences between them emerged as the Project progressed. Therefore, in order to simplify the presentation in this Chapter, most results for the Full Insurance treatment are compared with combined data for the Supervised and Integrated treatments.

The numbers of species found in the Full Insurance area relative to those in the Integrated and Supervised areas are shown in Table 10.1 for two major groups of polyphagous predators (ground beetles and money spiders). A list of all the species of ground beetles (Carabidae), rove beetles (Staphylinidae, excluding Aleocharinae) and money spiders (Linyphiidae) found during the Project is given in Appendix IV. A diverse range of species was present, including a larger than expected number of species that are usually associated with woodland. There were no consistent differences in numbers of species of either ground beetles or money spiders between treatment areas or between years (Table 10.1). The Full Insurance treatment regime does not appear to have caused an overall reduction in the numbers of species. Possible effects of the high input regime must be sought in changes in abundance of individual species rather than in the number of species present.

Changes in catches of ground-living predators

Changes in the numbers/activity of predators in the Full Insurance area, as a percentage of the average in the lower-input areas, are shown in Figure 10.1 for nine types of predators. This presentation obscures differences between fields, but allows broad trends to be identified. Variation between years was considerable for certain species, but is not revealed by Figure 10.1. However, it should not affect the conclusions drawn in this Chapter. The patterns will be described fully in future publications.

The species considered in Figure 10.1 may be classified into groups according to their risk of exposure to pesticides (determined primarily by their overwintering strategies and reproductive seasonality) and the ability of their populations to recover following adverse effects of exposure (which depends on sources of new colonisers and rates of dispersal). This classification is neither rigid nor complete, because certain predators do not fit easily into any category and the physiological

Table 10.1 **Numbers of species trapped in the centre of fields in the Full Insurance area as a percentage of those in the Integrated and Supervised areas.** (NE indicates that no estimate was made).

	Baseline Years		Treatment Years				
	1982	1983	1984	1985	1986	1987	1988
Ground beetles (Carabidae)	103	108.7	98	90.9	94.7	109.8	102.1
Money spiders (Linyphiidae)	NE	91.7	NE	NE	95.0	NE	116.7

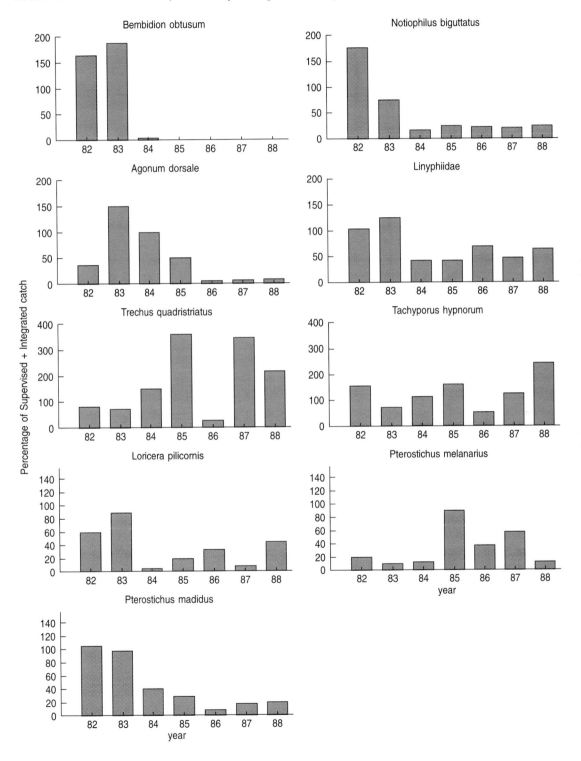

Figure 10.1 Changes in the pitfall-trap catches (representing numbers and activity) of selected invertebrate predators in the Full Insurance area at Boxworth, shown as a percentage of the average catch in the Supervised and Integrated areas.

susceptibility of particular species to pesticides is not taken into account. However, the scheme in Table 10.2 provides a framework for comparing the widely differing effects shown on various groups of predators.

Bembidion obtusum and *Notiophilus biguttatus* (group (i) in Table 10.2) both showed an immediate, major reduction in numbers/activity in the Full Insurance area relative to the baseline catches (Figure 10.1). These species are poorly dispersive, remaining in the centre of the fields as adults during the winter, at a time when the majority of broad-spectrum insecticides are applied, and when crop cover is sparse. Therefore the exposure of these ground beetles is high, and their ability to recolonise fields is poor. There was no recovery from the low level reached after the first year of the Full Insurance programme. Similar results were obtained for these species using suction sampling (see Chapter 9).

The ground beetle *Agonum dorsale* did not show an immediate adverse effect of the Full Insurance regime, and a major change from the very variable trap catches in the baseline years was first seen only in the third year after the start of the treatment phase of the Project (Figure 10.1). A large decline occurred in 1986, and there was no subsequent recovery. This spring-breeding species overwinters as an adult in field boundaries, recolonising cereal fields in spring. Since dispersal from the field boundary usually takes place two or three months after the final winter insecticide application, direct contact with an insecticide should be infrequent. The decline in 1986 took place after a late application of dimethoate in the Full Insurance area (Appendix II), in a year when crop cover was patchy. Failure of *Agonum* populations to recover may reflect poor dispersive ability, but could also be due to indirect effects of the Full Insurance treatment on alternative prey species (see Chapter 9).

The ground beetle *Loricera pilicornis* has a similar life-history to *A. dorsale*, overwintering in woodland and colonising cereal fields to breed in spring. Unlike *A. dorsale*, there was a major decline in the first treatment year, and numbers/activity remained generally low thereafter (Figure 10.1). This species is also active in autumn, and is a more specialist springtail predator than *A. dorsale* (see later). Therefore, it may have been more strongly affected by the Full Insurance treatment, especially by effects on its springtail prey.

The rove beetle *Tachyporus hypnorum* is representative of highly dispersive species which colonise fields from distant overwintering sources in spring in a similar way to aphid-specific predators. There was a reduction in catches of this species in the Full Insurance area in 1986, probably a result of the poor crop cover when summer insecticides were applied (see Chapter 9), but otherwise the index of numbers/activity was similar

Table 10.2 Dispersive ability, overwintering characteristics and vulnerability of four ecological categories of predators in cereal fields to pesticides used in the Boxworth Project.

Ecological group	Dispersive ability	Principal overwintering stage and habitat	Vulnerability to long-term effects of pesticides		Representative species
			Direct effects	Indirect effects	
(i)	Poor	Adult (mid-field)	High	High	*Bembidion obtusum* *Notiophilus biguttatus*
(ii)	Moderate	Adult (field boundary)	Low	High	*Agonum dorsale*
(iii)	High	Adult (non-crop habitat)	Low	Low	Tachyporinae, aphid-specific predators
(iv)	Moderate	Larva (mid-field)	Low	High	*Trechus quadristriatus,* *Pterostichus melanarius*

to, or greater than, that in the lower pesticide input areas.

The three ground beetles *Trechus quadristriatus, Pterostichus melanarius* and *P. madidus* are principally late summer or autumn breeders and moderate or poor dispersers. Since the majority of their populations overwinters as immature stages in the soil, they may be less exposed to the direct effects of pesticides even when crop cover is sparse. *P. melanarius* and *T. quadristriatus* catches showed no decrease; indeed, the numbers/activity of *T. quadristriatus* in the Full Insurance area were much higher than the baseline levels in several years (Figure 10.1), and a similar effect was noted using suction sampling (see Chapter 9). This increase is hard to explain in view of other evidence for a shortage of alternative prey (see Chapter 9), but a reduction in competition or predation may have played a role. In contrast, *P. madidus* showed a progressive reduction in numbers/activity relative to the baseline years (Figure 10.1). The reason for this difference is not known, but it is evident that species with apparently similar life-histories may experience different effects of pesticide treatments.

The money spiders (Linyphiidae) comprise a large number of species. The group as a whole showed a reduction in catches in the Full Insurance area relative to baseline levels (Figure 10.1). There was an early season fall in numbers/activity in the Full Insurance area due to pesticides applied in autumn and winter, followed by a rapid increase in all areas following summer immigration (see Figure 9.9). The money spiders therefore show both the vulnerability and high dispersive ability of groups (i) and (iii) in Table 10.2, having more marked within-season effects than between-seasons trends.

In summary, the monitoring study showed long-term decreases in catches of many of the important polyphagous predators in the Full Insurance treatment area. The effects of pesticide treatments are a function of species' physiological sensitivity and exposure. The major impact on polyphagous predators appeared to be due to treatments applied during winter and the timing of

such applications may be crucial. The ability to recover from the adverse effects of pesticides depends on dispersive ability (contrast group (i) predators with the highly-dispersive money spiders). Whereas indirect effects may have contributed to the patterns shown by catches of group (ii) predators (*Agonum*) and group (iv) predators (*Trechus and Pterostichus*), their importance is not clear from data on adults alone.

Indirect effects on predators

Availability of alternative prey

Polyphagous predators in cereal crops depend on alternative sources of prey before the arrival of cereal aphids in spring. Several studies have shown the importance of springtails in the diet of such predators (Griffiths, 1985; Chiverton, 1984). The effect of changes in the abundance of these prey was examined by studying the diet of *A. dorsale* and *L. pilicornis*.

Material for gut analysis by dissection was collected in pitfall traps between May and July (*A. dorsale*) and July and August (*L. pilicornis*). Solid remains in the guts were identified and assigned to the following groups: springtails, aphids, cecidomyiid midges (adults and larvae), other flies, mites, beetle larvae, earthworms and unidentified remains. In addition, the numbers of mature eggs in the ovaries were counted as an index of fecundity. Differences in dietary composition between treatments were examined using Analysis of Variance.

Figure 10.2 shows changes in the frequency of occurrence of ground-dwelling springtails (excluding the Sminthuridae) found in pitfall traps in the Full Insurance area relative to the Supervised and Integrated treatment areas. A decline in catches of springtails in the Full Insurance area is apparent also for the lucerne-flea *Sminthurus viridis* (family Sminthuridae) in suction samples (see Chapter 9), and for certain subterranean springtails such as *Folsomia quadrioculata* collected in soil cores (Chapter 11). Like the group (i) predator species (Table 10.1), ground-dwelling

Figure 10.2   Numbers of springtails (Collembola) trapped in the Full Insurance area as a percentage of the average in the Supervised and Integrated areas.

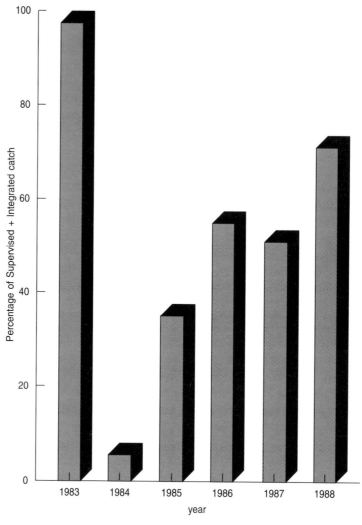

springtails are likely to be exposed to pesticides applied when the crop canopy is open. Some species of the Sminthuridae, which are active in the upper parts of the crop, are also likely to be exposed to fungicides and insecticides applied in spring and summer.

In the baseline years of the Project, there were no differences between the treatment areas in the occurrence of springtails in the guts of *L. pilicornis* or *A. dorsale*, although the frequency in the guts of *L. pilicornis* was greater, in line with its habits as a specialised springtail predator (eg

Bauer, 1982) (Figure 10.3). Data are shown in Figure 10.3 for those treatment years in which adequate numbers of the predators could be collected for gut analysis. In all cases there was a statistically significant lower frequency of springtails in the guts of beetles from the Full Insurance area compared with the other treatment areas, indicating that this important component of their diet was less easily obtained.

Long-term shortages of prey may cause a reduction in predators' fecundity, which could potentially result in a decrease in recruitment. The data presented in Table 10.3 suggest a lower fecundity for *L. pilicornis* in the Full Insurance area in 1988, but the variation in fecundity for *A. dorsale* makes interpretation of data for that species more difficult. Egg production measured by dissection of female beetles may not give a reliable indication of the true extent of effects on egg production in the field. However, this study has established clear links between pesticide use, dietary habits and availability of alternative prey, and has indicated a possible correlation between such changes and egg production in a specialist predator of springtails.

Colonisation of fields during spring migration

Colonisation of fields by species such as *A. dorsale* may be affected by pesticide use in several ways. The factors stimulating spring movement into fields

Figure 10.3 Frequency of occurrence of springtails in the guts of the ground beetles *Agonum dorsale* and *Loricera pilicornis*, shown by dissection.

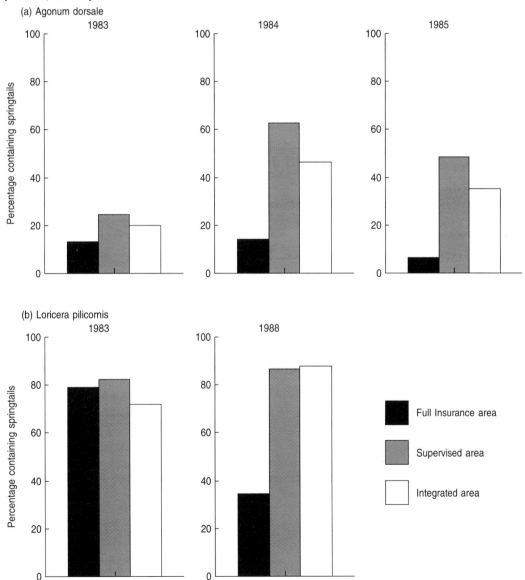

from hedgerow overwintering sites are not well understood, but it is likely that adequate vegetation cover is required. The elimination of weeds through herbicide use may be detrimental to some beetles (eg Vickerman, 1978) and thus reduced abundance of weeds might prevent or retard colonisation by such predators. Reduced prey availability might result in greater movement by predators (Chiverton, 1984) and so could lead to more rapid colonisation of fields. Alternatively, if predators aggregate at sites of high prey density a reduction of prey abundance in the fields might inhibit colonisation. These conflicting tendencies might also affect the total numbers of individuals moving into fields, as well as their rate of colonisation. However, without more detailed studies involving individually-marked animals, it would not be possible to differentiate this from direct mortality caused by pesticides.

Figure 10.4 illustrates, in a simplified form, changes in one measure of the rate of colonisation by *A. dorsale*, calculated as the interval between peak catches in the field boundary and the date when colonisation of the field appeared to be complete. There was a tendency for more rapid colonisation in the Integrated area (although this may have been due to the smaller fields with a greater amount of hedgerow per unit of field area). However, the rates of colonisation in the Supervised and Full Insurance areas remained similar throughout the study, and there were no obvious indications that either changes in prey abundance or the effect of Full Insurance herbicide treatments affected this component of the predator's behaviour.

Effects of the treatment regimes on predation

Introduction

The major importance of polyphagous predators in the cereal ecosystem lies in their predation of the grain aphid (*Sitobion avenae*) and rose-grain aphid (*Metopolophium dirhodum*) in summer, and of aphids transmitting Barley Yellow Dwarf Virus in autumn and winter. It was shown above (see also Chapter 9) that populations of many polyphagous predators were adversely affected by the Full Insurance treatment regime. In order to determine the implications of this for pest outbreaks, changes in predation on aphids were measured.

A range of techniques was used to measure levels of predation directly and to manipulate the numbers of predators experimentally. These aimed to reveal any ways in which the pesticide regimes might have interfered with the relationship between predators and prey.

Predator exclusion experiments

The principal technique used was to exclude predators from three small plots (approximately 80 m² in area) in each of the intensive study fields from April to August in 1983 to 1987. Pest and predation levels were then compared within exclusion and control plots in each treatment area. In this way the effectiveness of predator populations which survived the pesticide treatments could be measured by monitoring plots from which those predators had been excluded. To exclude predators, plastic bar-

---

Table 10.3   The mean number of eggs per gravid female of *Agonum dorsale* and *Loricera pilicornis* in each of the three treatment areas. For each year, means followed by the same letter are not significantly different.

	Agonum dorsale			*Loricera pilicornis*		
	Full Insurance	Supervised	Integrated	Full Insurance	Supervised	Integrated
1983	3.13 a	4.63 b	4.93 b	7.10 ab	5.16 a	7.88 b
1984	3.24 a	3.26 a	3.40 a	–	–	–
1985	3.84 a	5.16 b	4.02 ab	–	–	–
1988	–	–	–	5.24 a	6.59 b	7.41 b

Figure 10.4   Movement of *Agonum dorsale* into fields from hedgerow overwintering sites, 1983–88.

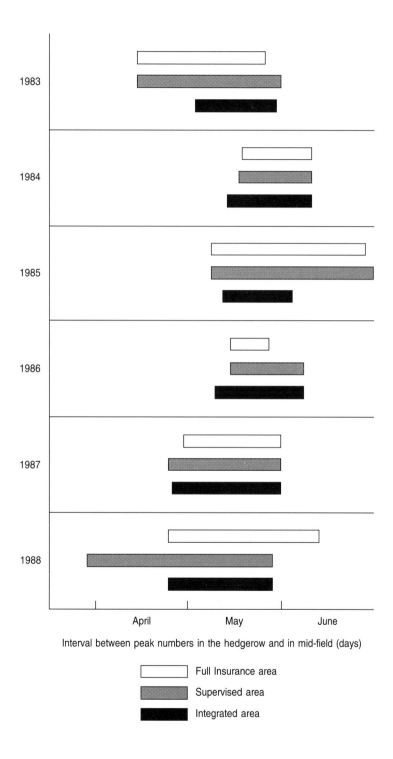

Interval between peak numbers in the hedgerow and in mid-field (days)

☐ Full Insurance area

▨ Supervised area

■ Integrated area

riers were erected, extending 15–30 cm below the soil surface and 30–45 cm above ground, and were supplemented with a soil insecticide treatment around the outside of the plot, and periodically within it early in the season. All exclusion and control plots were protected from direct overspraying of the summer aphicides during spray applications by a 2.2 m plastic windbreak. This allowed the continued development of natural aphid infestations within the plots, whereas the majority of the (more mobile) predator population was exposed to all pesticides. Within the plots, aphid and predator populations were measured using wheat tiller counts, soil samples or suction samples, and predator activity was measured using pitfall traps.

In 1988 five smaller circular exclusion plots (25 cm in diameter) were used in place of the larger barriers. These were located more widely over the field and were not protected from aphicide applications, in order to permit measurement of any subsequent population resurgence. Within these smaller exclusion plots and corresponding control plots, aphid numbers were counted on marked tillers throughout the summer.

The exclusion plots were effective in reducing numbers of ground-moving polyphagous predators in categories (i), (ii) and (iv) (Table 10.2); for example A. dorsale was reduced by 98–100% at its peak of activity. Group (iii) and other more dispersive species were reduced to a lesser extent; for example, money spiders were reduced by about 50% at the time of the peak in aphid populations. The effects of predator exclusion on aphid populations are presented in Figure 10.5, which shows the relative numbers of aphid-days per tiller in exclusion and control plots. In this simplified presentation of the six years' data, the roles of different predators at different stages of the aphids' population development cannot be distinguished. However, Figure 10.5 illustrates that in some years exclusion of predators allowed the development of a significantly higher aphid population.

Evidently, a simple comparison of aphid numbers in the three treatment areas without the information obtained from exclusion experiments would have given a misleading impression of the effect of pesticides on predation of aphids. An overall comparison of grain aphid populations in the three areas shows no consistent differences between the Full Insurance and lower input areas during the five treatment years. In fact, in the first two years of the treatment phase of the Project (especially in 1985), this species reached its highest levels in the lower pesticide input areas (Figure 9.5). It failed to reach high levels in the Full Insurance area probably because of the low overwintering population there. However, when comparisons are made between exclusion and control plots (Figure 10.5), it is apparent that in those years when grain aphids reached outbreak levels (1984, 1985 and 1988), predator exclusion did not lead to any appreciable further increase in aphid numbers in either lower or high pesticide treatment areas. In those years when aphids did not reach outbreak levels (1986 and 1987) predator exclusion had a greater effect in the Supervised and Integrated areas than in the Full Insurance area, indicating that in those years predators were more effective in the reduced input areas.

Rose-grain aphids generally arrived in cereal fields later than grain aphids and did not spend the winter on the crop. Hence the pesticides applied in winter would not be expected to affect numbers, and differences between the treatment areas might reflect more clearly the activity of predators. In the two years (1986 and 1987) when a heavy infestation by rose-grain aphids developed, populations were up to ten times higher in the Full Insurance area than in the other treatment areas. As for grain aphids, exclusion of predators had a significant effect on rose-grain aphid development in the Supervised and Integrated areas (Figure 10.5), but little effect in the Full Insurance area.

In summary, it appears from this study that the excluded predators had a measurable effect only in those years when aphid populations built up slowly, and did not reach outbreak levels. In all three years in which aphids exceeded spray threshold levels due to a rapid population build-up, comparison of exclusion and control plots showed

Figure 10.5 Effect of predator exclusion on populations of aphids. Columns show the aphid population levels (in accumulated aphid-days per tiller) in control plots as a percentage of those in exclusion plots. Shorter columns indicate a greater impact of predators on aphids.

(a) Grain aphid *Sitobion avenae*

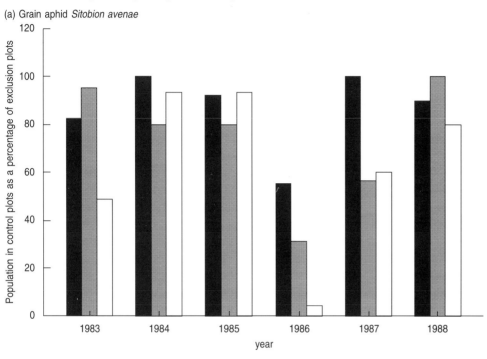

(b) Rose-grain aphid *Metopolophium dirhodum*

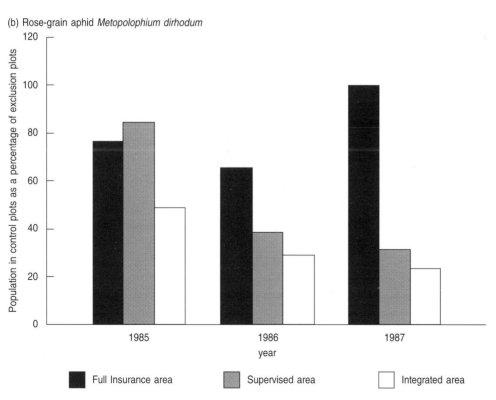

that predators did not have a major effect on aphid populations in any of the three treatment areas. In other years (1986 and 1987), the effect of predators on populations of both the grain aphid and the rose-grain aphid was less in the Full Insurance area.

Direct measurements of predation

Two methods were used to obtain a direct measure of predation rates in the different treatment areas. First of these was the use of *Drosophila* (fruit fly) pupae as artificial prey to measure predation at ground level (Speight & Lawton, 1976; Sotherton *et al.*, 1987). In the present study, sets of 12 pupae were presented on 2.5 cm² cards. These were placed at ground level in one field in each of the treatment areas for 24 hours on five or six occasions during summer in 1983 to 1987. In 1988 this method was used to measure predation rates in two fields in the Integrated area and three in the Supervised and Full Insurance areas.

Rates of predation of artificial prey are summarised in Figure 10.6. The fitted regression line represents the relationship between predation rates in the Full Insurance and lower input areas in the 1983 baseline year. The data points are means of predation for each sampling occasion during 1984–1988. If rates of predation were significantly reduced in the Full Insurance area relative to the other areas, then data points from post-treatment samples would be expected to fall well below the line. However, this was the case only for 1986 (Figure 10.6). During the first two years of treatments, predation was not reduced in the Full Insurance area, despite the very great reduction in one category of predator (Figure 10.1), whereas on three occasions there was a significantly higher rate of predation there, possibly related to the reduction in availability of alternative prey (see Figure 10.2 and below).

The second measure of predation pressure was made at crop foliage height. Data were obtained during summer in 1986–1988 by recording the rates of loss of grain aphids from artificially established colonies in exclusion and control plots.

Five colonies of five or more nymphs were established in each plot on fully expanded leaves near the crop canopy on two or three occasions during the summer. Rates of loss from the colonies were measured daily. These trials were intended to provide an independent relative measure of aphid predation rates in the three treatment areas in support of the measurements of natural aphid population development rates described above.

In Figure 10.7 predation is expressed as the proportion of colonies attacked (viz. with three or more aphids lost between sampling occasions) rather than as total numbers of aphids lost, in order to allow for the fact that an attack on a colony often resulted in dispersal of the remainder of the colony from the plant. Losses due to factors such as wind action were assumed to be similar in control and exclusion plots, and the results in Figure 10.7 represent losses due to attacks by predators only.

The higher losses of aphids observed in 1986 and 1987 (Figure 10.7) correspond to those years when a major effect of predation was shown for natural aphid populations in the predator exclusion experiments. The largest differences between lower- and high-input treatments were in 1986. Although the artificial colonies were exposed to predation for only two or three periods of up to one week, and therefore do not give a full picture of predation throughout the season, these results support the evidence from exclusion experiments and from artificial prey experiments (Figure 10.6) that the major effect on summer predation occurred in 1986.

Aphids in the diet of predators

On those occasions that a difference in predation on aphid populations was detected, the rate of predation was always lower in the Full Insurance area (Figures 10.5 and 10.7). Conversely, evidence from experiments using artificial prey showed an increase in rates of predation there on some occasions (Figure 10.6). Reduction in alternative prey availability in the Full Insurance area may have affected the response of surviving predators. In

Figure 10.6 Predation of artificial prey in summer in Full Insurance fields compared to Integrated and Supervised fields. Data points represent average losses of pupae per field on different sampling occasions. The regression line is calculated from data collected in the baseline years (1983, open circles).

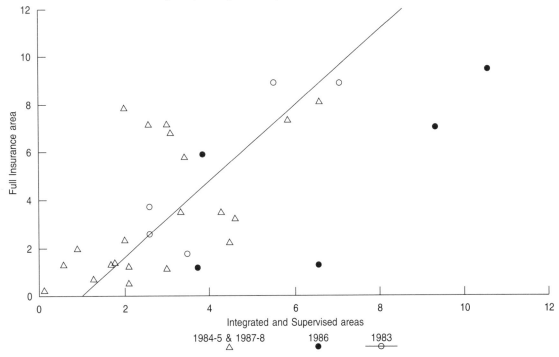

Figure 10.7 Proportions of artificial aphid colonies that were attacked in control plots in 1986–88.

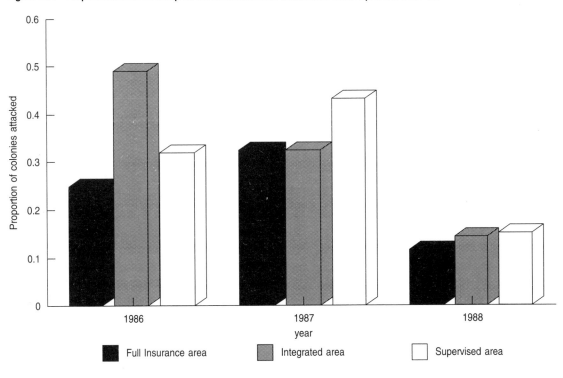

order to examine changes in the response of predators to their aphid prey, as well as to provide quantitative information on their rate of feeding on aphids, samples of polyphagous predators were taken throughout the Project for gut analysis using Enzyme Linked Immunosorbent Assay (ELISA). The technique used is described by Crook & Sunderland (1984).

Aphid antiserum, which reacts with remains of grain aphids and rose-grain aphids, was provided by the Institute of Horticultural Research (Littlehampton) and tests were made on a wide range of aphid predators with no solid gut contents, collected throughout the summer during 1983–1987.

Figure 10.8 presents the incidence of feeding on aphids in relation to their density for two groups of predators that regularly ate aphids, the ground beetle *Trechus quadristriatus* and the money spiders (the three most frequently tested species were *Erigone atra*, *Lepthyphantes tenuis* and *Bathyphantes gracilis*). More intensive feeding on aphids would be expected when aphids are more abundant. Figure 10.8 contrasts data from the Full Insurance and lower input areas over a wide range of aphid densities. For money spiders, aphid predation was similar in both areas but *T. quadristriatus* showed a higher frequency of predation on aphids in the Full Insurance area, especially at high aphid densities.

T. quadristriatus, which is a polyphagous predator, might be expected to eat various prey in proportion to their relative frequency on the ground. Where springtails are less abundant, as in the Full Insurance area, a higher proportion of aphids might be taken. In contrast, money spiders are more likely to take aphids in their webs in proportion to absolute abundance, not relative abundance. The frequency of aphids in their diet would therefore be independent of the availability of alternative prey such as springtails.

Predation of slugs in summer

Slugs are a significant pest problem in cereals grown on heavy soils such as that at Boxworth, and on two occasions during the study, virtually the entire Project area required a molluscicide application (Appendix II). The role of natural enemies, including polyphagous predators, in slug population dynamics is poorly understood. Ground beetles may have a significant effect in some agricultural systems (Altieri *et al*., 1982) and many polyphagous predators found in winter wheat are known to feed on the grey field slug (*Deroceras reticulatum*), which is a major pest (Tod, 1970). It was decided to investigate whether the role of polyphagous predators as slug antagonists was affected by the pesticide treatments at Boxworth.

Accurate population estimates of slugs are difficult to obtain, and sampling was difficult in the heavy clay soil. Therefore an indirect method was used to investigate predation, by monitoring slug activity in traps within exclusion and control plots. Where populations of ground beetles had been reduced, a rise in slug numbers might be expected, and should be reflected in an increase in the index of numbers/activity which slug traps provide. Towards the end of the growing season in each year from 1983 to 1987, between five and nine tiles (13×15 cm, each covering a bait containing methiocarb) were placed within each of the three predator exclusion and control plots in each treatment area. Since the exclusion barriers were usually erected in mid-April, ground-moving predators had been excluded from these plots for four months prior to monitoring slugs. The presence of slugs (mostly the grey field slug) under tiles was recorded immediately before harvest for three periods of two to three days in each year. Results for the trapping period having the greatest number of slugs in each year are presented in Table 10.4.

In eight out of 15 trials, significantly fewer slugs were trapped in control plots than in exclusion plots (Table 10.4), suggesting that the predators had contributed to a reduction in slug numbers and/or activity. Similar reductions were apparent in control plots (to which predators had access) in both high and lower pesticide input areas, even four years after the commencement of contrasting treatments (Table 10.4). Variation in the efficiency of

Figure 10.8 Relationships between aphid density and the proportion of (a) Linyphiidae (money spiders) and (b) *Trechus quadristriatus* that contained aphid remains.

(a) Linyphiidae (money spiders)

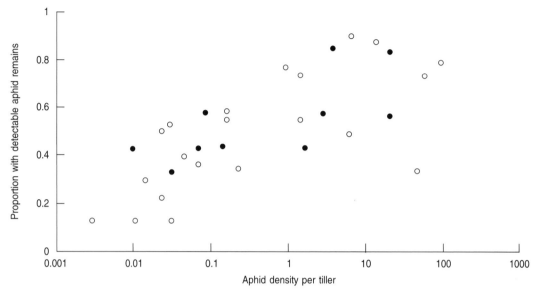

(b) Trechus quadristriatus

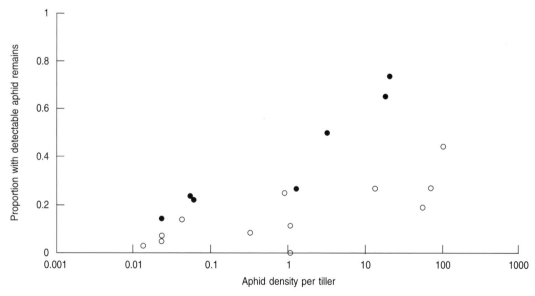

Supervised and Integrated areas ○

Full Insurance area ●

predator exclusion might have obscured some effects of the pesticide treatments on predation of slugs. However, there is no evidence that the impact of predator exclusion on slug activity during summer was any different in the Full Insurance and lower input areas. Therefore it appears that there was no measurable effect on predation of slugs caused by the Full Insurance treatment. Tod (1970) showed a positive correlation between the body size of ground beetle species and the frequency with which they fed on slug material. The larger ground beetles at Boxworth tend to be late summer and autumn breeding species (category (iv) in Table 10.2) which may have been less severely affected by the Full Insurance treatment.

Predation of artifical prey in winter

Predators might have an important role in reducing populations of aphids in autumn and winter, when aphids may migrate into fields and spread Barley Yellow Dwarf Virus (BYDV) within the crop. Historically, BYDV has not been a problem at Boxworth, and winter aphid populations were low during this study. As a result, and also because predator exclusion barriers could not be set up during winter owing to other farming operations, it was not possible to measure aphid predation rates directly during winter. However, there may be a reduction in numbers and/or activity of the predators that are found in the crop in winter (Figure 10.1). Artificial prey (see above) were used to determine whether the reduction in predator activity was accompanied by a measurable reduction in predation rate.

Winter predation was measured in only one year (1987–1988), in two fields in the Full Insurance area, and in the six fields of the lower-input areas. In that year a molluscicide treatment was necessary in most fields in the Supervised and Integrated areas (Appendix II). Methiocarb pellets were incorporated with the seed at drilling in the Integrated fields, and were broadcast (as in the Full Insurance treatment) in those Supervised fields which required treatment. Therefore, winter predation rates could be compared between the Full Insurance area and two methiocarb application methods in the lower input areas as well as a no-molluscicide treatment.

In this study twenty cards carrying *Drosophila* pupae were placed within selected fields in each treatment area for 48–72 hours on three dates during winter and early spring. The results presented in Figure 10.9 are an illustration for one sampling period only; other data gave a similar picture (Burn, 1988). Two fields received the Full Insurance pesticide programme (pair 1 in Figure 10.9). The Supervised and Integrated fields had a similar insecticide history up to 1987, but in 1987 pair 2 (Supervised) received a broadcast application of methiocarb pellets, pair 3 (Integrated) received the molluscicide incorporated into the soil at drilling, and pair 4 (Integrated and Supervised) received no molluscicide treatment. Between-fields variation, and lack of randomisation within the original Project design, prevent a valid statistical comparison between these treatments, since no baseline measurement of overwinter predation was made. However, the results (Figure 10.9) suggest that the Full Insurance treatment reduced the level of predation on artificial prey in winter relative to predation rates in lower-input areas (field pairs 3 and 4 in Figure 10.9). The data also indicate that a single molluscicide application within the Supervised treatment area had a significant impact.

Table 10.4 Mean numbers/activity of the grey field slug in control plots as percentages of those in exclusion plots.

Significant differences between numbers trapped in control and exclusion plots within each treatment area (t-test) are indicated by * $p < 0.05$, ** $p < 0.01$, *** $p < 0.001$.

Year	INTEGRATED	SUPERVISED	FULL INSURANCE
1983	67% *	24% *	74% *
1984	200%	100%	100%
1985	33%	120%	11% **
1986	200%	44% **	50%
1987	33% *	31% ***	41% ***

The method of application was apparently crucial. A reduction in predation was observed only where slug pellets had been broadcast. During the study, poorly-dispersive predators in category (i) (Table 10.2) formed the major part of the surface-dwelling fauna, and it seems likely that the reduction in numbers/activity in this group (Figure 10.1) was largely responsible for the reduced predation in the Full Insurance area. It is not possible to determine the consequences of reduced predation for the development of pest populations, but since a high proportion of the aphids transmitting BYDV are likely to be found at or near ground level during winter, it seems probable that the observed reduction in predation could have a major effect.

Rates of parasitism of aphid populations

A wide range of Hymenoptera has been identified as important parasites of cereal aphids, and they

Figure 10.9   Average numbers of pupae destroyed per card in four pesticide treatments during winter (see text for details). Columns with the same letter are not significantly different.

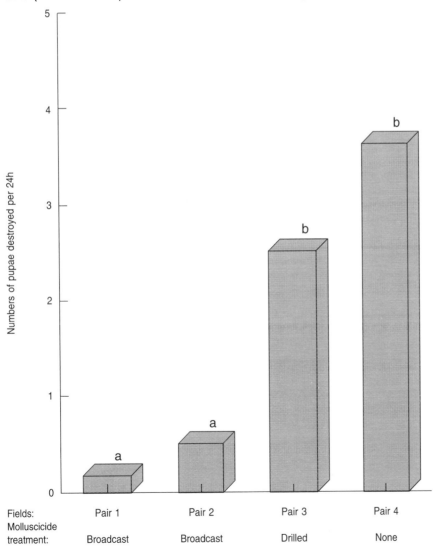

frequently have a major effect at an early stage during the development of aphid populations (Chambers *et al.*, 1983). Fungal parasites are usually more important later in the aphid population cycle, but are often too late to substantially reduce aphid population growth (Chambers *et al.*, 1983). Although not necessarily specific to cereal aphids, both types of parasites are found in fields only after an aphid population has started to develop, and rates of parasitism are closely linked to the density of the host species. For this reason, the effects of pesticides are likely to be apparent only within a season. Any between-seasons effects, and hence long-term trends, will probably be masked by reinvasion each year. Moreover, as described earlier (Chapter 9), overwinter survival of grain aphids, *S. avenae*, was usually lower in the Full Insurance area than in the other two treatment areas, because of aphicide applications in autumn. This is likely to have affected the numbers of parasitoids overwintering in the Full Insurance area irrespective of any long-term effects on their populations.

Results from suction sampling for the aphid parasitoids are presented in Chapter 9. This section describes rates of parasitism of *S. avenae* by entomophagous fungi and hymenopteran parasites, measured in the field during visual counts of aphid populations. Although assessments in field counts cannot provide an accurate measure of the true level of parasitism, counts of mummies (aphid remains after parasitism) and infected aphids do give a relative measure of parasitism in different fields, and lend support to independent counts of hymenopteran parasites in suction samples.

Figure 10.10 presents rates of parasitism measured at the aphid population peak in the intensive study fields, where the plots in which aphid counts were made had received no aphicide during the summer.

There was wide variation between years. Peak levels of parasitism by hymenoptera tended to occur in the lower-input areas in those years when aphid populations reached an early peak. Similar proportions of parasitised aphids were found in all treatment areas in the baseline year and

in 1984, 1986 and 1988, but there were apparent differences between areas in 1985 and 1987. During 1985 aphid population development in the Full Insurance intensive study field was slow, and the peak was reached at about Growth Stage 90, allowing both hymenopteran and fungal parasitism to build up beyond that in the Supervised and Integrated areas, where peaks were reached at Growth Stage 84. In 1987, parasitism levels were higher in the Integrated than in the Supervised or Full Insurance areas. Reasons for this are not clear.

In general, no obvious long-term treatment effects were apparent from measurements of parasitism in the field. There was substantial between-years and between-fields variation, and rates of parasitism appeared to be related to aphid population development rather than a direct response to effects of the treatments.

Discussion

This study has shown significant differences between treatment areas in the pitfall-trap catches of invertebrates. The unreplicated experimental design of the Project demands caution in ascribing those differences to the pesticide treatments. By comparing trends with baseline year measurements it was possible to demonstrate that many effects were indeed due to the treatments, but in the case of predation on aphids, variation between years limited the value of the pre-treatment measurements. In such cases interpretation relies more upon the body of evidence arising from several different methods of measuring predation, which may indicate the probable mechanisms which led to the effects observed.

The possibility exists that husbandry or pesticide treatments in surrounding fields may have affected population densities within the experimental areas. Indeed, reductions in populations of predators, parasites or prey in one treatment area might possibly interfere with populations in the other areas. As far as possible, the design of the experiment attempted to ensure that surrounding activities were not likely to interfere with the exper-

Figure 10.10 Rates of parasitism of grain aphids by (a) hymenopteran parasites and (b) fungal parasites.

(a) Parasitism by Hymenoptera

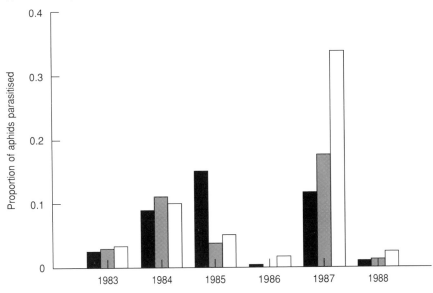

(b) Parasitism by fungi

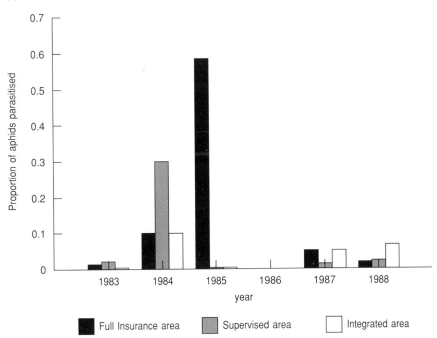

imental programme (see Chapter 2). It is improbable that any such impact would be demonstrated on group (iii) (longer-distance migrants) or group (i) (sedentary) species. By concentrating studies of locally migrating species (groups (ii) and (iv)) on those field boundaries that were not adjacent to surrounding farms, any influences of surrounding land not within the treatment area should have been minimised.

Reductions in the activity and/or abundance of many polyphagous predators under the Full Insurance regime were clear, although for some species these were only apparent after several years of contrasting treatments. However, catches of a few species in the Full Insurance area increased relative to the lower pesticide input areas, at least during part of the year. The size and persistence of such effects were influenced by both the dispersive ability of the predators, and their exposure to pesticides, but in general the most severe effects were due to pesticides applied during winter when the crop cover was thin and therefore the risk of exposure was high. The study of predation rates during winter indicated a possible severe effect of methiocarb, which merits further investigation.

Shorter-term effects were also evident, especially in 1986 when crop cover was uneven, and there were late applications of dimethoate and triazophos against yellow cereal fly. It was demonstrated that longer-term reductions in predator numbers might occur through changes in food availability. Further work is needed to establish relationships between density of prey, fecundity and population size in polyphagous predators. Short-term population decreases were demonstrated for highly dispersive species, but the experimental scale was insufficient to reveal any general decline in such species.

Treatment effects on predation of aphids were only identified through the use of predator-exclusion barriers. During years when aphid population development was slow, predation on aphids was reduced in the Full Insurance area relative to the lower-input areas. Predation on artificial prey was greatly reduced in the Full Insurance area during a brief study carried out in winter. The effect of high pesticide inputs on predation of artificial prey in summer was more variable. Predation was reduced in 1986, corresponding to the severe effects on predators observed in that year. In other years, rates of predation on artificial prey were occasionally higher in the Full Insurance area, indicating the caution needed in predicting predation levels from knowledge of predator populations alone.

This study has shown that a high pesticide input can have long-term effects on a variety of predators. Although predation on aphids was reduced in the Full Insurance area, there was no apparent effect on the frequency of outbreaks of *S. avenae* there. Nevertheless, it is possible that in some years (for example in the Integrated area in 1983) natural enemies may have held aphid densities below the spray threshold. Also, it must be accepted that the potential of predators to control aphids might have been greater in the complete absence of pesticides than in the Supervised and Integrated fields. In contrast to *S. avenae*, *M. dirhodum* twice (in 1986 and 1987) reached densities 10 times higher in the Full Insurance area than in the lower-input areas, probably as a result of differences in predation by natural enemies. In agricultural systems where pest outbreaks are sporadic or the effectiveness of predators is variable (both appear to be the case for cereal aphids), studies of even five years may not be sufficient to detect any effect of pesticide treatments on pest resurgence.

Although the results of manipulation experiments may be difficult to interpret if there is variability in pre-treatment populations, they do have the benefit of measuring the combined effects of both changes in numbers and feeding behaviour of predators in response to the density of their prey. In the present study there were effects on predation when pest resurgence was not apparent.

Several components of the Full Insurance regime probably contributed to population declines observed in the species monitored, but the evidence suggests that one or two broad-spectrum compounds applied in winter may have a particularly severe effect, depending on the timing of application relative to the activity of sensitive species. It is only in a relatively long-term study such as the Boxworth Project that infrequent events, such as deviations from the usual timing of pesticide applications, are likely to occur, with possible profound effects on non-target organisms.

Acknowledgements
I should like to thank the following for invaluable assistance in the field: Mrs M Free, P Halfpenny, D Farrington, S Ameer, U Banda Ekanayeke, J Mauremootoo and M Bryan. I am indebted to the following for helpful discussions and help with data interpretation: Dr T H Coaker, Dr N Galwey and M Bryan.

References

Altieri, M A, Hagen, K S, Trujillo, J & Caltigirone, L F (1982). 'Biological control of *Limax maximus* and *Helix aspersa* by indigenous predators in a daisy field in central coastal California.' *Acta Oecologia/Oecologia Applicata* **3**, 387–390.

Bauer, T (1982). 'Predation by a carabid beetle specialised for catching Collembola.' *Pedobiologia* **24**, 169–179.

Burn, A J (1987). 'Cereal crops.' In: Burn, A J, Coaker, T H & Jepson, P C (Eds). *Integrated Pest Management*. Academic Press. pp. 209–256.

Burn, A J (1988). 'Assessment of the impact of pesticides on vertebrate predation in cereal crops.' *Aspects of Applied Biology* **17**, 279–288.

Burn, A J (1989). 'Long-term effects of pesticides on natural enemies of cereal crop pests'. In: Jepson P C (Ed.) *Pesticides and Non-Target Invertebrates*. Intercept Press. pp. 177–193.

Chambers, R J, Sunderland, K D, Stacey, D L & Wyatt, I J (1983). 'Control of cereal aphids in winter wheat by natural enemies: aphid-specific predators, parasitoids and pathogenic fungi.' *Annals of Applied Biology* **108**, 219–231.

Chiverton, P A (1984). 'Pitfall trap catches of the carabid beetle *Pterostichus melanarius* in relation to gut contents and prey densities, in insecticide treated and untreated spring barley.' *Entomologica Experimentalis et Applicata* **36**, 23–30.

Crook, N E & Sunderland, K D (1984). 'Detection of aphid remains in predatory insects and spiders by ELISA.' *Annals of Applied Biology* **105**, 413–422.

Dixon, A F G (1987). 'Cereal aphids as an applied problem.' *Agricultural Zoology Reviews* **2**, 1–57.

Frampton, G K (1988). 'The effects of some commonly-used foliar fungicides on Collembola in winter barley: laboratory and field studies.' *Annals of Applied Biology* **113**, 1–14.

Griffiths, E (1985). 'The Feeding Ecology of the Carabid Beetle *Agonum dorsale* in Cereal Crops.' Unpublished Ph.D. Thesis, University of Southampton.

Kendall, D A, Smith, B D, Chinn, N E & Wiltshire, C W (1986). 'Cultivation, straw disposal and Barley Yellow Dwarf Virus infection in winter cereals.' *Proceedings of the 1986 British Crop Protection Conference, Pests and Diseases* **3**, 981–988.

Powell, W, Dean, G S & Bardner, R (1985). 'Effects of Pirimicarb, Dimethoate and Benomyl on natural enemies of cereal aphids in winter wheat.' *Annals of Applied Biology* **106**, 235–242.

Shires, S W (1985). 'A comparison of the effect of cypermethrin, parathion-methyl and DDT on cereal aphids, predatory beetles, earthworms and litter decomposition in spring wheat.' *Crop Protection* **4**, 177–193.

Sotherton, N W, Moreby, S J & Langley, M G (1987). 'The effects of the foliar fungicide pyrazophos on beneficial arthropods in barley fields.' *Annals of Applied Biology* **111**, 75–87.

Speight, M R & Lawton, J H (1976). 'The influence of weed cover on the mortality imposed on artificial prey by predatory ground beetles in cereal fields.' *Oecologia* (Berlin) **23**, 211–223.

Tod, M E (1970). 'The significance of predation by soil invertebrates on field populations of *Agriolimax reticulatus (Gastropoda: Limacidae)*.' Unpublished Ph.D. Thesis, University of Edinburgh.

Vickerman, G P (1978). 'The arthropod fauna of under-sown grass and cereal fields.' *Scientific Proceedings of the Royal Dublin Society (A)* **3**, 273–283.

Vickerman, G P (1980). 'Important changes in the numbers of insects in cereals.' *The Game Conservancy Annual Report* **11**, 67–72.

Vickerman, G P and Sunderland, K D (1977). 'Some effects of Dimethoate on arthropods in winter wheat.' *Journal of Applied Ecology* **14**, 767–777.

Changes in the soil fauna at Boxworth

11

Geoffrey K Frampton, Stephen D Langton, Peter W Greig-Smith
and **Anthony R Hardy**
(MAFF Central Science Laboratory)

Introduction

Many of the invertebrates inhabiting cereal fields live in the soil, at least for part of their life cycle. The soil fauna includes earthworms (Lumbricidae), proturans (Protura), symphylids (Symphyla), and many species of mites (Acari) and springtails (Collembola), which are well-adapted to a subterranean lifestyle. Some beetles and flies spend their larval life in the soil.

Springtails are among the most abundant of the soil fauna, and they formed the focus of this study. They are prey for a variety of predatory arthropods including ants, bugs, earwigs, flies, mites, centipedes, beetles and spiders, many of which occur in the diet of farmland birds (Chapter 15). Some ground beetles, rove beetles and money spiders are natural enemies of cereal aphids and other crop pests, and springtails may be an important alternative prey for these predators when numbers of cereal aphids are low (Chapter 10).

The aim of this study was to determine whether the soil fauna at Boxworth was affected by the pesticide regimes used in the Project. Springtails are known to be influenced by a variety of insecticides (eg Edwards & Thompson, 1973) and fungicides (Frampton, 1988). However, there have been no long-term studies of the effects on these insects of repeated applications of different pesticides. If pesticides cause perturbations of populations of springtails or other invertebrates, there could be knock-on effects on beneficial invertebrates, through changes in their food supply.

Other groups of invertebrates found in soil samples (such as mites) were also examined in case they were affected by pesticides. In addition, sampling for soil nematodes was carried out in June 1988 to assess the numbers of pest species (Hancock, 1989).

Sampling was carried out both on a field-scale and in the replicated plots. Field-scale comparisons between treatments are realistic in scale because pesticides are usually applied to whole fields, but they lack the replication necessary for orthodox statistical analysis. For soil animals with very limited powers of dispersal, the replicated plots should provide a good indication of effects likely to occur at a field-scale. Accordingly, statistical analysis of any treatment differences observed in the replicated plots may help to indicate whether differences between the field-scale treatments were real rather than due to chance variation.

Methods

The soil fauna was sampled by taking soil cores from fields and plots using an auger. The cores were 5 cm in diameter and were taken to a depth of 10 cm. Each core was transferred to the laboratory in a 6 cm-diameter sealed aluminium canister, groups of which were kept in chilled storage boxes.

Extraction of the fauna from cores commenced within 48 hours of sampling, using a modified version of the high gradient extractor described by Macfadyen (1961). An extractor at Brooms Barn Experimental Station was used initially but identical extractors were constructed at the Central Science Laboratory, Tolworth, and became available for use in August 1984. An assessment of the relative efficiencies of the two sets of apparatus was made in August and September 1984. Thereafter the soil fauna was extracted from cores using only the extractors at Tolworth.

The extraction apparatus

Each extractor consisted of an array of Tullgren funnels set in a vertical temperature gradient, created by 15-watt light bulbs above the funnels and cold water piped around their bases. This arrangement relies on avoidance of unfavourable stimuli by soil animals to expel them from soil cores. The cores were located in sieves above the funnels, and dried slowly in the temperature gradient, causing many animals to move downwards to more favourable conditions (of lower temperature and higher humidity). If the drying period is sufficiently long (seven days in this study), some of the soil animals eventually fall out of the cores and can be collected in vials below the funnels. Although this method does not extract all the animals in a soil core, it provides a consistent method whereby differences between soil cores may be identified, if the cores are subjected to the same drying conditions.

Invertebrates were preserved in 99% ethanol, then identified and counted under a binocular microscope. Most were identified to order or family, but where possible springtails were identified to species. Table 11.1 lists the invertebrate taxa extracted from soil cores using this method.

Relative efficiency of extracting the soil fauna

There were changes in the efficiency of extracting springtails and mites from soil cores on two occasions: with the change from the use of the extractor at Brooms Barn to those at Tolworth, between September 1984 and March 1985; and with the installation of a 'flow cooler' to increase the temperature gradient of the Tolworth extractors, between August 1986 and March 1987. The efficiency of the extractors at Tolworth relative to the one at Brooms Barn was assessed in August and September 1984 by extracting springtails and mites from duplicate pairs of soil cores. For springtails this was initially c. 70%, and increased to 117% after the flow cooler was installed; for mites it varied considerably between dates (48%–68% in August and September 1984) and was not increased by the installation of the flow cooler (45% in March 1987). Analysis of the results takes account of these changes by comparing the proportions rather than the numbers of each species or

Table 11.1 The variety of invertebrates in soil cores taken from Boxworth Project fields from 1983 to 1988. The 14 species or groups of springtails are indicated by asterisks. Their classification follows Kloet & Hincks (1964).

Species or group	Average number extracted from each soil core (based on c. 2000 cores)
Mites (order Acari)	42.57
*Isotoma spp. (family Isotomidae)	7.12
*White blind springtails (family Onychiuridae)	6.19
*Folsomia quadrioculata (family Isotomidae)	6.15
*Isotomiella minor (family Isotomidae)	2.67
*Hypogastruridae (springtail family)	2.20
Symphylids (order Symphyla)	0.95
Pot worms (family Enchytraeidae)	0.79
*Neelidae (springtail family)	0.69
Fly larvae (order Diptera)	0.61
*Lepidocyrtus spp. (family Entomobryidae)	0.54
*Pseudosinella alba (family Entomobryidae)	0.52
Beetle larvae (order Coleoptera)	0.45
*Sminthuridae (springtail family)	0.37
Booklice or psocids (order Psocoptera)	0.33
Thrips (order Thysanoptera)	0.29
Pauropods (order Pauropoda)	0.29
Millipedes (class Diplopoda)	0.28
Bugs (order Hemiptera)	0.25
Spiders (order Araneae)	0.22
Bristletails (order Diplura)	0.18
Adult flies (order Diptera)	0.16
*Isotomodes productus (family Isotomidae)	0.14
Adult beetles (order Coleoptera)	0.14
Woodlice (order Isopoda)	0.12
Centipedes (class Chilopoda)	0.12
*Pseudosinella decipiens (family Entomobryidae)	0.10
Proturans (order Protura)	0.08
Earthworms (family Lumbricidae)	0.07
*Entomobrya spp. (family Entomobryidae)	0.03
*Isotomurus palustris (family Isotomidae)	0.02
*Moss springtails, Heteromurus nitidus (Entomobryidae)	0.01

group extracted from soil cores from each of the treatment areas.

Field-scale sampling

Sampling commenced in the spring of 1983, the second of the 'baseline' years. Soil cores were taken on three occasions in each year, in March or April, May or June and July or August. On each occasion, soil cores were taken from three fields in the same 'triplet group' (see Chapter 2).

In each of these fields two transects were marked out parallel to a hedgerow; one at 75 m from the hedgerow, the other at 150 m. The location of the transects was the same on each sampling occasion. In 1983 five soil cores, and in subsequent years ten, were taken along each transect at 10-m intervals.

Sampling in the replicated plots

The 24 rectangular plots comprising this experiment were located in Shackles Aden (a field in the Full Insurance area), and are described in Chapter 2. The layout, which was randomized, gave eight plots each of the Full Insurance, Supervised and a 'Minimum Input' treatment, which consisted only of the minimum number of herbicide applications deemed necessary to keep the plots weed-free; this replaced the field-scale Integrated treatment, which could not be recreated in the plots.

Sampling commenced in March 1984, the first of the treatment years. Except in June 1984, when only one soil core per plot was taken (because the soil was very sticky), two cores, 2 m apart in the centre of each plot, were taken on each sampling occasion. In most years, soil cores were taken on three occasions in the spring and summer, between March and September.

Although soil cores contain mostly subterranean fauna, some surface-dwelling species may also be trapped. Conversely, suction samples contain mostly surface-dwelling invertebrates but some species in the uppermost layers of the soil may be collected with a powerful suction sampler, especially if the soil is very dry. These two methods, which can provide an insight into invertebrates' vertical distribution, were compared by taking matched soil cores and suction samples from the same plots on 3 September 1987. The suction samples were taken using a Dietrick vacuum insect sampler (D-vac) (see Chapter 9).

Two suction samples, each comprising five randomly-placed 10-second sub-samples, were taken from each of the plots in two of the four blocks (ie from 12 of the 24 plots). Each sample was sealed in a polythene bag then transferred in a chilled storage box to the laboratory at Tolworth where it was frozen for storage. Inorganic material was removed from thawed samples by mixing them with a saturated salt solution; floating organic material was separated by sieving and then was preserved in 99% ethanol. Invertebrates were sorted from the organic material under a binocular microscope and were identified and counted in the manner described for those extracted from soil cores.

Results

Of 32 species or groups of fauna examined (Table 11.1) only the springtails and mites were sufficiently numerous to show clear differences in their populations between the field-scale or replicated plot treatments.

Trends in the proportions of springtails or mites in the fauna of each treatment area might indicate long-term effects of pesticides. These were examined at a field-scale by plotting, for each of 40 sampling occasions, the number in Full Insurance soil cores relative to the Supervised and Integrated samples, and the number of Supervised relative to Integrated soil cores (Figure 11.1). Trends in the replicated plots over 13 sampling occasions were also examined for comparison with the field-scale results. For each of the 13 sampling occasions, statistically significant differences between the plot treatments were identified using Analysis of Variance with a square root transformation (Figure 11.2). For the field-scale, trends were identified using regression analysis, while in the replicated plots a more rigorous test involving analysis of contrasts over time was used.

Figure 11.1 Trends in the relative proportion of springtails and mites in different field-scale treatments. (a)–(e) show the numbers in Full Insurance soil cores as a percentage of the numbers in all soil cores; (f) shows the numbers in Supervised soil cores as a percentage of the combined numbers in Supervised and Integrated soil cores. Lines indicate statistically significant regression equations. The total springtails and mites showed no statistically significant trends.

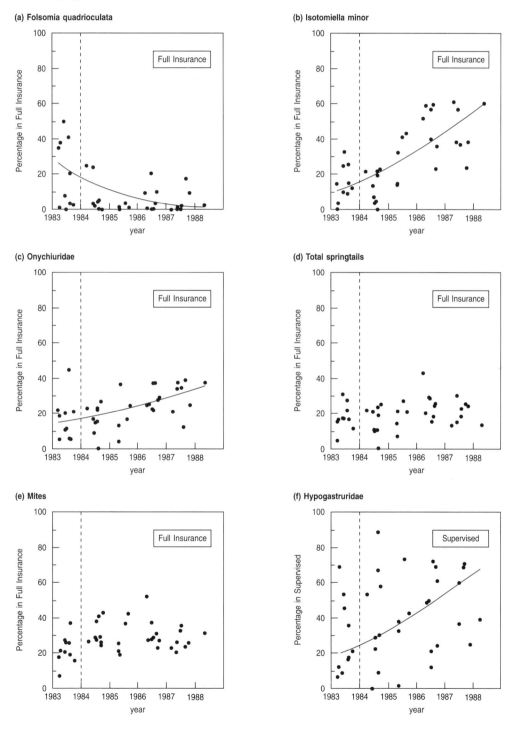

Figure 11.2 Relative proportions of springtails and mites extracted from the three replicated plot treatments on each of 13 sampling occasions. Asterisks indicate dates on which differences between the Full Insurance and the other regimes were statistically significant (*P<0.05, **P<0.01, ***P<0.001). Comparisons are omitted when there were too few insects for analysis.

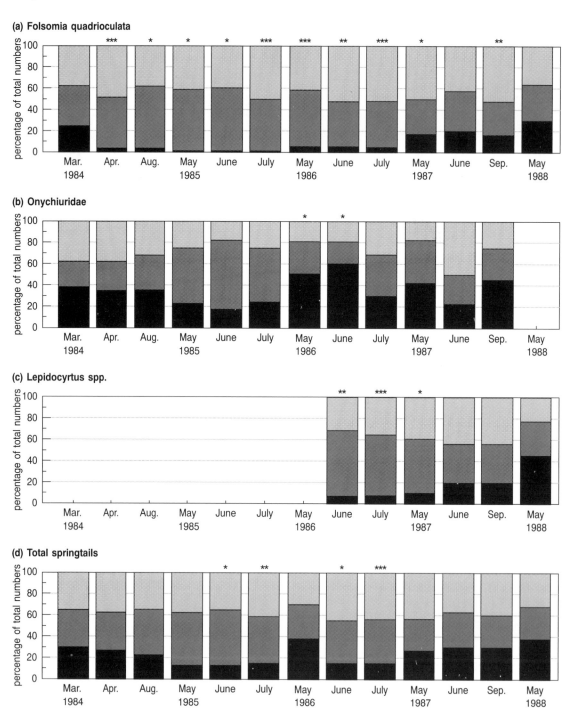

(a) Folsomia quadrioculata

(b) Onychiuridae

(c) Lepidocyrtus spp.

(d) Total springtails

Plate 1   *The Full Insurance area at Boxworth EHF, looking towards Grange Wood and Backside field, with Grange Piece and Shackles Aden in the foreground.*

Plate 2   *The Supervised area at Boxworth EHF, viewed from the west, showing Knapwell field to the left of Thorofare Spinney, with Thorofare and Top Pavements fields to the right.*

Plate 3   *The Integrated area at Boxworth EHF, showing Extra Farm in the centre of the picture, surrounded by Extra Close West and Extra Close East fields in the foreground, with Eleven Acre Extra and Bushes & Pits in the background.*

Plate 4 *Slugs are among the most damaging pests at Boxworth.*

Plate 5 *One of the major weeds on the wheat crops was cleavers* (Galium aparine).

Plate 6   *The distribution and persistence of pesticide residues was measured after spraying with insecticides in summer.*

Plate 7   *Some field margins at Boxworth contain trees that provide nesting sites for songbirds.*

Plate 8   *Occurrence of weeds and other plants was studied in the hedges and headlands of the Project fields.*

Plate 9 *'D-vac' suction samples were used to collect invertebrates from the wheat crops.*

Plate 10 *Cores of soil were taken regularly from the fields, allowing populations of soil invertebrates such as springtails to be studied.*

Plate 11 *Insects and spiders that are active on the ground can be collected by the use of sunken pitfall traps.*

Plate 12 *The beetle* Agonum dorsale *is a predator that spends the winter in field margins, migrating into wheat fields in the spring.*

Plate 13   *The lucerne-flea is abundant in cereal crops.*

Plate 14   *Money-spiders (Linyphiidae) include species that are significant predators of pests in the crop.*

Plate 15   *Ground-beetles (Carabidae) include predatory species of a wide variety of ecological habits which may influence their vulnerability to the effects of pesticides.*

Plate 16   *Parent tree sparrows were observed foraging and delivering food to their young, by teams of observers in the crop.*

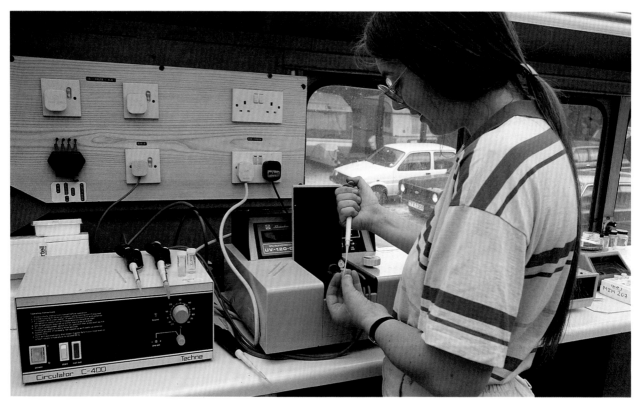

Plate 17   *Blood samples taken from nestling birds were analysed in a mobile laboratory at Boxworth to detect effects of pesticides on enzyme activity.*

Plate 18   *Detailed studies of tree sparrows* (Passer montanus) *were carried out in order to explore the effects of aphicides on birds.*

Plate 19   *Infra-red detectors at the entrance-holes of tree sparrow nest boxes provided a continuous automatic record of the birds' visits.*

Plate 20 *The species of small rodent most commonly found in the fields at Boxworth was the wood mouse* (Apodemus sylvaticus).

Plate 21 *Rabbits can be attracted to traps baited with chopped carrots.*

Plate 22 *Longworth live-traps were placed in the crop to catch small mammals.*

Figure 11.2–*continued*

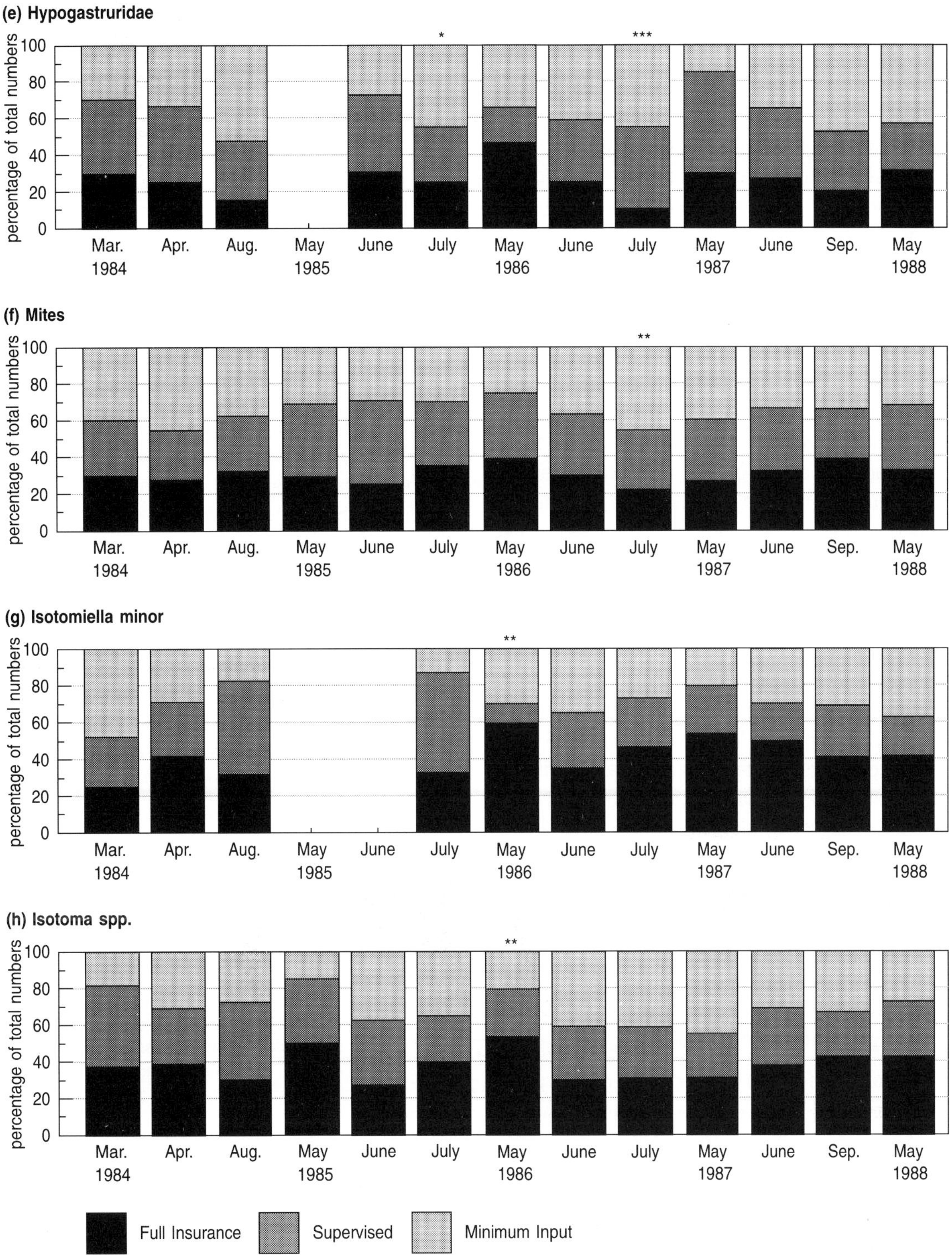
(e) Hypogastruridae
*
***
percentage of total numbers
100
80
60
40
20
0
Mar. 1984
Apr.
Aug.
May 1985
June
July
May 1986
June
July
May 1987
June
Sep.
May 1988
(f) Mites
**
(g) Isotomiella minor
**
(h) Isotoma spp.
**
Full Insurance
Supervised
Minimum Input

Field-scale trends

On the field-scale, *Folsomia quadrioculata, Isotomiella minor* and the springtail family Onychiuridae showed changes during the course of the Project (Figure 11.1). The proportion of *F. quadrioculata* from the Full Insurance area decreased from 1983 onwards and remained low throughout the remainder of the Project, with no evidence of a recovery, although it is not possible to identify precisely when the decline started (Figure 11.1a). In contrast, *I. minor* and the Onychiuridae (Figure 11.1b and 11.1c) showed an increase from 1983 onwards, this being particularly pronounced for *I. minor*. There was evidence for a significant, though less pronounced, decline in the Sminthuridae in Full Insurance fields. This might reflect adverse effects of the Full Insurance regime on the lucerne-flea *Sminthurus viridis* (Chapter 9), although this is primarily a surface-dwelling species and relatively few occurred in soil cores (Table 11.1). None of the other species or groups examined showed any clear trends in the Full Insurance fields and this was also true for the total springtails (which comprised the 14 species or groups listed in Table 11.1) and mites (Figure 11.1d and 11.1e). Only for the springtail family Hypogastruridae was there evidence of a statistically significant trend indicating a difference between the Supervised and Integrated areas, with a relative increase in the proportion of individuals from the Supervised area during the treatment phase of the Project (Figure 11.1f). However, as the figure shows, there was great variation between samples.

Springtails and mites in the replicated plots

In the replicated plots nine species or groups of springtails and mites showed statistically significant differences between treatment regimes. Most of these were between the Full Insurance area and the other treatments (Figure 11.2). Excluded from Figure 11.2 are the Sminthuridae, whose numbers were sufficient for analysis only on one of the 13 sampling occasions (June 1986), when significantly

fewer of these springtails were extracted from Full Insurance soil cores than from Supervised or Minimum Input cores.

Numbers of *F. quadrioculata* were lower in soil cores from Full Insurance plots on all 13 sampling occasions, the difference being statistically significant for 10 of these (Figure 11.2a). *Lepidocyrtus* spp., the Hypogastruridae, the total springtails, and mites, were also less numerous in Full Insurance soil cores on several sampling occasions but relatively few of these differences were significant (Figure 11.2 c–f). In contrast, *I. minor* was often more numerous in Full Insurance than in Supervised or Minimum Input soil cores (Figure 11.2g). For *Isotoma* spp. and the Onychiuridae, the proportions of individuals which were from Full Insurance plots varied between sampling occasions, with no consistent pattern (Figure 11.2b and h), though an overall trend in the Onychiuridae was significant (see below).

Significant differences between Supervised and Minimum Input plots are given in Table 11.2. *F. quadrioculata* and mites were less numerous in Supervised plot soil cores whereas the reverse was true for the Onychiuridae, *Lepidocyrtus* spp. and *I. minor*, though no group showed significant differences on more than one sampling occasion.

Four groups of springtails, *F. quadrioculata*, *Lepidocyrtus* spp., the Onychiuridae, and the total springtails, showed progressive changes during

Table 11.2 Differences in the relative numbers of springtails and mites extracted from Supervised and Minimum Input plots.

Species or group	Sampling occasion	Treatment from which most were extracted	Significance of difference
Folsomia quadrioculata	Sept. 1987	Minimum Input	$P < 0.05$
Mites	July 1984	Minimum Input	$P < 0.05$
Onychiuridae	June 1985	Supervised	$P < 0.01$
Lepidocyrtus spp.	June 1986	Supervised	$P < 0.05$
Isotomiella minor	July 1985	Supervised	$P < 0.05$

the Project, identified using an analysis of contrasts over time.

As on the field-scale (Figure 11.1a), the proportion of *F. quadrioculata* extracted from Full Insurance plots decreased initially in 1984. Hardly any individuals of this species were found in Full Insurance soil cores in 1985, but from 1986 onwards, in contrast to the field-scale results, the proportion increased; by 1988 it was similar to that recorded in 1984 (Figure 11.2a).

The Onychiuridae showed considerable variation between sampling occasions, with some suggestion of a complicated pattern; this appears to obscure a significant trend indicating an overall increase from 1984 to 1988 in the proportion from Full Insurance plots (Figure 11.2b).

Numbers of *Lepidocyrtus* spp. were very low until June 1986 but thereafter there was an increase in the proportion extracted from Full Insurance plots, which exceeded the proportion from Supervised or Minimum Input plots in 1988 (Figure 11.2c). *Lepidocyrtus* spp. were the only springtails showing a significant trend in the relative proportion extracted from Supervised and Minimum Input plots. There was a decrease in the proportion of these species extracted from Supervised plots from 1986 onwards (Figure 11.2c).

The trend for the total springtails in the Full Insurance plots was similar to that for the most abundant species, *F. quadrioculata*; there was an initial decrease followed by a recovery. The trend also reflects high proportions of the Hypogastruridae, *I. minor* and *Isotoma* spp. in Full Insurance plots in May 1986 (Figure 11.2d).

Springtails and mites in soil cores and suction samples

The different sampling biases of soil cores and suction samples provide information on the vertical distribution of springtails and mites. Knowledge of this may give an insight into the mechanisms of any observed pesticide effects on the soil fauna.

Information from the matched samples collected on 3 September 1987, together with data from c. 2000 soil cores collected during the course of the Project (Table 11.1) was compared with results obtained from three years' sampling of cereal fields in southern England (using 884 suction samples and 1330 pitfall traps; Frampton, 1989) to group the springtails according to their vertical distribution (Table 11.3).

Springtails' morphological characteristics (such as the size of the antennae, eyes and springing organ) may also reflect their vertical distribution (eg Gisin, 1943). The morphological characteristics of *F. quadrioculata* suggest that it is shallower-living than the Onychiuridae or *I. minor*, but sampling indicated that it was rarely found above ground (Table 11.3).

Both subterranean and surface-dwelling springtails showed trends suggesting pesticide effects (Table 11.3). Mites, which were extracted in large numbers both from soil cores and suction samples, are excluded from Table 11.3 because the total represents many different species which may differ in their vertical distribution. As a group,

Table 11.3 Vertical distribution of springtails in cereal crops and the effects of the Boxworth pesticide regimes. '√' indicates a trend suggesting a pesticide effect.

	Fields	Plots
Surface-dwelling		
Sminthuridae	√	
Lucerne-flea (Chapter 9)*	√	
Lepidocyrtus spp.*		√
Entomobrya spp.		
Surface-dwelling & subterranean		
Hypogastruridae	√	
Isotoma spp.		
Pseudosinella alba		
Pseudosinella decipiens		
Mostly subterranean		
Folsomia quadrioculata	√	√
Wholly subterranean		
Onychiuridae	√	√
Neelidae		
Isotomiella minor	√	

*also found on cereal plants

mites did not show any trends which might be indicative of pesticide effects.

Large numbers of Sminthuridae were found in suction samples, and the identification of individual species was thought to be worthwhile. Numbers of the lucerne-flea (*Sminthurus viridis*), *Sminthurinus elegans* and the garden springtail (*Bourletiella hortensis*) were noticeably lower in samples taken from Full Insurance plots than from Supervised or Minimum Input plots; these differences were statistically significant for the lucerne-flea and *S. elegans*. Too few Sminthuridae were extracted from soil cores to permit analysis. None of the other species or groups of springtails in suction samples showed significant differences between the treatments.

Discussion

Identification of treatment effects

On each sampling occasion, all samples were handled identically. Therefore, the differences described above in the proportions of springtails or mites extracted from different treatment regimes should reflect real population differences. However, it cannot be assumed that they necessarily reflect effects of pesticides.

The decrease of *F. quadrioculata* under the Full Insurance regime was clearly evident in both fields and plots. The field-scale pattern seems to represent a long-term effect of the Full Insurance regime, rather than a response to a specific pesticide event, as there was no evidence of recovery at any time during the treatment phase of the Project. The reason for the apparent recovery in the Full Insurance plots after 1986 is unclear; if it represents a response to a specific pesticide application, a similar pattern might have been expected on the field-scale.

In contrast to *F. quadrioculata*, the proportion of *I. minor* and the Onychiuridae in the Full Insurance fields increased during the treatment phase of the Project. There was a statistically significant trend in the proportion of Onychiuridae

in Full Insurance plots which mirrored the even greater field-scale increase. However, the marked field-scale increase in the proportion of *I. minor* in the Full Insurance fields was not seen in the replicated plots. For both these groups, on the few occasions when the differences between plot treatments were statistically significant (May and June 1986 for Onychiuridae; May 1986 for *I. minor*), numbers were highest in the Full Insurance plots.

The increase in the proportion of Onychiuridae in the Full Insurance plots could be explained by the effects of pesticides on predation or competition. For example, the numbers of some predatory ground beetles, such as *Bembidion obtusum* and *Pterostichus* spp. (whose larvae are largely subterranean) were considerably lower in Full Insurance than Supervised and Integrated fields (Chapters 9 and 10). Predatory mites might also have been affected by the Full Insurance regime; increases in populations of *I. minor* in an arable field were seen in an earlier study after populations of predatory mites were reduced by DDT (Edwards, Dennis & Empson, 1967).

Populations of two abundant springtail species, the lucerne-flea (Chapter 9) and *F. quadrioculata*, were adversely affected by the Full Insurance regime, but it is not known if these species compete with the Onychiuridae for food or space; this seems unlikely in the case of the lucerne-flea, which feeds on plants and lives above ground.

For *I. minor*, the disparity between the field-scale and replicated plot results suggest that the increase in the occurrence of this species in Full Insurance fields was not caused by the pesticide regime. Indeed, there was no marked change at the start of the treatment phase of the Project to suggest otherwise. However, it is possible that if there was an effect of pesticides on predation of springtails, this might have been obscured in the replicated plots (but not on the field-scale) by rapid dispersal of mobile predators, overcoming any differences in predation due to the treatments.

Some pesticides might have had favourable indirect effects on some species. In other studies, increases in numbers of springtails have been observed after herbicide applications, caused, it seems, by an increased rate of litter input to the soil (eg Conrady, 1986). However, some herbicides may also have direct adverse effects on springtails (eg Edwards & Stafford, 1979). There was no evidence to suggest that applications of herbicides used in the Project caused such favourable or adverse effects, though it is conceivable that cumulative effects of successive applications may have contributed to the overall effect of the Full Insurance regime.

Also, the location of the replicated plots in the Full Insurance area, where populations of some predators were lower than in the other treatment areas (see Chapters 9 and 10), would have made any effects of pesticides on predation more difficult to detect. Therefore, the possibility that the increase in the proportion of *I. minor* in Full Insurance fields was caused by indirect effects of the pesticide regime on predation cannot be ruled out.

For *Lepidocyrtus* spp. there was no obvious pattern in the proportion in Full Insurance fields, though a reliable trend indicated an increase in the Full Insurance plots. Numbers of *Lepidocyrtus* spp. were very low in all fields until 1986, after which they increased, but there was much variation between fields and this could explain the lack of any obvious effects of the treatment regimes on the field-scale.

There were sporadic significant differences in the proportions of *Isotoma* spp., the Hypogastruridae and mites in different plot treatments (Figures 11.2 f–h). There were also significantly fewer of the surface-dwelling lucerne-flea and *Sminthurinus elegans* in suction samples taken from Full Insurance plots. These differences, which might represent the transient effects of specific pesticide applications, or increased susceptibility on particular occasions, indicate that a wider range of species was affected by the Full Insurance regime than the long-term trends alone suggest. Only the most abundant of the groups sampled (Table 11.1) showed obvious responses to pesticides and it seems likely that more effects would have been detected if some of the rarer

animals had been sampled more efficiently. For example, the lucerne-flea, which was not efficiently sampled by soil cores, was abundant in suction samples, and showed major effects of the Full Insurance regime (Chapter 9).

Variation in the susceptibility of springtails and mites to pesticides might occur if sparse crop cover allowed greater than usual penetration of pesticides on some occasions, or if pesticide residues were washed into the soil by rainfall. It is notable that for most species, the majority of significant differences between plot treatments was in 1986 (Figure 11.2), a year in which crop cover was exceptionally thin.

Only the Hypogastruridae showed a reliable difference between the two reduced-input regimes, suggesting that there was a transient increase in occurrence in Supervised fields between 1985 and 1987. It is not known if this increase was caused by pesticides. In the replicated plots, the only consistent difference between Supervised and Minimum Input plots was shown by *Lepidocyrtus* spp.; this indicated a decrease in the proportion of these springtails in Supervised plots. In very few comparisons were differences between numbers in Supervised and Minimum Input plots statistically significant (Table 11.2).

Life cycles and vertical distribution

The springtails and mites described above spend their entire life cycle in arable crops, so it seems reasonable that surface-dwelling species such as the lucerne-flea are likely to be exposed to most of the pesticides applied to the Project fields, whereas subterranean springtails such as the Onychiuridae might be afforded some protection. However, the results of this study show that both surface-dwelling and subterranean springtails may be sensitive to the effects of the Full Insurance pesticides.

Potential effects of the pesticides used at Boxworth

Other studies have shown that springtails may be susceptible to some of the pesticides used at

Boxworth. Dimethoate has been shown to reduce numbers of surface-dwelling springtails, including the lucerne-flea, in winter barley (Frampton, 1988). This broad-spectrum organophosphorus insecticide was used at Boxworth only in May 1986. Most of the sporadic significant effects of the Full Insurance regime were seen in May–June 1986 (Figure 11.2). However, there is no direct proof of a connection. Organophosphorus insecticides tend to be detrimental to springtails (eg Madge, 1981), and the routine applications of triazophos and demeton-S-methyl to Full Insurance fields could well have contributed to the overall effects of the Full Insurance regime.

Many species of springtails eat various kinds of fungi, so fungicides could affect them indirectly via their food supply. Three of the fungicides used in the Boxworth Project, propiconazole, triadimenol and carbendazim, have shown adverse effects on some surface-dwelling springtails (including the lucerne-flea) in wheat plots, though these effects were brief and sporadic (Frampton, 1989).

It is clear that at least some of the pesticides used at Boxworth were potentially harmful to springtails and mites. However, none of the observed effects of the Full Insurance regime, or differences between the Supervised and Integrated regimes, could be traced definitely to individual pesticide applications.

Conclusions

The Full Insurance regime had long-term effects on populations of some springtails, whilst for others the effects were transient. Overall, the changes were varied: *F. quadrioculata* were adversely affected whereas the Onychiuridae and *Lepidocyrtus* spp. appeared to benefit in the long-term. The mechanisms for these effects are not known, though the beneficial effect on the Onychiuridae might reflect a lower predation pressure in the Full Insurance regime. *I. minor* seemed also to be favoured in the long-term by the Full Insurance regime, but the evidence for this was circumstantial.

Discrete guilds of mites were not studied separately, but there was no evidence that mites as a group experienced any long-term effects of the Full Insurance regime. Other groups of soil fauna were too rare in samples to allow analysis.

A number of springtail groups, and mites, exhibited short-term responses to the pesticide regimes. Most of these were in 1986, perhaps influenced by exceptionally poor crop cover in that year. These transient effects, like the field-scale effects, were varied: some groups were adversely affected by the Full Insurance regime whilst a few appeared to benefit. Although seemingly unimportant in comparison with long-term effects, brief within-season reductions in numbers of springtails could be important if they occur at times when other prey for beneficial predatory arthropods are scarce.

Relatively few significant differences in springtail populations were observed between the Supervised and Integrated areas, reflecting the broad similarity in the pesticide applications which these areas received. The only long-term effect seen in the lower-input regimes was an increase in *Lepidocyrtus* spp. in Supervised relative to Minimum Input plots. The underlying reason for this is not clear.

The markedly different responses of different species or groups to the Full Insurance regime at Boxworth make it difficult to determine the importance of the overall effect on the fauna in cereal fields. Surface-dwelling springtails are usually regarded as beneficial insects because they are known to be important in the diet of beneficial predators, such as money spiders and ground beetles which are antagonists of pests (eg Sunderland, 1975, 1986). A reduction in numbers of surface-dwelling springtails could, therefore, have undesirable effects on their predators (Chapter 10). Predators which specialize in eating springtails, such as some ground beetles (Bauer, 1982), are particularly at risk. On the other hand, some subterranean springtails, notably the Onychiuridae, are pests of a variety of crops because they may attack plant roots (eg Getzin, 1985). An increase in these springtails, like that caused by the Full Insurance regime at Boxworth, is likely to be undesirable, particularly if the use of fungicides increases the springtails' propensity to attack plant roots by reducing the amount of fungal material available as alternative food.

It is unwise to extrapolate the consequences of the Full Insurance regime at Boxworth to other farming situations but it is clear that some components of the soil fauna may be adversely affected in the long-term by the continued use of a prophylactic pesticide programme.

Summary

The soil fauna at Boxworth was examined using soil cores. Springtails and mites were the only groups sufficiently numerous to show effects of the pesticide regimes, but there was no evidence for long-term effects on mites as a group.

Effects of the Full Insurance regime on springtails were mixed, and affected both surface-dwelling and subterranean insects. *Folsomia quadrioculata* was adversely affected in the long-term whereas the family Onychiuridae and *Lepidocyrtus* spp. appeared to benefit from the Full Insurance regime.

The implications of these effects are unclear but they could be undesirable if, as was the case at Boxworth, they favour pest springtails such as the Onychiuridae and lower populations of beneficial species such as *Lepidocyrtus*.

Acknowledgements

We thank the following Hatfield Polytechnic students, each of which contributed a year's research to the study of the soil fauna at Boxworth: Jem Smith (1982–1983), Les Reeves (1983–1984), Helen Shortridge (1984–1985), Dawn Harratt (1985–1986), Helen Dickens (1986–1987), Philip Williamson (1987–1988) and Frank Wimpress (1988–1989). L Allen-Williams kindly agreed to supervise the students, A Dewar and J Cooper provided help with the use of the Brooms Barn extractor and H Gough assisted with advice on the identification of springtails.

References

Bauer, T (1982). 'Predation by a carabid beetle specialised for catching Collembola.' *Pedobiologia* **24**, 169–179.

Conrady, D (1986). 'Okologische Untersuchungen uber die Wirkung von Umweltchemikalien auf die Tiergemeinschaft eines Grunlandes.' *Pedobiologia* **29**, 273–284.

Edwards, C A, Dennis, E B & Empson, D W (1967). 'Pesticides and the soil fauna: effects of aldrin and DDT in an arable field.' *Annals of Applied Biology* **60**, 11–22.

Edwards, C A & Thompson, A R (1973). 'Pesticides and the soil fauna.' *Residue Reviews* **45**, 1–79.

Edwards, C A & Stafford, C J (1979). 'Interactions between herbicides and the soil fauna.' *Annals of Applied Biology* **91**, 125–146.

Frampton, G K (1988). 'The effects of some commonly-used foliar fungicides on Collembola in winter barley: laboratory and field studies.' *Annals of Applied Biology* **13**, 1–14.

Frampton, G K (1989). 'Effects of some commonly-used foliar fungicides on springtails (Collembola) in winter cereals.' Unpublished PhD. Thesis. University of Southampton.

Getzin, L W (1985). 'Chemical control of the springtail *Onychiurus pseudarmatus* (Collembola: Onychiuridae).' *Journal of Economic Entomology* **78**, 1337–1340.

Hancock, M (1989). 'Soil Nematodes.' In: *The Boxworth Project 1988 Annual Report*. Ministry of Agriculture, Fisheries and Food. p. 79.

Kloet, G S & Hincks, W D (1964). *A check-list of British insects, 2nd ed. Part 1: Small orders and Hemiptera.* Royal Entomological Society, London. pp. 1–11.

Macfadyen, A (1961). 'Improved funnel-type extractor for soil arthropods.' *Journal of Animal Ecology* **22**, 65–77.

Madge, D S (1981). 'Influence of agricultural practice on soil invertebrate animals.' In: Stonehouse, B. (Ed.) *Biological husbandry. A scientific approach to organic farming.* Butterworths, London. pp. 79–98.

Sunderland, K D (1975). 'The diet of some predatory arthropods in cereal crops.' *Journal of Applied Ecology* **12**, 507–515.

Sunderland, K D (1986). 'Spiders and cereal aphids in Europe.' *International Organisation for Biological Control WPRS Bulletin* **1987/X/1**, 82–102.

Populations and diet of small rodents and shrews in relation to pesticide usage

Ian P Johnson, John R Flowerdew and **Robert Hare**
(Department of Applied Biology, University of Cambridge)

Introduction

A variety of small mammal species occurs at Boxworth. Field voles *Microtus agrestis* occupy grassy areas, while the woods hold species such as the weasel *Mustela nivalis*, bank vole *Clethrionomys glareolus*, shrews *Sorex* spp. and the wood mouse (or long-tailed field mouse) *Apodemus sylvaticus*. To varying degrees all these species are found in arable farmland. However, the wood mouse is the only species which is abundant in fields and the only one which is present there throughout the year (Green, 1979; Havers, 1989; Pelz, 1989). For these reasons the work on small mammals concentrated on wood mice, the populations of which were monitored regularly throughout the Project. In addition, a detailed study was made of the diet of shrews in the different pesticide treatment areas.

The wood mouse has been the subject of many studies in woodland areas, but relatively little work has been carried out on arable farmland, although more wood mice may live in this habitat than in woodland (Pollard & Relton, 1970). In woodland, wood mice show a clear annual cycle of numbers (eg Flowerdew, 1985; Montgomery, 1989). Highest population density occurs in early winter, at the end of the breeding season. Numbers then decline through the winter and early spring, to reach a minimum in late spring or early summer. As breeding commences, young animals are recruited into the population and density gradually increases.

Green's (1979) study of a population of wood mice on arable farmland in Suffolk revealed a similar pattern of population density changes through the year. However, studies in fields adjacent to woodland revealed that movement between habitats may alter this annual cycle of numbers. Wood mice tend to move from woodland into fields during the summer and breed there Bergstedt, 1966; Havers, 1989). Following harvest in the autumn, most mice return to woodland. This movement between habitats is likely to occur only over a relatively short range (Flowerdew, 1976).

As well as an annual cycle of population density, there may be large variations between years. For example, Green (1979) reported peak population densities in a three-year study ranging from 8.4 to 17.5 mice per hectare. The extent of the decline over winter appears to be strongly linked to food availability, at least in woodland (Watts, 1969). Green (1979) considered that this was likely to be true also for populations on arable land.

The work at Boxworth was directed at detecting any differences in population density, age structure or diet that were associated with the contrasting pesticide regimes. In addition, palatability trials were carried out in the laboratory to investigate whether shrews and wood mice will feed on invertebrates that have been exposed to pesticides.

Studies of wood mouse populations

Methods

Numbers of wood mice at Boxworth were monitored by regular trapping using Longworth live traps (Chitty & Kempson, 1949). Animals caught in the traps were marked either by toe clipping (up to 1986) or by fur-clipping, so that recaptured animals could be recognised. They were then weighed and their sex and reproductive condition recorded, after which they were released at the point of capture.

Traps for catching mice were baited with wheat and hay was provided as a bedding material. Casters (fly pupae) were also provided in case shrews were caught, because these animals are unable to survive overnight without feeding. Traps were checked each morning of the trapping session.

In 1982, the first of the 'baseline' (pre-treatment) years of the Project, preliminary surveys of field populations were carried out in May-June and October. One line of twenty points was set in each of the ten study fields, with a spacing of 15 m between points and two traps were set at each point. Traps were set to catch for four nights in each trapping session.

During the remainder of the Project two distinct procedures were followed, namely 'grid trapping' and 'line trapping'. Grid trapping was carried out in the three intensive study fields, one in each of the treatment areas (see Chapter 2). In each of these three fields traps were set out in a grid of 8×8 points at approximately 20 m intervals, with two traps at each point. Again the traps were set to catch for four nights in each session.

Because carrion crows were liable to raid Longworth traps in the winter, trapping was restricted to the period from May to November. Three trapping sessions (May, July—August and October—November) in each year revealed the annual population minimum in spring (which resulted from over-winter losses before breeding) and the subsequent changes during the breeding season up to the time just after harvest.

Several methods are available for calculating population densities of animals captured in a grid of traps (e.g. Green, 1979). In this study the number of different individuals captured during a trapping session was taken as the minimum number of animals present within the area of the grid. This may over- or under-estimate actual density, but it provides a means of comparing population densities in similar habitats, and avoids problems caused by low capture success.

The grid trapping procedure is very labour-intensive, with over 1500 trap-nights required to monitor the intensive study fields in each session. The line trapping method was devised to provide a less intensive means of sampling all the Project fields, including those with trapping grids. The method is based on lines of ten points at 20 m intervals, with two traps per point, operated for two nights. Trap lines were randomly assigned to tractor wheelings at the start of the Project, with the number of trap lines proportional to the field area, and varying from three to eight trap lines per field (approximately one trap line per 2 ha). As with grid trapping, the traps were checked daily. All Project fields could be sampled within a two-week period (four trapping sessions, each occupying three days).

Trapping success is influenced by several environmental factors (Flowerdew, 1988). Wood mouse activity is reduced by moonlight (Wolton, 1983) and by cold, wet conditions (Gurnell, 1976). Because of the short trapping sessions, results from individual trap lines may be strongly influenced by the weather. To minimise these effects, the trap lines in each field were not all used on the same nights (but each was operated for two days and nights).

Line trapping provides a less direct index of wood mouse numbers than grid trapping. However, there was a strong correlation between the two sets of results (Flowerdew, 1988) which indicates that, at the range of densities encountered, the trap lines provide a reliable means of monitoring wood mouse populations over large areas.

Wood mouse population changes

The wood mouse was the commonest small mammal in the cereal fields, although population density varied greatly during the course of the Project, both within and between years.

The aim of the monitoring studies during the two baseline years (1981–1983) was to detect any inherent differences in small mammal populations between the three prospective treatment areas which might be related to the proximity of non-crop habitats, for example. In early summer 1982 small mammal numbers were very low at Boxworth. No wood mice were captured in the arable fields and only five were caught in the hedgerows during May and June. The absence of wood mice from the cereal fields is presumed to be due to very poor survival over the severe 1981–82 winter. Results from the Mammal Society's National Small Mammal Survey conducted at approximately the same time indicated that small mammal numbers were exceptionally low over the whole of the British Isles. The second survey, in October 1982, showed a substantial increase in wood mouse numbers at Boxworth; 127 animals were caught in the fields and a further 135 in traps set in hedgerows and woodland. This increase in numbers indicates that survival of juveniles born early in the breeding season must have been unusually good, with a high proportion subsequently breeding successfully later in the year.

During the course of the Project, including the baseline years, fewer wood mice were caught using the line-trapping method in the Full Insurance area than elsewhere. This difference in trapping success (measured as the proportion of 'trap-nights' catching mice) was statistically significant in 1983 and 1985 (Table 12.1). As this difference was evident before the treatment phase of the Project began, it is unlikely to be related to pesticide usage.

Although numbers were generally lower in the Full Insurance area, the trends in population density during the course of the Project were very similar in all three treatment areas, as shown by the results of grid trapping in the three intensive study fields (Figure 12.1). Numbers of mice at Boxworth were exceptionally low in 1984, 1985 and early 1986. No trapping was carried out in the autumn of 1986, but by May 1987 the wood mouse population had increased considerably. The mean density for mice in the intensive study field trapping grids was 20.5 mice per hectare, the highest recorded since the beginning of the Project and close to the maximum previously recorded on farmland (Green, 1979). The very low numbers of wood mice in the fields at Boxworth in August 1986 indicate that immigration must have been an important factor in the subsequent population increase, coupled with improved overwinter survival. In 1988 there were again good numbers of wood mice in the cereal fields, although densities were lower than in 1987.

It is difficult to explain the fluctuations in wood mouse numbers at Boxworth since 1983. The results of the National Small Mammal Survey in deciduous woodland indicate that there was a general decrease in winter populations of wood mice during 1984–1986. However, the national woodland figure for summer 1987 was the lowest since 1982, a result strikingly at odds with the population changes that occurred at Boxworth. Exceptionally good overwinter food supplies in 1986–1987, from a heavy mast crop in the local woodland, may account for the high population densities in 1987. What is clear, however, is that

Table 12.1 Numbers of wood mice caught along trap lines in October. No data were collected in 1986; 'NS' denotes that the difference between areas was not statistically significant. Note that the trapping technique used in 1982 differed from that which was used subsequently (see text for details).

		1982	1983	1984	1985	1987
	No. of trap-nights	480	920	1000	1000	760
Full Insurance area	No. of mice caught	43	17	14	1	63
	No. per 100 trap-nights	8.96	1.85	1.40	0.10	8.29
	No. of trap-nights	912	1360	1360	1360	1040
Combined Supervised and Integrated areas	No. of mice caught	84	63	24	11	102
	No. per 100 trap-nights	9.21	4.63	1.77	0.81	9.81
Significance of difference between areas (Chi-squared test)		NS	$P<0.001$	NS	$P<0.05$	NS

Figure 12.1   Numbers of wood mice trapped in 2-hectare trapping grids in the three intensive study fields, 1983–88.

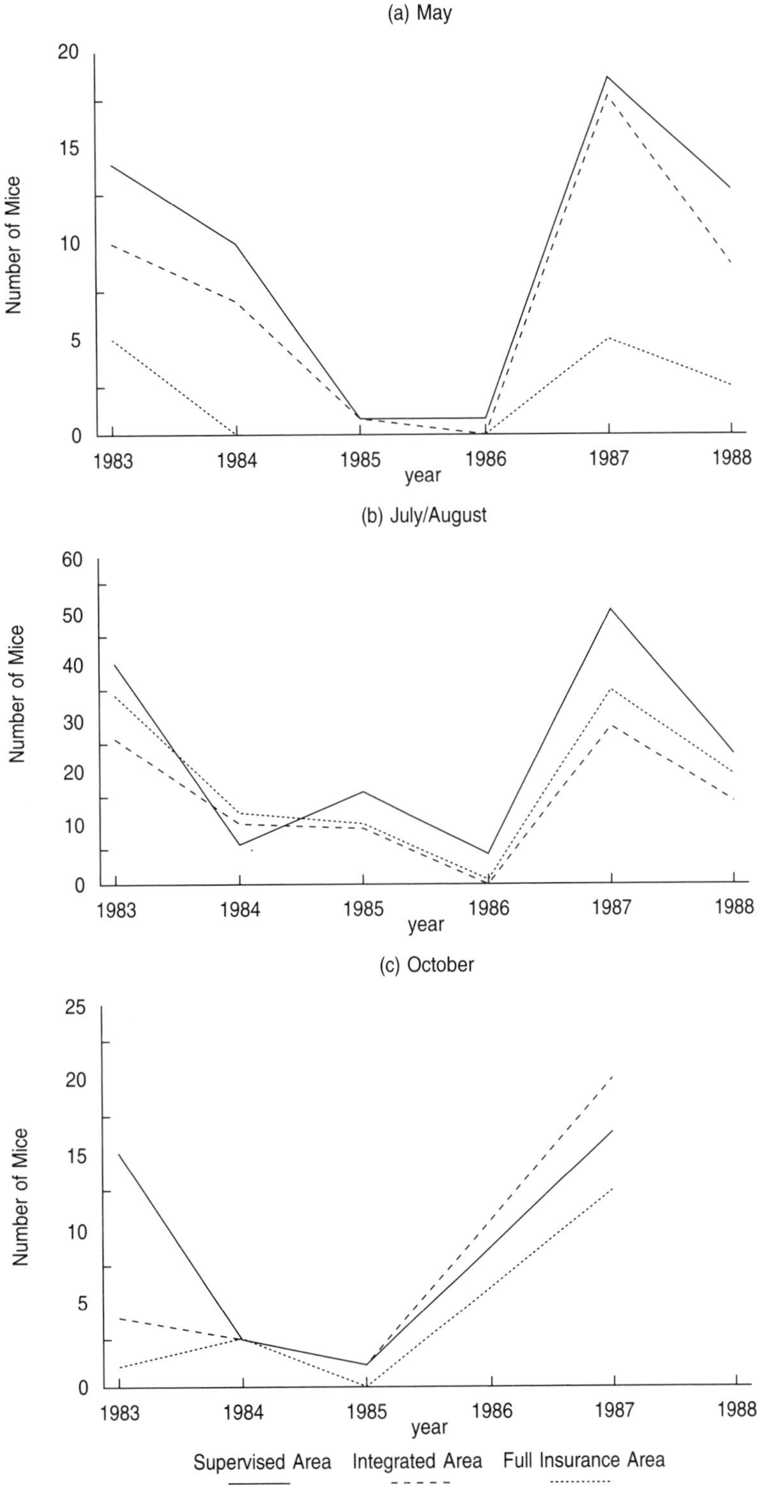

(a) May

(b) July/August

(c) October

Supervised Area   Integrated Area   Full Insurance Area

Figure 12.2   Changes in the population density of wood mice between spring (May), summer (July-August) and autumn (October) trapping sessions in Grange Piece.

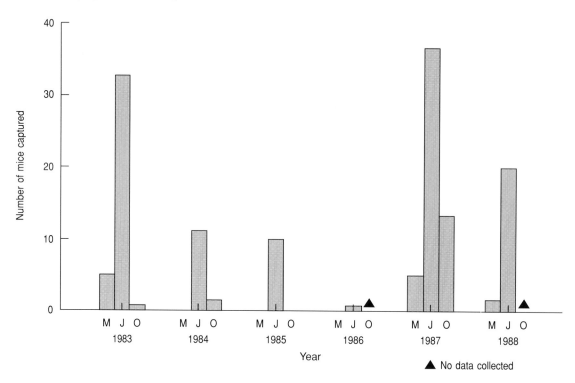

trends in wood mouse numbers were broadly similar in all three treatment areas at Boxworth and therefore the variation between years cannot be ascribed to the effects of the different pesticide treatments.

The annual cycle of wood mouse populations at Boxworth was also somewhat different to that described in woodland studies. The results obtained from grid trapping showed very similar fluctuations in numbers in all three intensive study fields (Figure 12.1). Figure 12.2 shows the results obtained in Grange Piece, in the Full Insurance area. In each year, numbers of wood mice captured in the grids increased from May to July or August, as would be expected as juveniles are recruited into the population. From other studies (see above) numbers would be expected to continue to rise to a peak sometime in the winter, and then decline over the late winter and spring, until the start of the next breeding season, when they would again rise.

However, at Boxworth numbers of mice in the fields fell sharply between the summer and the autumn (October–November) trapping sessions. This decline in mouse numbers can be attributed to the physical disturbance of harvesting, ploughing and preparing the seed bed. In addition, evidence gathered in the course of this Project (see pp. 150–151) shows that a molluscicide applied when drilling seed in the autumn had a dramatic effect on field populations of wood mice.

Another practice which appears to affect wood mice is that of straw burning in the fields. Table 12.2 shows the results of trapping wood mice after harvest in the treatment fields and Eleven Acre Childerley (a field outside the Project area), in relation to the method of straw disposal. The results showed a significantly higher capture success in those fields where straw had been baled and the stubble burned, than on those fields where the straw had been burned with the stubble. This

Table 12.2 Catches of wood mice compared between fields (a) where both straw and stubble were burned and (b) where only the stubble was burned. The proportion of mice caught in fields where both straw and stubble were burned was significantly lower than in fields where only the stubble was burned (Chi-squared test, $P < 0.001$).

	Number of trap-nights	Number of mice caught	Number of captures/100 trap-nights
(a) Fields in which both straw and stubble were burned	1720	57	3.3
(b) Fields in which straw was baled then stubble was burned	520	35	6.7

implies that the burning of straw and stubble has a detrimental effect on wood mice, compared with burning the stubble alone.

Other small mammal species recorded in the fields

The bank vole, *Clethrionomys glareolus*, is generally found in woodland, scrub and hedgerow habitats, having a preference for thick cover (Southern & Lowe, 1968), and does not usually move far into fields (Pollard & Relton, 1970). In most years at Boxworth no more than one or two individuals were captured in any field. However, in 1984 summer breeding populations were established in some of the Project fields. Between five and eleven individuals were captured on trap lines set in Thorofare, Backside, Grange Piece, Pamplins and Extra Close. The largest number of voles recorded was in Knapwell field, where 26 were captured on the trapping grid in August 1984. These animals included adult males, pregnant females and a number of juveniles, including two very young animals weighing less than 10 g. As would be expected, no bank voles were captured in the fields after harvest in the autumn. The reasons for the movement of bank voles into the fields in 1984 are

not clear, and this unusual occurrence was not repeated during the remainder of the Project.

The field vole, *Microtus agrestis*, was occasionally caught in the fields. This species usually occurs in rough grassland (Evans, 1977) and when recorded at Boxworth it was restricted to areas with dense ground cover, such as patches infested with blackgrass, *Alopecurus myosuroides*.

The house mouse, *Mus musculus*, was captured in very small numbers every year. During the Project it was recorded from each of the ten fields, but was most numerous in Extra Close where it was regularly trapped during the summer months. The farm buildings adjacent to Extra Close were known to have a large population of house mice, providing a source of immigrants into the neighbouring fields.

The harvest mouse, *Micromys minutus*, was the rodent species least frequently encountered in the study. However, the low capture frequency is unlikely to reflect its true abundance at Boxworth, since it spends much of the summer climbing in the vegetation and is consequently harder to trap than more terrestrial species. Signs of harvest mouse activity were recorded in all three treatment areas in 1983, and some animals were caught in Longworth traps in 1987 and 1988. Harvest mice were also occasionally recorded in insect pitfall traps.

The weasel, *Mustela nivalis*, was the only carnivore captured, and no more than four individuals were recorded in any one year.

The common shrew, *Sorex araneus*, and the pygmy shrew, *S. minutus*, were captured regularly, but in small numbers. The common shrew was the more frequent of the two species, but the distribution of captures in insect pitfall traps indicated that population densities for both species were significantly higher in the hedgerows than in the fields (p. 153).

Effects of molluscicide on wood mouse populations

The molluscicide used in the Project fields was 'Draza', a pelleted formulation containing 4%

methiocarb. The pellets have a cereal base which is potentially attractive to mice, and laboratory feeding studies have shown that wood mice are susceptible to poisoning by methiocarb pellets (Tarrant & Westlake, 1988). Methiocarb was applied at Boxworth in the autumn, to prevent slug damage to the newly sown crops. The standard method of application was to broadcast pellets on the field surface. However, in 1987 and 1988 some fields were treated by drilling the molluscicide with the seed (an alternative approved means of application). Both methods used methiocarb at a rate of 5.5 kg of pellets per hectare.

The first indication that methiocarb applications might be having a harmful effect on field populations of wood mice came in the autumn of 1982, when farm workers reported seeing a number of dead mice in the fields shortly after methiocarb had been applied. The effects of methiocarb applications on wood mice were investigated in detail in 1983, 1987 and 1988 (Johnson *et al.*, 1991).

In 1983 all the fields in the Supervised and Integrated areas reached ADAS thresholds for molluscicide application. Methiocarb was also applied routinely to the Full Insurance area fields.

Line trapping was used in five fields in three trapping sessions during September—October 1983. Methiocarb was applied to two of the fields

before the second trapping session, and to the remaining three fields between the second and third sessions. Trapping was also carried out in an untreated field adjacent to the Project area in the second and third trapping sessions to provide a further control.

In the few days immediately following the broadcast application of methiocarb, there was a large and statistically significant reduction in mouse numbers (Table 12.3). No mice were caught in the two fields sampled 2–4 days after molluscicide was applied. However, following this rapid decline, numbers increased again, and after a week trapping success was not significantly different to that recorded before treatment (Chi-squared test, $P > 0.05$).

Similar numbers of mice were caught in the first and third trapping sessions in the methiocarb-treated fields (28 and 29 respectively; Table 12.3) but only one of those released prior to the application was recaptured in the third session. This sharp reduction in survival is highlighted by a comparison between captures in sessions two and three in a control field next to the Full Insurance area where no methiocarb was applied, and the Project fields where methiocarb was used (Knapwell, Grange Piece and Extra Close). In the control field, 5 of the 6 mice released in session two were recaptured in session three, whereas in the other

Table 12.3 Trapping success of wood mice (the number caught per 100 trap-nights) in five fields before and after surface-broadcast applications of methiocarb slug pellets.

		Before methiocarb application	2–4 days after application*	7–27 days after application
(1) Pamplins and	No. of trap-nights	157	160	159
Top Pavements	No. caught	7	0	12
(2 trap-lines per field)	No. per 100 trap-nights	4.5	0	7.5
(2) Knapwell, Grange Piece and				
Extra Close	No. of trap-nights	478	–	474
	No. caught	21	–	17
(4 trap-lines per field)	No. per 100 trap-nights	4.4	–	3.3

*Significantly fewer mice were caught 2–4 days after methiocarb application than either before or 7–27 days after the application (Chi-squared test, $P < 0.02$).

three fields, none of 20 were recaptured (Fisher Exact test, P<0.001). It was also noted that the mouse populations in methiocarb-treated fields consisted mainly of juvenile mice (ie mice weighing 15 g or less and therefore likely to be less than three months old; Flowerdew, 1972).

More detailed information on the survival of individual mice was obtained in 1988. In the autumn, grid trapping was carried out in Backside field before and after the application of methiocarb. Shackles Aden field was trapped at the same time to provide a control (with no molluscicide treatment). Significantly fewer mice were recaptured in Backside than in Shackles Aden in the second trapping session (Table 12.4), indicating that survival was reduced following the application of methiocarb.

Table 12.4 Recaptures of marked wood mice in a field broadcast with methiocarb pellets and in a control (untreated) field. The proportion of mice recaptured in the molluscicide-treated field was statistically significantly smaller than in the control field (Chi-squared test, P<0.005).

	Field	
	Backside (treatment)	Shackles Aden (control)
Initial catch	29	68
Recaptures	5 (17%)	29 (43%)

These results suggest that methiocarb killed most of the mice present in the fields within two or three days of application, and that the population then increased quickly due to migration into the field of juvenile mice from outside the treated area.

The findings above refer only to surface applications of methiocarb. The alternative recommended method of application, drilling the pellets with the seed when sowing the next season's crop, uses pellets at the same rate as broadcast applications, but is thought by some to give less effective slug control. Work carried out in 1987 and 1988 indicated that the effects of methiocarb pellets on

wood mice were greatly reduced when the pellets were drilled rather than broadcast (Johnson et al., 1991).

An experiment carried out in 1987 compared mouse populations in untreated fields with those in fields where methiocarb pellets were drilled. Wood mice were trapped in three Project fields before and after drilled applications of methiocarb. In each case trapping was also carried out in an untreated control field on approximately the same dates. Results (Johnson et al., 1991) showed that mouse survival over the course of the experiment did not vary significantly between treated and untreated fields (Chi-squared test, P>0.05). Neither did the total number of mice captured, nor the proportion of juveniles in the population change significantly in any of the fields.

In 1988 a direct comparison was made of the impact of drilled and broadcast methiocarb on wood mice in Taylor's field, adjacent to the Project area. Approximately half the field was drilled with seed and methiocarb pellets, while the remaining area was drilled with seed and subsequently pellets were broadcast. A grid of traps was set in each area for a period of ten days following the broadcast application. Trapping revealed a significant difference in the proportion of adult and juvenile mice in the two areas: 88% of mice (14/16) captured in the broadcast area were juveniles, compared with only 40% (6/15) in the drilled area (Chi-squared test, P<0.02). This result is consistent with the earlier work which indicated that the proportion of juveniles in the mouse population rose following broadcast applications of methiocarb, but not drilled applications.

These trapping studies have shown clearly a dramatic short-term effect on field populations of wood mice caused by broadcasting slug pellets. However, monitoring has shown that there was no detectable long-term effect on mouse populations (see Figure 12.1). The application of slug pellets at Boxworth coincides with the peak in the annual cycle of numbers of wood mice, when many juveniles are likely to be dispersing (Green, 1979). Thus, any depletion in field populations in the autumn is likely to be relatively short-lived. Applications at other times of year, especially in early summer, could have a much more serious effect on local mouse populations. The impact of methiocarb is also likely to be affected by such factors as the available food supply for mice, the area treated and the weather. However, the work at Boxworth has clearly shown that the detrimental effect of methiocarb applications can be greatly reduced by changing the method of application, by drilling rather than broadcasting the slug pellets.

In this study field populations of mice recovered rapidly following methiocarb applications. It may be that applications on substantially larger fields, or in areas lacking woods and copses (which are likely to hold populations of mice capable of reinvading the treated areas), or at different times of year when densities are usually low (for example on potatoes in summer), might have a longer-lasting impact on wood mice.

Palatability of methiocarb-poisoned prey

The work on molluscicide pellets was prompted by the presence of dead mice following an application of methiocarb in 1982, but it has since proved very difficult to find dead animals at Boxworth. This may be due partly to the very low numbers of mice present at Boxworth in 1984–1986, although population densities were much higher in 1987 and 1988. One consequence of this is that the route of poisoning of wood mice by methiocarb is still not clear. Laboratory studies (Tarrant & Westlake, 1988) have shown that mice will ingest methiocarb pellets. However, in the field it may be that mice are poisoned secondarily through feeding on other animals, such as slugs and earthworms, which have themselves eaten the pellets. Of ten mice killed in breakback traps in the Project fields, the stomachs of seven contained identifiable invertebrate remains: insect remains were found in six stomachs and mollusc remains in four. The possibility of secondary poisoning via the diet was investigated by offering earthworms and slugs, poisoned with methiocarb, to captive wood mice.

Wood mice were captured at Boxworth and kept in the laboratory for at least three days prior to being offered slugs, *Deroceras reticulatum*, that had been killed by feeding on methiocarb pellets in the laboratory. A standard pelleted mouse diet remained available to animals when they were offered poisoned slugs.

Six animals were offered newly poisoned slugs for a 24 hr period, with five other animals as controls. The six mice ate an average of 5.9 g of slugs over 24 hrs (a range of 1.3–10.4 g). One of the control animals died in this time, but all animals fed with the slugs were still alive five days after treatment.

These results suggest that although wood mice might eat dead and moribund slugs they find after the application of methiocarb, this would not itself poison the mice, unless they take slugs in large numbers. Green's (1979) study suggests that molluscs are not a major component of the diet of wood mice on arable land. Since methiocarb-poisoned slugs are likely to be available to mice for only a short period, it seems that the disappearance of wood mice at Boxworth, following methiocarb applications, is more likely due to the mice eating the molluscicide pellets directly, rather than eating contaminated prey.

Studies of shrews

Shrew distribution

The common shrew and the pygmy shrew were both encountered frequently at Boxworth. During the course of the Project many shrews were accidentally caught in pitfall traps set for insects. The pattern of these captures provides an indication of shrew activity in different habitats and in different years.

Pitfall trapping was carried out chiefly between April and August in 1982–1988. There was no clear evidence of an effect of pesticide treatments on shrew captures in the different treatment areas. However, the trapping results reveal differences between years in shrew abundance. In most years shrews were captured in between 4% and 8% of pitfall traps set, but in two years (1982 and 1986) less than 1% of traps caught shrews. In these two years wood mouse numbers at Boxworth were also very low and the low populations of both species were likely to be a consequence of severe winter weather.

As well as giving an index of shrew numbers in the different treatment areas, pitfall traps have revealed the relative densities of shrews in field margin or hedgerow habitats and in the open fields in the different treatment areas. In both the Full Insurance and the Supervised and Integrated areas approximately 40% of pitfall trap captures were in traps set in the hedgerows or within 10 m of the field boundary, although the proportion of pitfall traps set in these areas was much smaller. This indicates that hedgerow or field margin habitats supported a much higher density of shrews than open fields. However, some shrews evidently managed to breed in these open field habitats. There

was a tendency for relatively more shrews to be caught in the open fields than in the hedgerows as the Project progressed (Figure 12.3). Forty-two percent of common shrew captures in 1983 were in open field habitats, rising to 57% in 1984, and during 1985–1988 over 70% of captures were in open fields rather than hedgerows. This may be because cultivation was closer to the hedges later in the Project.

Shrew diet

Work on populations of invertebrates has shown considerable variation between the three pesticide treatment areas (Chapter 9). Common and pygmy shrews feed almost exclusively on invertebrates (Rudge, 1968; Churchfield, 1982). The diet of shrews in the study fields was investigated by examination of the gut contents of a total of nearly 300 shrews found dead in insect pitfall traps (Chapter 10) during the course of the Project.

Examination of gut contents and particularly of faecal samples is biased towards those prey

Figure 12.3 Distribution of common shrews in insect pitfall traps between fields and hedgerows, shown by the percentages of captures that were made in these habitats each year.

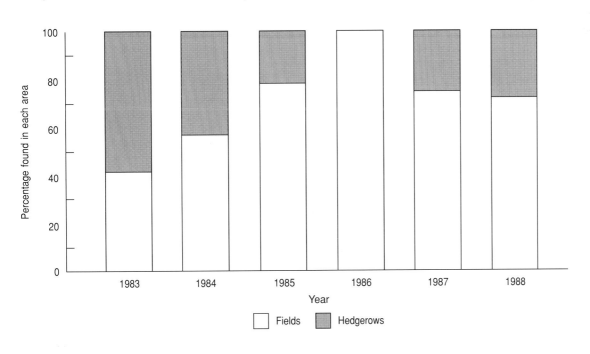

items which leave the most readily identifiable remains. Thus, small soft-bodied insects such as springtails may be under-recorded in comparison with larger tougher-bodied insects such as beetles. In addition, more digestible prey may have a shorter transit time through the gut and therefore were less likely to be encountered in a study of this sort.

For these reasons, the results shown here are not likely to indicate the exact composition of the shrews' diet. They should, however, provide an indication of the main components of the diet, and also provide a means of comparing the relative frequency of particular items in the diet in the different treatment areas.

Of 229 shrews examined during the treatment years (1984–88), the guts of 210 (92%) contained identifiable remains of invertebrate prey. The dietary composition for common and pygmy shrews at Boxworth is given in Table 12.5. Results are shown for each of the pesticide treatment areas for the period 1984–1988. A wide variety of invertebrates was recorded from the shrew guts examined. Prey ranged in size from springtails, thrips and aphids to slugs and leatherjackets.

Common shrew diet

Beetles were a major component of the diet of common shrews at Boxworth, with over 40% of the guts examined containing remains of adult beetles. Most beetles were carabids (ground beetles), but staphylinids (rove beetles), curculionids (weevils) and chrysomelids (leaf beetles) also occurred. Mollusc remains, generally the grey field slug, *Deroceras reticulatum*, and aphids were each found in approximately 30% of guts examined. Molluscs and aphids were found equally commonly in shrews from all three treatment areas, indicating that regular pesticide applications to control these pests in the Full Insurance area were not affecting their availability as prey items for shrews. However, the fourth major prey type, fly larvae or pupae (chiefly leatherjackets), did occur at different frequencies in the different treatment areas. Significantly more leatherjackets were present in adult shrews captured in the Integrated area than in the Full Insurance area (Chi-squared test, $P < 0.01$). It was also noted that leatherjackets occurred more frequently in adult, overwintered shrews than in immature animals captured in their year of birth ($P < 0.001$). Because of the relatively small number

Table 12.5   Diet of the common shrew and pygmy shrew at Boxworth, as revealed by analysis of the gut contents of dead shrews (1984–1988). Figures show the percentage of guts examined that contained the item.

Item	Common Shrew			Pygmy Shrew		
	Full Insurance area	Supervised area	Integrated area	Full Insurance area	Supervised area	Integrated area
Adult beetles	43	47	40	60	88	92
Molluscs	34	27	27	0	25	0
Aphids	23	36	25	40	13	33
Leatherjackets	16	23	40	0	13	0
Adult Diptera	18	16	13	30	38	17
Caterpillars	16	14	8	30	13	25
Beetle larvae	4	8	8	0	0	0
Arachnids	0	0	0	20	0	0
Hymenoptera	2	2	2	0	13	0
Springtails	0	0	2	10	0	0
Other items	5	5	4	0	0	0
Centrorhynchus sp.*	12	41	23	50	63	25
No. of Guts examined	83	64	52	10	8	12

(*parasite, not part of diet)

of shrews captured at Boxworth in 1981–1983, before the three pesticide treatment regimes began, it is unclear how the frequency of leatherjackets in the diet of common shrews in each treatment area has changed since then. However, pitfall-trap catches of leatherjackets indicated that they were much reduced in the Full Insurance area compared with the Supervised and Integrated areas (A J Burn, personal communication).

The only beetle identified to species was the ground beetle *Notiophilus biguttatus*. This species tended to occur more frequently in shrews in the two lower-input areas than in the Full Insurance area, reflecting its relative abundance in the different areas (Chapter 9).

A prey type notably absent from the gut contents examined was earthworms. Rudge (1968) identified these as a major component of the diet of common shrews, and in the laboratory worms were eaten by captive shrews. The soil at Boxworth is a heavy clay and the absence of earthworms in the diet probably reflects the rather low numbers of worms present at Boxworth, and possibly the shrews' difficulty in capturing worms in the heavy soil.

Pygmy shrew diet

The few pygmy shrew guts examined did not reveal any significant differences between treatment areas, but the sample sizes were probably too small to detect minor differences. However, the results suggest that the diet of the pygmy shrew differs markedly from that of the common shrew. Significantly more guts of pygmy shrews contained remains of adult beetles (24/30 pygmy shrew guts compared with 87/199 common shrew guts; Chi-squared test, P < 0.001). Pygmy shrews tended to take rather smaller beetles than the common shrews, although there was some overlap (for example, both species of shrew took the ground beetle *Notiophilus biguttatus*). Generally, pygmy shrews took smaller food items than common shrews. Significantly fewer mollusc and leatherjacket remains were found in the guts of pygmy shrews (Chi-squared tests: both P < 0.02). How-

ever, remains of adult flies, caterpillars and aphids appeared in similar proportions in the guts of both species.

Shrew parasites

Analysis of shrew guts at Boxworth revealed the presence of an encysted acanthocephalan parasite, probably *Centrorhynchus aluconis*, in both common and pygmy shrews (see Table 12.5). Infection was restricted to adult, overwintered individuals, and was significantly more common in males than in females. Fewer adult male shrews from the Full Insurance area were infected, compared with animals from the Supervised area (Chi-squared test, P < 0.001). This distribution of the parasite is likely to relate to the abundance of its primary host (an unknown invertebrate) in the different treatment areas. However, the relative rarity and abundance of the parasite in the different treatment areas cannot be ascribed to the effects of the pesticide treatments, since in 1981–1983, before the treatments began, parasites occurred less frequently in shrews from the prospective Full Insurance fields.

Summary

Populations of wood mice in the fields declined each year following husbandry activities such as harvesting, straw burning and ploughing. Overall, there was no evidence for any long-term effects of pesticide usage on wood mice at Boxworth. However, broadcast applications of molluscicide pellets had a serious effect on wood mouse populations, killing most of the mice within a few days, but immigration of juvenile mice from untreated areas allowed a rapid recovery in numbers. Broadcast applications of molluscicide at other times of year or in fields without adjacent woodland, where recolonisation of treated areas by juvenile mice is slow or absent, could have longer-lasting effects than those observed. The harmful effects of the molluscicide on wood mice were greatly reduced by drilling it with the seed instead of surface-broadcasting the pellets.

There was no evidence of long-term effects of the different pesticide treatments on populations of common shrews, estimated from pitfall-trap captures. However, there were changes in the shrews' diet in the different areas. Leatherjackets were less frequent in the stomachs of shrews caught in the Full Insurance area than in the Supervised area, reflecting the abundance of leatherjackets in the different areas. However, the occurrence of the grey field slug and of aphids in the diet was similar in shrews from all three areas.

The results of this study indicate that variation in the use of pesticides (at least over the range considered) will probably have little long-term effect on populations of small mammals, at least in areas where alternative untreated habitats, such as hedgerows and woods, persist. Probably of greater significance to small mammals are cultural practices such as straw burning, the reduction of distance between crop and hedge, and the removal of hedgerows.

References

Bergstedt, B (1966). 'Home ranges and movements of the rodent species *Clethrionomys glareolus* (Schreber), *Apodemus flavicollis* (Melchior) and *Apodemus sylvaticus* (Linne) in Southern Sweden.' *Oikos* **17**, 150–157.

Buckner, C H, Sarazin, R & McLeod, B B (1977). 'The effects of the fenitrothion spray program on small mammals.' In: Roberts J R, Greenhalgh R & Marshall W K (Eds). Proceedings of a symposium on fenitrothion: the long-term effects of its use on forest ecosystems. *National Research Council of Canada. Publication No. 16073*, pp 337–390.

Chitty, D & Kempson, D H (1949). 'Prebaiting small mammals and a new design of live trap.' *Ecology* **30**, 536–542.

Churchfield, S (1982). 'Food availability and the diet of the common shrew, *Sorex araneus*, in Britain.' *Journal of Animal Ecology* **51**, 15–28.

Evans, D (1977). 'Field Vole.' In: Corbet G B & Southern H N (Eds). The Handbook of British Mammals (2nd edition). Blackwell Scientific Publications, Oxford. pp. 185–193.

Flowerdew, J R (1976). 'The effect of a local increase in food supply on the distribution of woodland mice and voles.' *Journal of Zoology, London* **180**, 509–513.

Flowerdew, J R (1985). 'The population dynamics of wood mice and yellow-necked mice.' *Symposium of the Zoological Society of London* **55**, 315–338.

Flowerdew, J R (1988). 'Methods for studying populations of wild mammals.' In: Greaves M P, Smith B D & Greig-Smith P W (Eds) *Field Methods for the Study of Environmental Effects of Pesticides*. BCPC Monograph No. 40 pp. 67–76.

Green, R E (1979). 'The ecology of Wood mice (*Apodemus sylvaticus*) on arable farmland.' *Journal of Zoology, London* **188**, 357–377.

Gurnell, J (1976). 'Studies on the effects of bait and sampling intensity on trapping and estimating Wood mice, *Apodemus sylvaticus*.' *Journal of Zoology, London* **178**, 91–105.

Havers, S J (1989). *The ecology of wood mouse populations in contrasting farmland habitats*. Unpublished PhD thesis, University of Southampton.

Johnson, I P, FLowerdew, J R & Hare, R (1991). 'Effects of broadcasting and of drilling methiocarb molluscicide pellets on field populations of wood mice, *Apodemus sylvaticus*.' *Bulletin of Environmental Contamination and Toxicology* **46**, 84–91.

Montgomery, I W (1989). 'Population regulation in the wood mouse, *Apodemus sylvaticus*. I. Density dependence in the annual cycle of abundance.' *Journal of Animal Ecology* **58**, 465–475.

Pelz, H-J (1989). 'Ecological aspects of damage to sugar beet seeds by *Apodemus sylvaticus*.' In: Putman R J (Ed). *Mammals as Pests*. Chapman & Hall, London. pp. 34–48.

Pollard, E & Relton J (1970). 'Hedges V: A study of small mammals in hedges and cultivated fields.' *Journal of Applied Ecology* **7**, 549–557.

Rudge, M R (1968). 'The food of the common shrew *Sorex araneus* L. (Insectivora: Soricidae) in Britain.' *Journal of Animal Ecology* **37**, 565–581.

Southern, H N & Lowe, V P W (1968). 'The pattern of distribution of prey and predation in tawny owl territories.' *Journal of Animal Ecology* **37**, 75–97.

Tarrant, K A & Westlake, G E (1988). 'Laboratory evaluation of the hazard to wood mice, *Apodemus sylvaticus*, from the agricultural use of methiocarb molluscicide pellets.' *Bulletin of Environmental Contamination and Toxicology* **40**, 147–152.

Wolton, R J (1983). 'The activity of free-ranging wood mice, *Apodemus sylvaticus*.' *Journal of Animal Ecology* **52**, 781–794.

Exposure of rabbits to aphicides

13

Kenneth A Tarrant and **Helen M Thompson** (MAFF Central Science Laboratory, Tolworth)

Introduction

The rabbit (*Oryctolagus cuniculus*) is very abundant in the UK in many types of farmland, but at the start of the Boxworth Project there was no reason to suppose that rabbits are especially vulnerable to the effects of pesticides.

During the early years of the Project, rabbit runs were found to extend into the fields from the boundaries and droppings were found well beyond the headlands. There was rabbit damage to the growing wheat plants, and in summer rabbits were frequently disturbed during daylight hours in the mature crop. It became apparent that rabbits were using the crop to a considerable extent, and it seemed possible that they might be exposed directly to some pesticide applications, particularly the aphicides used in early summer. This possibility was investigated by trapping rabbits in the fields for biochemical assays and histological examinations that would identify the animals' exposure to demeton-S-methyl applications. Similar approaches were used in the studies of birds at Boxworth (Chapters 15 and 16).

A preliminary study was carried out in the summer of 1986; live-traps baited with carrot were placed in several fields, before and after the application of demeton-S-methyl. Samples of blood were taken from the ear veins of the rabbits caught, which were then released. Analysis of these samples revealed an apparently elevated blood level of the liver enzyme glutamate oxaloacetate transaminase (GOT) after demeton-S-methyl application. This enzyme may be released into the blood when liver cells are subjected to stress such as the toxic action of a pesticide (Tarrant, 1988).

In order to investigate this apparent effect, and other possible changes in enzyme levels associated with liver damage, further trapping was carried out in the summers of 1987 and 1988, and the rabbits were sacrificed for laboratory examination.

Collection and analysis of samples

Baited live-traps in the fields in 1987 captured four rabbits before spraying, and five after spraying. In 1988, three 'control' rabbits were caught (well after the first demeton-S-methyl spray, but before the second), and a total of eight was captured after spraying (six within 24 hours, one after 48 hours, and one after 60 hours).

These animals were subjected to an autopsy to reveal any gross pathological damage. Liver tissue was removed, and sections were cut and stained for microscopical examination. With the aid of an image analysis system, the size of liver cells and the presence of binucleation (both of which are indicative of a physiological response to poisoning) were measured.

To assess effects on enzyme activity, brain acetylcholinesterase (AChE) and blood serum cholinesterase (ChE) were assayed, using methods described by Thompson *et al*. (1988), and GOT was measured by the method of Bergmeyer & Bernt (1963). As indicated above, exposure to an organophosphorus pesticide should produce an increase in GOT, but would be expected to reduce the measured activity of AChE and ChE, by inhibition of these enzymes.

Results

Table 13.1 summarises the results of the biochemical and histological studies, showing the values obtained for rabbits trapped after demeton-S-methyl application relative to those trapped before treatment. The sample sizes are too small to permit statistical comparisons between rabbits that were exposed to the chemical and 'control' animals. However, previous work suggests that enzyme activity of about 20% more or less than the control is likely to be a real difference, particularly if there are also changes in histology.

Table 13.1 Biochemical and histological evaluation of rabbits trapped after demeton-S-methyl sprays at Boxworth in 1987 and 1988, relative to control samples collected before spraying. No assays for serum ChE or GOT were carried out in 1987. Normal values of the binucleation and cell count indices are c. 12 and c. 18

		% of control values				
Date of capture	Time after aphicide spray	Serum ChE	Brain AChE	Serum GOT	binucleation	cell counts
1 July 87	24 h	–	78	–	31.6	16.4
15 July 87	48 h	–	96	–	16.5	16.7
15 July 87	48 h	–	68	–	13.5	16.8
16 July 87	72 h	–	107	–	2.5	16.9
16 July 87	72 h	–	90	–	5.3	18.2
15 June 88	24 h	81	74	86	7.2	17.1
15 June 88	24 h	77	98	86	6.1	19.6
15 June 88	24 h	103	55	93	9.1	20.8
17 June 88	24 h	203	98	214	5.6	15.0
17 June 88	24 h	109	92	236	40.5	17.6
20 July 88	24 h	35	117	93	9.9	17.1
16 June 88	48 h	35	55	136	7.1	16.1
18 June 88	60 h	24	68	200	20.2	16.2

In 1987, levels of brain AChE activity in rabbits exposed to demeton-S-methyl ranged from 68% to 107% of mean control values. In 1988, activity of this enzyme in the rabbits trapped after 24 hours was 55% to 117% of controls, and enzyme inhibition was observed in the two animals trapped later (55% and 68% at 48 and 60 hours).

There was also inhibition of serum ChE in both years. The degree of inhibition was greater in the rabbits trapped 2–3 days after application (35% and 24% of the mean control activity) than in all but one of those trapped after 24 hours.

GOT levels were measured in 1988, and revealed evidence of increases in half of the rabbits trapped after the application of demeton-S-methyl.

Liver histology measurements showed less obvious differences. However, there was an increase in cell size in those animals which had the greatest ChE inhibition. Counts of binucleated liver cells showed an increase in a few animals, including one rabbit caught 60 hours after spraying. These differences were not statistically significant when the groups of pre- and post-spray animals were compared. In contrast, animals which were diseased did have significantly altered cell sizes and binucleation counts, which suggest that disease effects would have masked any changes due to pesticide exposure. Minor variations might also be expected, due to differences in the age of the animals.

Overall, liver morphology in the disease-free rabbits was normal, with the exception of occasional minor diffuse areas of inflammation and other mild symptoms of damage. No irreversible damage was observed, apart from one rabbit which showed typical lesions associated with coccidiosis.

Conclusions

This short study demonstrated that rabbits present in the fields after demeton-S-methyl spraying experienced short-term effects on certain blood and brain enzymes. The strongest evidence of effects was seen in the animals trapped after 48 and 60 hours, which suggests that exposure was accumulated during time spent in the fields over the first few days after spraying. The biochemical results were supported by histological evaluation that showed only minor liver cell changes, with no evidence of long-term damage. It is likely that the potential for transmission of disease (coccidiosis and myxomatosis) presents a greater hazard to rabbit populations than does exposure to aphicide sprays.

References

Bergmeyer H U & Bernt E (1963). 'Glutamate oxaloace-tate transaminase.' In: Bergmeyer H U (Ed.), *Methods of Enzymatic Analysis 1st Edition.* Verlag Chemie, Weinheim. pp. 837–842.

Tarrant K A (1988). 'Histological identification of the effects of pesticides on non-target species.' In: Greaves M P, Greig-Smith P W & Smith B D (Eds) *Field Methods for the Study of Environmental Effects of Pesticides.* British Crop Protection Council, Monograph 40. pp. 313–317.

Thompson H M, Walker C H & Hardy A R (1988). 'Avian esterases as indicators of exposure to insecticides; the factor of diurnal variation.' *Bulletin of Environmental Contamination and Toxicology* **41,** 4–11.

Population density and breeding success of birds

14

Mark R Fletcher*, Sian A Jones*, Peter W Greig-Smith†, Anthony R Hardy # and Andrew D M Hart† (MAFF Central Science Laboratory, Tolworth*, Slough # and Worplesdon†

Introduction

Population densities of many farmland birds can be assessed by mapping the number of territories held by breeding birds in an area. This is based on observing birds' territorial behaviour, such as aggressive displays or singing, during repeated visits to a study area. If the registrations of territorial behaviour follow stringent guidelines, which include the minimum number of visits required in each year, then year-to-year changes in the density of a particular species may be identified by comparing the estimated numbers of territories obtained in different years.

This study aimed to identify the effect on farmland birds of any changes in the quality of the environment which might have resulted from pesticide use, such as depletion of invertebrate prey, or alterations to the habitat. Census methods are unreliable for within-year comparisons so it was not possible to identify the direct effects of individual pesticide applications. Censuses were carried out each year to investigate changes in bird populations and also to identify any consistent differences between the treatment areas.

Monitoring concentrated on eleven resident species which are typical of farmland birdlife and were present on the farm at the beginning of the Project: blackbird (*Turdus merula*), blue tit (*Parus caeruleus*), chaffinch (*Fringilla coelebs*), dunnock (*Prunella modularis*), great tit (*Parus major*), robin (*Erithacus rubecula*), song thrush (*Turdus philomelos*), starling (*Sturnus vulgaris*), tree sparrow (*Passer montanus*), wren (*Troglodytes troglodytes*) and yellowhammer (*Emberiza citrinella*). These species breed in woods or hedges close to the crops at Boxworth, so they could be affected directly by pesticide use either through spray drift or by feeding in pesticide-treated crops. It was not feasible to monitor populations of woodpigeons

(*Columba palumbus*), carrion crows (*Corvus corone*), rooks (*C. frugilegus*), jackdaws (*C. monedula*), magpies (*Pica pica*), game birds (partridges, *Perdix perdix* and *Alectoris rufa*, and pheasants, *Phasianus colchicus*) nor birds of prey such as the kestrel (*Falco tinnunculus*), because their large ranges may have extended well beyond the area of study. Skylarks (*Alauda arvensis*), although numerous and likely to be exposed to pesticides (see Chapter 16), were difficult to census in fields without extending the visit times greatly and were excluded from monitoring, as were migrant birds such as the whitethroat (*Sylvia communis*), whose numbers varied greatly in each year.

In addition to population censuses, the breeding performance of some species was studied, to examine possible influences of the pesticide regimes. Pesticides might affect breeding success by direct poisoning, by reducing the availability of food, or by affecting birds' behaviour, which in turn could influence their susceptibility to predation. Such effects of pesticides on the breeding success of a particular species might help to explain any pesticide-related effects observed on that species at the population level.

Methods

As birds are highly mobile, it was decided that the Supervised and Integrated areas, which were adjacent, shared common boundaries, contained a similar mosaic of habitats and received similar pesticide inputs (Appendix II) should be considered together as a single 'lower-input' area similar in size to the Full Insurance area (See Figure 2.1).

The method used to census birds was the 'Common Bird Census' (International Bird Census

Committee, 1969). To achieve adequate coverage of the Full Insurance and Supervised + Integrated areas, a total of ten visits, each lasting between three and four hours, was made to each area by one observer at intervals of seven to ten days in spring and early summer. On each visit, a pre-determined route was walked to give complete coverage of field margins and adjacent woods. Territorial activity such as singing, aggression and breeding behaviour was recorded for each bird species on large-scale field maps. Birdsong and other territorial behaviour tends to be more evident soon after dawn than at other times, and has usually declined by mid-morning, so most census visits were made in the morning. However, in each year it was usual to make one or two visits near dusk so that the presence of species which tend to sing more in the evening (such as the song thrush) might be detected. Weather conditions influenced the exact timing of census visits, since birds tend not to sing in heavy rain (although they often give a burst of singing after showers), and because drizzle may impair visibility. After the last census visits in each year, a summary map was prepared for each species, showing the number and location of territories, deduced from clusters of accumulated observations of territorial behaviour.

For the study of birds' breeding success, about 160 nest boxes were put up early in the Project in each of the treatment areas. A further 60 were added in the Supervised + Integrated area and 84 in the Full Insurance area in the winter of 1986–1987. About half of the original nest boxes had large holes suitable for starlings. The remainder had smaller holes and were used by blue tits, great tits and tree sparrows. The nest boxes were placed at heights of between 1.5 and 4 metres above ground in the hedgerows surrounding the Project fields and on trees in adjacent woodland.

From the beginning of the breeding season (the end of March) onwards, all nest boxes were checked at weekly intervals for the presence of eggs. When eggs were found, further frequent checks were made until the young had fledged, or the nest failed. Details of the species, the number of eggs or young birds which were present, and their stage of development, were recorded when each box was checked. Young birds were ringed so that those which returned to breed in the nest boxes in subsequent years could be identified. The information collected from the nest box study was used to compare the breeding success of the four hole-nesting species in the Full Insurance and Supervised + Integrated areas.

Population density

The total numbers of breeding territories for the eleven bird species studied during the Boxworth Project are shown in Figure 14.1. The initial increase observed in both the Full Insurance and the Supervised + Integrated areas was caused by the occupation of nest boxes put up at the start of the Project.

During the course of the Project there were several habitat changes which affected the numbers of bird territories. A hedgerow across Knapwell field was removed in the winter of 1982–1983, a stubble fire on a neighbouring farm burnt through a hedgerow beside Top Pavements in August 1983, dead elm trees were removed at several times from large areas of Grange Wood and Thorofare Wood, where there was some subsequent replanting, and there was removal of scrub along a stream beside Backside field during the autumn of 1983.

Because bird density fluctuated from year to year, an index of the relative numbers of territories in the Full Insurance and the Supervised + Integrated areas was used to examine differences which might be related to the use of pesticides. The index was obtained by expressing the number of territories of all eleven species in the Full Insurance area as a percentage of the total number present that year. Thus, a value of 50% would indicate that birds used the two treatment areas equally. Because of the unreplicated design of the Project (Chapter 2), it is not possible to test statistically the differences in bird density between areas. However, trends from 1984 to 1988 can be

Figure 14.1 Total numbers of breeding territories of eleven common bird species in the treatment areas at Boxworth, 1982–1988.

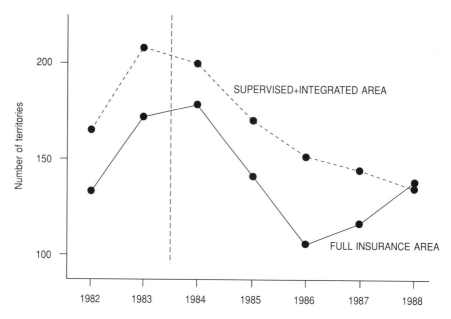

Figure 14.2 Distribution of territories of eleven bird species between the Full Insurance area and the Supervised + Integrated area in each year of the Project.

Total of all species

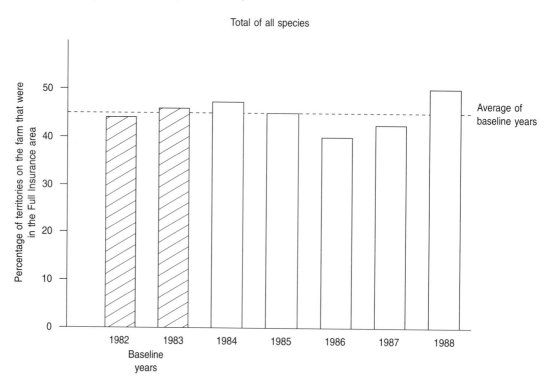

examined by testing for changes in the proportions of territories that were in the Full Insurance area.

The index for the eleven species remained fairly constant over the course of the Project, including the baseline years (Figure 14.2). This indicates that, taking all these species together, there was no change in birds' use of the two treatment areas.

When examined separately, five of the species studied (tree sparrow, blue tit, great tit, starling and wren) showed clear changes in numbers during the course of the Project, but these were apparently not related to the effects of pesticides.

Tree sparrow numbers increased rapidly at the beginning of the Project, as a result of the provision of nest boxes. After reaching a maximum in 1984 in both treatment areas, there was a rapid decline towards the end of the Project. The reason for this is not known. A similar national decline of tree sparrows was observed between 1977 and 1985 in the Common Bird Census results, during which the population appears to have fallen to about a quarter of its initial size (Summers-Smith, 1989). Other workers have observed that after initial occupation of nest boxes, local populations then decline, often to extinction, over a period of a few years.

Tree sparrows were dominant over other species with which they were in competition for the smaller nest boxes and often displaced other occupants, particularly blue tits and great tits. As the density of tree sparrows declined in the latter part of the Project the number of nest boxes taken up by blue tits and great tits increased (Figure 14.3). The specific nesting requirements of tree sparrows are such that until the nest boxes were put up there were no suitable nesting sites for them at Boxworth. Not all the boxes were used. At most 89% of the nest boxes were occupied (in 1984), which suggests that the availability of nesting sites was not the major factor limiting the tree sparrow population during the Project (although it must be recognised that some nest boxes may not have been in suitable places for occupation by tree sparrows).

A large decrease in the number of wren territories occurred in both treatment areas during 1986. This was probably caused by the prolonged cold winter of 1985–86 and reflects a general decline in the number of wren territories recorded throughout the country in the Common Bird Census results (Marchant et al., 1990).

Only starlings showed a statistically significant change in the distribution of their territories between the two treatment areas. There was a steady decline in the number of starling territories in the Full Insurance area relative to the Supervised + Integrated area from about 1984 to 1987, with a recovery in numbers in 1988 (Figure 14.4). Although this pattern is consistent with an adverse effect of the Full Insurance programme on food supplies, there was no obvious change in the use of pesticides that would account for the recovery in starling numbers in 1988. Biochemical studies (Chapter 16) revealed some sub-lethal effects of pesticides on this species at Boxworth, but these were transient and were detected outside the breeding season. Therefore, factors other than pesticides were examined as possible reasons for the change in distribution of breeding starlings.

If starlings were affected chiefly by the habitat changes which took place on the farm during the Project, one would expect declines in the numbers of breeding birds in both areas, since habitat changes occurred in both areas in most years. An excess of nesting sites was available in both treatment areas, so limitations on the availability of suitable nesting sites did not differ between the areas. Although there was predation on adult starlings by weasels (Mustela nivalis), kestrels (Falco tinnunculus) and tawny owls (Strix aluco), this occurred to a similar extent in both areas.

Two changes that occurred during the Project could explain the relative decline in starlings in the Full Insurance area. These were a decline in several groups of invertebrates in that area and a reduction in the number of cattle kept on the farm during the treatment phase of the Project.

Starlings breeding in the Supervised +

Figure 14.3 Occupation of small nest boxes at Boxworth by tree sparrows, blue tits and great tits.

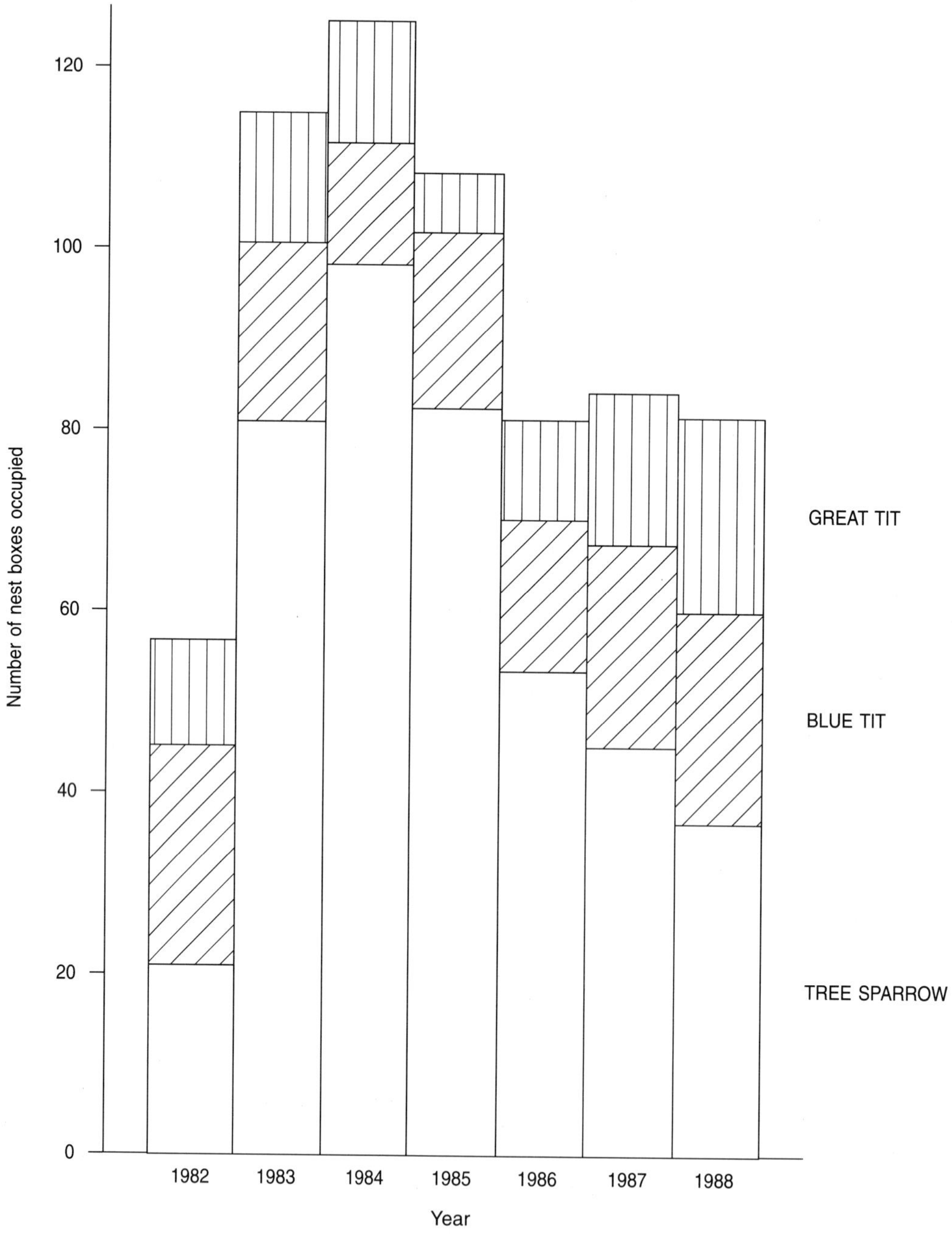

GREAT TIT

BLUE TIT

TREE SPARROW

Number of nest boxes occupied

Year

Figure 14.4   The distribution of starling territories between the Full Insurance area and the Supervised + Integrated area, 1982–1988.

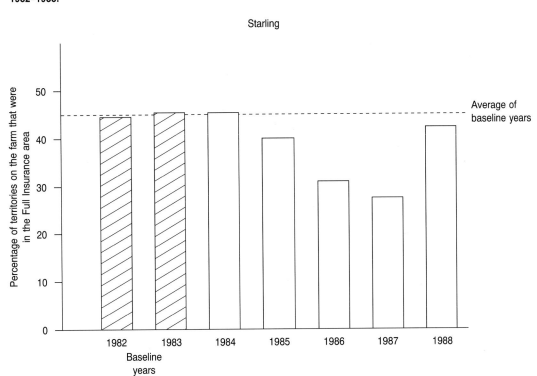

Integrated area foraged widely outside the Project area, but most of those in the Full Insurance area fed close to cattle kept on pasture near Grange Wood, the main breeding site for starlings in that part of the farm. Particularly in dry weather, they obtained insects by probing the softer ground where the cattle had urinated and defaecated. Changes in the location and number of cattle on the farm might account in part for changes in numbers of starlings in the Full Insurance area. However, since a major reduction in the number of cattle did not occur until 1987, this cannot explain the initial decline in the number of birds.

Starlings feed on a wide variety of invertebrates, of which leatherjackets (crane-fly larvae) are particularly important during the breeding season (eg Feare, 1984). After 1984, pitfall trap catches of leatherjackets showed a marked decline

in the Full Insurance area relative to the Supervised + Integrated area, and there were reductions in several other groups of invertebrates that contribute to the diet of starlings (Chapter 9). This suggests that the number of leatherjackets available to starlings (which may be as little as 1% of the total leatherjacket population; Tinbergen & Drent, 1980), decreased in the Full Insurance area after 1984.

Thus, none of the factors considered above provides a full explanation for the changes observed in the distribution of breeding starlings. Their initial decline in the Full Insurance area is consistent with an indirect effect of high pesticide inputs acting through the birds' prey populations, and this may have contributed to the trend for them to increasingly favour the Supervised + Integrated area for nesting.

Breeding performance

The aim of comparing nesting success between the Full Insurance and Supervised + Integrated areas was to discover any differences that could have been caused, directly or indirectly, by pesticide usage. These might be seen in breeding 'effort' (the number of broods in the season, or clutch size) or breeding 'success' (the proportions of eggs that hatch, and of nestlings that survive). However, care is needed in interpreting any observed differences in breeding performance, because these may be caused by other factors, such as predation and weather. Differences between the treatment areas observed during the treatment years of the Project must also be examined in the light of any differences present in the baseline years.

Studies of breeding were carried out on the four principal hole-nesting species: tree sparrow, starling, great tit and blue tit.

Tree sparrow

There are few suitable natural nest sites for tree sparrows at Boxworth, and no breeding pairs were recorded on the farm before 1982. However, the birds readily use nest boxes, and a substantial population built up during the baseline years of the Project.

Because it is likely that the entire local population used nest boxes, the results obtained by regular checks of boxes give a complete record of breeding performance for this species. Table 14.1 lists the total numbers of breeding pairs, and of young tree sparrows successfully reared each year in the two areas. There were huge fluctuations from year to year in the numbers of young reared per pair, and in some years there were large differences between the areas. However, these were not consistent, and did not suggest any obvious trends through the period of the Project.

In order to investigate breeding success in more detail, the first, second and third broods of each season were examined separately. First broods should give the clearest indication of any major differences in breeding prospects between the two treatment areas, because later nests represent a mixture of second and third broods and re-nesting by failed breeders, and performance may also be confounded by seasonal changes in birds' behaviour. However, in most respects, results were similar in first and later broods.

Figure 14.5 illustrates the outcome of tree sparrows' first broods, showing the average clutch size, the average percentage of eggs laid that hatched, and the average proportion of nestlings that survived to leave the nest. There is no clear pattern in any of these measures of breeding

Table 14.1 Breeding performance of tree sparrows in each year of the Project, measured by the average number of young reared per breeding pair, in each area.

	Supervised + Integrated area			Full Insurance area		
	Total fledged young	Number of breeding pairs	Number fledged per pair	Total fledged young	Number of breeding pairs	Number fledged per pair
1982	34	6	5.7	78	15	5.2
1983	149	38	3.9	256	44	5.8
1984	204	40	5.1	323	58	5.6
1985	118	32	3.7	162	53	3.1
1986	83	24	3.5	129	32	4.0
1987	102	18	5.7	103	29	3.6
1988	21	12	1.8	64	24	2.7

Figure 14.5 Success of tree sparrows' first broods in the treatment areas at Boxworth, shown by average clutch sizes, percentages of eggs that hatched, and percentages of chicks that survived to leave the nest each year. Bars above the columns are standard errors.

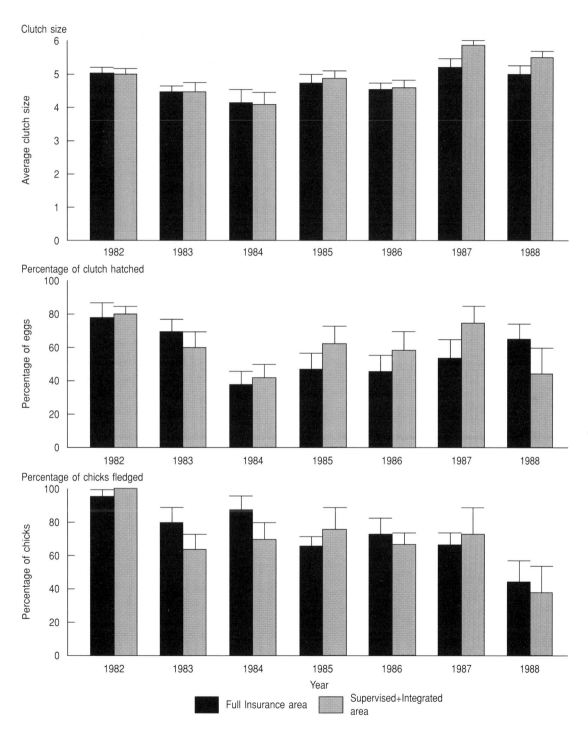

success, nor any consistent difference between the treatment areas, although there was an apparent steady decline in the survival of nestlings from 1982–1988. The growth and condition of nestlings, measured by their weight and wing length on the 10th day after hatching, also showed no obvious difference between the treatment areas (Hart *et al.*, 1990).

When nests are scored as 'successful' or 'unsuccessful' in producing at least one fledged young bird, there is evidence of a difference between the areas (Table 14.2). The percentage of nests that failed was lower in the Full Insurance area in the baseline years, but increased more in that area during the treatment years. Although there was much year to year variation in success, this difference between areas was greater than would be expected by chance alone. This result reflects the combined effects of predation, desertion, and other causes of total nest failure. Although pesticides might have influenced these events indirectly, Table 14.1 shows that the tree sparrow population's overall production of young was not consistently reduced.

Table 14.2   The percentage of first broods of tree sparrows and starlings that failed to produce at least one fledgling as a result of predation, desertion or starvation.

	Supervised + Integrated area	Full Insurance area
Tree sparrow		
1982–83	41% (41 nests)	18% (55 nests)
1984–88	45% (108 nests)	52% (174 nests)
Starling		
1982–83	21% (116 nests)	9% (104 nests)
1984–88	36% (238 nests)	46% (164 nests)

The causes of nesting failure are shown in Figure 14.6. In the baseline years, there was disproportionately more predation (nests in which the remains of chicks were found, or all the chicks disappeared) in the Supervised + Integrated area. However, this difference was not evident during the 1984–88 period, when predation accounted for

more than half the known causes of nest failures. More detailed examination of partial brood losses (Hart *et al.*, 1990) confirmed that nest losses due to causes other than predation were no more frequent in the Full Insurance area.

If pesticides were causing more nests in the Full Insurance area to fail (eg by making nestlings more susceptible to predation), the occurrence of second and third broods there might be high, to replace early failures. In contrast, an indirect effect of pesticides on invertebrate food supplies might cause an opposite trend, if poorer feeding conditions in the Full Insurance area permitted fewer pairs to raise a second or third brood. The results showed no clear patterns in the incidence of later broods. While this does not rule out the possibility of complex effects of the pesticide regimes, it does not offer any evidence for the existence of such effects.

Starling

Whereas tree sparrows only nested in boxes, there were some natural nest sites available for starlings at Boxworth, and not all nests could be monitored. However, a large majority of the population used the nest boxes during the Project. Therefore, although data on the total production of young from nests in boxes are not fully representative of the population's performance, they should identify any major differences between areas that corresponded to the introduction of the contrasting pesticide programmes.

Table 14.3 shows the average number of young birds produced by each breeding pair of starlings. As for tree sparrows, there was great variation from year to year, and no consistent difference between the areas in the treatment years of the Project.

For starlings' first broods there were no clear patterns or differences between the treatment areas in the percentages of eggs that hatched, or of chicks that survived to leave the nest (Figure 14.7). There was a tendency similar to that detected for tree sparrows, for an increase in the percentage of

Figure 14.6 Fate of tree sparrow nests in the two treatment areas in the baseline years (1982–1983) and the treatment phase of the Project (1984–1988).

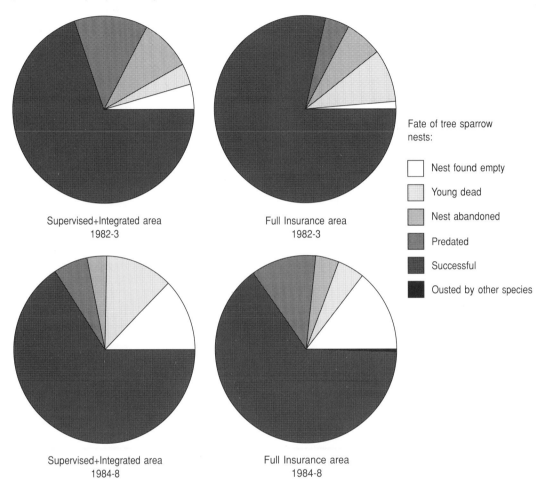

Supervised+Integrated area
1982-3

Full Insurance area
1982-3

Supervised+Integrated area
1984-8

Full Insurance area
1984-8

Fate of tree sparrow nests:

☐ Nest found empty

▨ Young dead

▨ Nest abandoned

▨ Predated

■ Successful

■ Ousted by other species

Table 14.3 Breeding performance of those starlings at Boxworth that bred in nest-boxes.

	Supervised + Integrated area			Full Insurance area		
	Total fledged young	Number of breeding pairs	Number fledged per pair	Total fledged young	Number of breeding pairs	Number fledged per pair
1982	128	50	2.6	148	46	3.2
1983	208	66	3.2	250	58	4.3
1984	83	59	1.4	99	55	1.8
1985	129	45	2.9	145	32	4.5
1986	140	52	2.7	28	28	1.0
1987	181	41	4.4	48	17	2.8
1988	91	41	2.2	70	32	2.2

Figure 14.7   Breeding performance of starlings in the treatment areas at Boxworth, shown by average clutch sizes, percentages of eggs that hatched, and percentages of chicks that survived to leave the nest each year. Bars above the columns are standard errors.

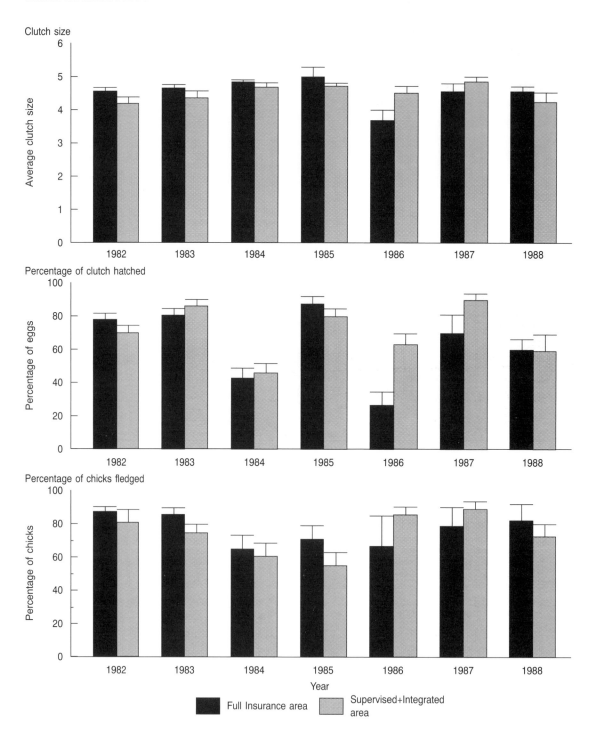

unsuccessful nests in the Full Insurance area relative to the Supervised + Integrated area (Table 14.2). However, statistical analysis showed that the variation from year to year was so great that the average difference between areas was no more than might occur by chance.

The main cause of nest failure was predation, which increased between the baseline period and the treatment years, but the relative importance of other causes were very similar in both areas (Figure 14.8).

The number of pairs of starlings that reared both first broods and second broods varied greatly, from 0% of pairs (in 1986) to 53% (in the Full Insurance area in 1983). However, it was generally equal to or higher in the Full Insurance area than that in the Supervised + Integrated area. This difference was present in the baseline years, and is probably a reflection of differences in the quality of the habitat for starlings (eg the presence of cattle close to the Full Insurance fields), rather than a result of pesticide usage.

Figure 14.8   Fate of starling nests in the two treatment areas in the baseline years (1982–1983) and the treatment phase of the Project (1984–1988).

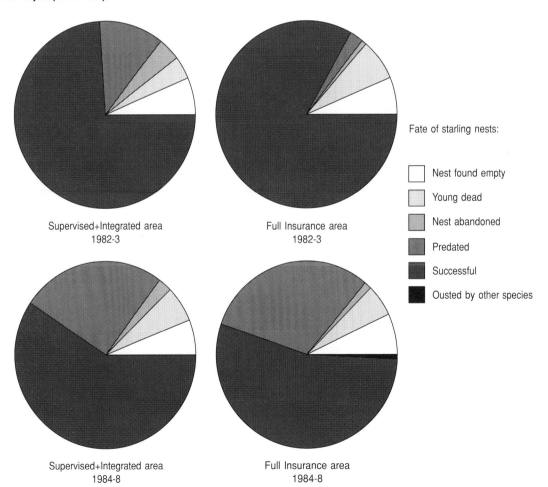

Supervised+Integrated area
1982-3

Full Insurance area
1982-3

Fate of starling nests:

☐ Nest found empty

▨ Young dead

▨ Nest abandoned

▨ Predated

■ Successful

■ Ousted by other species

Supervised+Integrated area
1984-8

Full Insurance area
1984-8

Figure 14.9 Breeding performance of blue tits in the treatment areas at Boxworth, shown by average clutch sizes, percentages of eggs that hatched, and percentages of chicks that survived to leave the nest each year. Bars above the columns are standard errors.

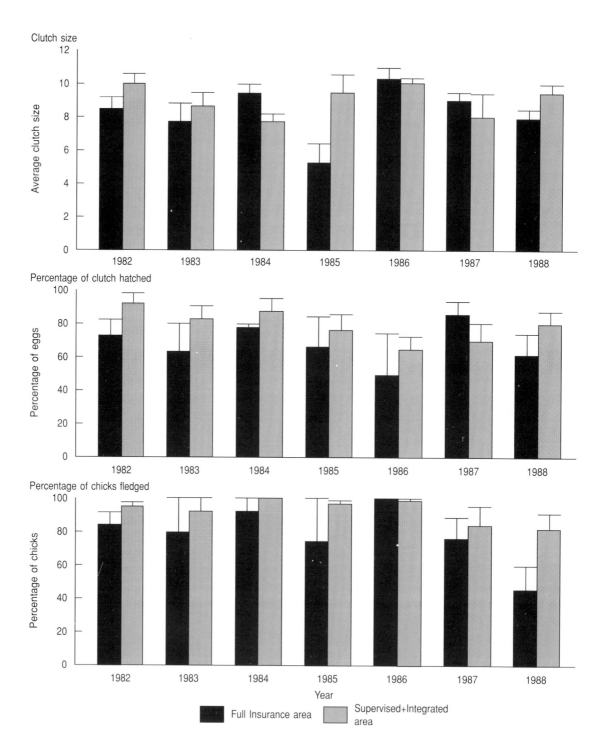

Figure 14.10 Breeding performance of great tits in the treatment areas at Boxworth, shown by average clutch sizes, percentages of eggs that hatched, and percentages of chicks that survived to leave the nest each year. Bars above the columns are standard errors.

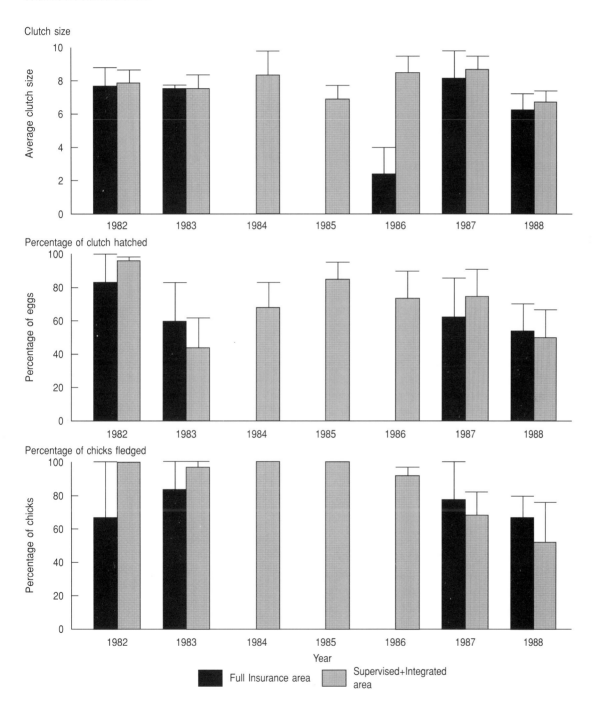

Blue tits and great tits

Some of the small nest boxes of the type used by tree sparrows were adopted by blue tits or great tits. However, the numbers of nests of these species were too few for a detailed analysis of breeding success. There were no indications of any clear differences between the two areas in the breeding success of blue tits (Figure 14.9) or great tits (Figure 14.10). For both species, predation was the major cause of nesting failure.

Discussion

The value of the information that can be gathered by simple, regular checks of the contents of nest-boxes was limited at Boxworth by the great variation in the breeding performance of the species studied. The statistical tests carried out may have failed to detect some differences in breeding that could be biologically important. However, the scale of the study was sufficient to have revealed any major adverse effects of the Full Insurance regime, at least for tree sparrows and starlings. The results show no patterns in population density or breeding performance that clearly match what might be expected as a result of adverse pesticide effects, and overall production of young was not reduced in the Full Insurance area. This suggests that populations of tree sparrows and starlings were robust enough to overcome any effects on their invertebrate food supply, or sub-lethal debilitation of breeding birds.

Summary

Five out of eleven farmland bird species exhibited notable changes in their populations during the course of the Project but of these only starlings showed a statistically significant decline in numbers in the Full Insurance area relative to the Supervised + Integrated area. Although the reason for the change in starlings' distribution is not fully understood, there was evidence for a decline in invertebrates caused by the high pesticide input regime, and there were changes in the numbers and location of cattle which could have affected starlings in the Full Insurance area. For all other species, the population changes could be explained by factors other than the effects of the differing pesticide treatment regimes.

Examination of the breeding performance of four hole-nesting species revealed no clear evidence of any changes, or differences between areas, that could be accounted for by effects of the high pesticide inputs in the Full Insurance area, either due to direct toxicity or to indirect depletion of food supplies. Although the possibility exists that some adverse effects of the Full Insurance regime were not detected, it can be concluded that the breeding success of birds was not seriously affected by the Full Insurance programme.

References

Feare, C J (1984). *The Starling*. Oxford University Press, Oxford. pp. 315.

Hart, A D M, Fletcher, M R, Greig-Smith, P W, Hardy, A R, Jones, S A & Thompson, H M (1990). 'Le Projet Boxworth: effets de regimes de pesticides contradictoires sur les oiseaux de terre arable.' In: *Relations entre les traitements phytosanitaires et la reproduction des animaux*, Annales de l'ANPP. No. 2. pp. 225–232.

International Bird Census Committee (1969). 'Recommendations for an international standard for a mapping method in bird census work'. *Bird Study* **16**, 248–255.

Marchant, J H, Hudson, R, Carter, S P & Whittington, P (1990). *Population trends in British breeding birds*. British Trust for Ornithology, Tring. pp. 300.

Summers-Smith, J D (1989). 'A history of the status of the Tree sparrow *Passer montanus* in the British Isles'. *Bird Study* **36**, 23–31.

Tinbergen, J & Drent, R (1980). 'The starling as a successful forager'. In: Wright E N, Inglis I R and Feare C J (Eds). *Bird Problems in Agriculture*. British Crop Protection Council. pp. 83–97.

Effects of summer aphicides on tree sparrows

Andrew D M Hart*, **Helen M Thompson#**, **Mark R Fletcher#**,
Peter W Greig-Smith*, **Anthony R Hardy†** and **Stephen D Langton#**
(MAFF Central Science Laboratory,
Worplesdon*, Tolworth# and Slough†)

Introduction

Of all the birds breeding around the cereal fields at Boxworth, tree sparrows are among the most likely to be exposed to summer insecticides. At the time that aphicides were applied to wheat in June and July, tree sparrows were gathering invertebrates from the fields to feed to their growing nestlings in nest boxes around the edges of the crop. This species therefore provides an opportunity to examine some of the potentially most severe consequences of pesticide use for birds.

The studies undertaken at Boxworth were designed to help address three questions: (1) to what extent were members of the tree sparrow population exposed to aphicides? (2) what were the effects of such exposure on individual birds? (3) were these effects ecologically significant?

To these ends, a detailed investigation was mounted to examine foraging behaviour and nest attendance by adult tree sparrows, food brought to nestlings, and nestling growth. Changes in the activity of certain blood and brain enzymes which are liable to be inhibited by the insecticide were also studied. Additional information on the success of nests in producing healthy young birds, and on

the density of the breeding population, were available from the bird studies described in Chapter 14.

The summer aphicide applications that occurred in each year of the Project are listed in Table 15.1. Demeton-S-methyl was used twice annually in the Full Insurance area, usually with a 14-day interval between applications, and occasionally in the Supervised area. Pirimicarb was used in the Integrated area in years when aphid thresholds were exceeded.

In principle, studies of nests in the Full Insurance area around the time of spraying should be comparable with those in the Supervised + Integrated area at the same time. However, the spread of breeding dates, combined with the declining population of tree sparrows (Chapter 14), meant that there were often insufficient nests at the time of spraying to permit comparisons between the treatment areas. An alternative approach was adopted, of examining changes in behaviour, diet and biochemistry from just before to just after the applications of demeton-S-methyl. This also carries the advantage of dealing with variation between nests and between individual nestlings, by letting

Table 15.1 Dates of applications of aphicides in the three treatment areas at Boxworth. Demeton-S-methyl was used in the Full Insurance and Supervised areas and pirimicarb in the Integrated area.

	Full Insurance area		Supervised area	Integrated area
	First spray	Second spray		
1984	18–19 June	3–6 July	23–28 June	24 June
1985	18 June	10 July	–	–
1986	27 June—1 July	14 July	–	–
1987	29 June	13 July	–	–
1988	16 June	18–20 July	13 June	23 June

each act as its own 'control'.

As for many aspects of the Project, the unreplicated design of the treatment areas (Chapter 2) prevents detailed statistical comparisons between areas. In addition, the small size of the tree sparrow population meant that it was often not possible to achieve sample sizes adequate to detect small differences. For this reason, many of the statistical tests reported are of low power, and the results must not be interpreted too strongly. All the statistical tests used were two-tailed.

Nestling diet

Although tree sparrows include some plant material in their diet, they are primarily insectivorous during the period when they are rearing nestlings. Casual observations at the beginning of the Project showed that the birds fed regularly in wheat fields, and therefore might be sensitive to changes in the availability of invertebrate prey, both as a progressive decline under the Full Insurance regime, and as a short-term change at the time of particular pesticide applications. Also, if parent tree sparrows gather food from treated fields shortly after spraying, they are liable to provide a contaminated diet to their young. With these possible effects in mind, a study of the diet of nestling tree sparrows was carried out.

Methods

Early in the Project, the composition of the birds' diet at Boxworth was surveyed by placing soft collars around the necks of nestlings, temporarily preventing them from swallowing food. This allowed meals fed by the parent birds to be removed from time to time, and examined, before taking off the collars. Although this procedure did not result in any harm to nestlings, it caused disturbance that sometimes reduced the parents' subsequent visits to the nest. The method was therefore abandoned, in favour of collecting faecal sacs from nestlings when inspecting nests and weighing the chicks. Identification of invertebrate remains in faeces is a widely-used technique of diet analysis (e.g. Davies, 1976; Moreby, 1988), and allowed a large number of samples to be collected without any additional disturbance.

Faecal samples were stored in a solution comprising 90% ethanol and 10% glycerol for later examination. It was not always possible to determine which, or how many of the nestlings had produced the material, and all the samples collected at a nest inspection were combined. Therefore the results provide a picture of relative proportions of different types of prey, rather than a quantitative estimate of how much nestlings ate.

Prey remains were identified and counted by teasing apart the faecal sacs under a microscope, and comparing fragments of insects' legs, wings, mouthparts and other pieces to a reference collection of the species concerned. Many beetles, particularly ground beetles (Carabidae), rove beetles (Staphylinidae) and ladybirds (Coccinellidae), were identified to species, but many other invertebrates were assigned only to families or broader categories. The numbers of fragments were then converted to the corresponding numbers of prey animals they represent, and expressed as a proportion of the total in the sample.

There are possible biases in the results obtained by analysis of faeces, because soft-bodied animals are likely to be under-represented, and prey species may pass through the bird's gut at different rates. Subject to this limitation, the data should be sufficient to detect any major changes in diet composition at the time of aphicide applications.

Results

Table 15.2 illustrates the range of food items fed to nestlings, based on the contents of neck-collar samples collected in June and July of 1983 and 1984. At that time of year, the diet was heavily dominated by Hemiptera (almost entirely aphids) although, because of their small size, they formed a

Table 15.2 Food items delivered to nestling tree sparrows in June—July 1983 and 1984. The figures indicate the numbers of each category in 36 neck-collar samples.

springtails	(Collembola)	28
earwigs	(Dermaptera)	1
thrips	(Thysanoptera)	1
plant bugs & aphids	(Hemiptera)	
	—Aphididae	519
	—Psyllidae	1
	—Cicadellidae	1
beetles	(Coleoptera)	
	—Carabidae	45
	—Staphylinidae	33
	—Elateridae	5
	—Chrysomelidae	6
	—Curculionidae	6
	—Coccinellidae	15
	—others	6
scorpion-flies	(Mecoptera: Panorpidae)	1
flies	(Diptera)	
	—Tipulidae	6
	—Muscidae	6
	—Syrphidae	1
	—others	3
butterflies & moths	(Lepidoptera)	32
sawflies	(Hymenoptera: Cephidae)	2
woodlice	(Isopoda)	2
spiders & harvestmen	(Arachnida)	
	—Linyphiidae	7
	—Tetragnathidae	2
	—Opiliones	1
	—Lycosidae	2
	—others	1
slugs	(Mollusca)	6
wheat grains		12

smaller proportion of the biomass fed to chicks. Beetles were the second most numerous items, but the species recorded were largely from different families in the two years. In 1984 a large proportion was ladybirds (Coccinellidae), which usually are only rarely found in the diet of farmland birds. In both years the ground beetles (Carabidae) that were eaten included larvae of *Notiophilus biguttatus* and *Loricera pilicornis*, indicating that tree sparrows took prey from the ground as well as from the wheat plants. Large caterpillars were the only items in some samples and on average they formed a major contribution to the total biomass.

Although invertebrates constituted over 97% of the food items identified, whole wheat grains were present in 19% of the samples collected in 1983 and 44% of those in 1984. Evidently, tree sparrows are not entirely insectivorous, even when feeding their young.

Examination of faecal samples revealed a broadly similar variety of prey items (Figure 15.1), although there was much variation between samples from different nests, and at different times from the same nest. This probably reflects the proximity of nests to different foraging sites, the age of nestlings (which may be fed large items only

Figure 15.1   Composition of the diet of nestling tree sparrows, shown by the results of examining faecal pellets collected from seven nests in the summer of 1986.

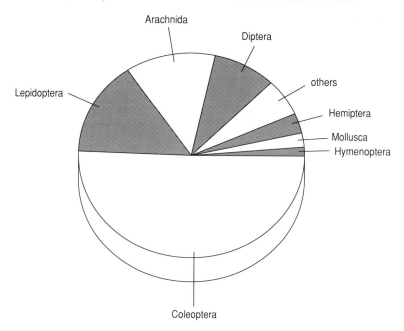

Figure 15.2   Changes in the occurrence of aphids and beetles in the diet of five broods of tree sparrow nestlings from three days before to three days after the first aphicide application in 1988.

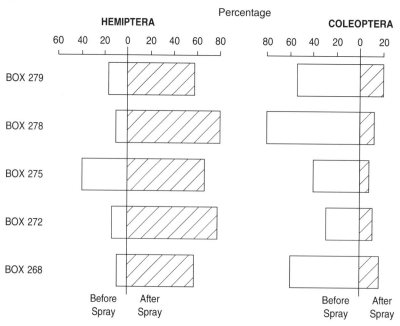

as they grow older), possible feeding specialization by the parent birds, and a large element of chance variation. Many samples contained insects (eg certain caterpillars) that occur only in hedgerows and other non-crop habitats, as well as aphids, demonstrating the flexibility of birds' foraging habits.

Because of this variability in the tree sparrows' diet and the fact that in most years samples were collected opportunistically when nests were visited for other reasons, it is difficult to interpret the results in detail. Sufficient material for a statistical comparison of diet before and after aphicide application was obtained only for 1988, when an effort was made to collect frequent samples from particular nests.

Figure 15.2 summarises the main changes in the diet of nestlings in five nests in the Full Insurance area, from three days before to three days after the first spraying of demeton-S-methyl on 16 June. The proportion of the diet made up by aphids greatly increased after spraying, and the proportion of beetles, particularly ground beetles (Carabidae) declined. Both these patterns were highly significant (Analysis of Variance, $P<0.001$). The proportion of spiders also fell slightly, but numbers were low both before and after spraying, and the change could have been due to the large increase in aphids. There was a tendency for samples collected in the morning to contain fewer ground beetles than the afternoon samples. Also, there were differences between nest boxes in the degree of change after spraying. This is to be expected because birds from different nest boxes varied in the proportion of their foraging time that was spent in the treated fields (page 180). A similar analysis of samples collected around the time of the second spray, in mid-July, did not show a consistent increase in the proportion of aphids.

In order to assess the generality of these patterns, data collected in other years were combined, by scoring each nest monitored before and after a spray as either 'increasing' or 'decreasing' in the proportions of aphids and of beetles. Although this is a crude evaluation of changes, it revealed a pattern that was consistent with the

1988 data. Most nests in the Full Insurance area showed an increase in Hemiptera and a decrease in Coleoptera (Table 15.3), although in some cases the differences were small.

Table 15.3 Numbers of tree sparrow nests at which the proportions of aphids and of beetles in the diet increased or decreased from before to after spraying of demeton-S-methyl in the Full Insurance area. Data collected in 1988 are excluded (see Figure 15.2). For both groups, the change in the Full Insurance area is significant (Fisher exact tests, $P<0.01$).

	Full Insurance area	Supervised + Integrated area
Aphids (Hemiptera)		
proportion increased	8	0
proportion decreased	0	4
Beetles (Coleoptera)		
proportion increased	0	5
proportion decreased	8	0

Too few data were gathered within 3 days of pirimicarb spraying in the Integrated area to permit a similar analysis. However, comparison of samples collected within two weeks before and two weeks after the spraying in 1984 suggested a similar change, with an increase of aphids in the diet of tree sparrows after the pirimicarb spray.

The implication of these results is that after spraying, tree sparrows concentrated their foraging on aphids, which would be killed or immobilized by the pesticide, and hence perhaps were more easily gathered. A drop in the proportion of beetles might then be simply a consequence of specializing on aphids. However, it is also possible that beetles became less available after spraying, and their decline in the diet was a direct effect. The data do not permit these possibilities to be distinguished.

Whatever the reason, it appears that during the first few days after an aphicide spray, tree sparrows feed to their nestlings primarily those insects most likely to carry residues of the pesticide, raising the possibility of dietary poisoning.

Foraging destinations

Information on the destinations taken by adult tree sparrows on foraging flights from their nests was collected for two purposes. First, it indicated the time that individual birds spent foraging in wheat fields, which was likely to affect the degree to which they and their young were exposed to insecticide. Second, it was undertaken to investigate whether the distribution of foraging destinations changed after an insecticide application, as might be expected if the spray was aversive to the birds, or altered the numbers or availability of their prey.

Methods

Observations were made at nest boxes for periods of 30 minutes. The parent tree sparrows were watched every time they flew away from the nest. It was rarely possible to watch a bird continuously until it returned to the nest, so each trip was classified according to the bird's location and direction of flight when it was lost to view. The categories used were: 'crop' (seen to land in one of the Project fields or its margin, or lost to view while flying on a route normally taken to the crop), 'non-crop' (seen to land in, or flying towards, areas away from the crop), and 'unknown' (neither of the above). Note that these categories refer to initial destinations, and could be misleading if the birds habitually visited both crop and non-crop destinations within a single trip. On a number of occasions at several nest boxes, birds were observed to return directly to the nest box from the place to which they first flew. However, usually they were lost to view immediately after arriving at their initial destination.

Observers worked in pairs, so that both parents could be watched even if they left the nest simultaneously. The observers were stationed about 100 m from the nest, to avoid disturbing the birds, and watched through telescopes and binoculars.

Observations were made at 10 nest boxes in the Full Insurance area between 12 and 22 July 1988. Demeton-S-methyl was applied to winter wheat adjacent to the nest boxes on 18 July (Backside) and 19 July (Grange Piece). There were no nests suitable for observation at this time in the Supervised and Integrated areas. As far as was possible, each nest was observed every day for one period in the late morning, and another in the afternoon. One or two additional periods of observation per day were completed at selected nests, close to the time of spraying.

Results

The frequency with which birds flew to the crop, as a proportion of all known destinations, is shown in Figure 15.3 for six breeding pairs of tree sparrows. Overall, 18% of visits were to 'unknown' destinations. There were large differences between the pairs in their use of the crop. For example, birds from nest box 260 rarely visited the crop, while those from nest box 278 used it almost exclusively.

Use of the crop changed during the period of study, in a variety of different ways. For example, foraging in the crop was most common during the middle of the period for birds from nest box 275, but was least common at that time for birds from nest box 260.

For birds from nest boxes 256, 260, 278 and 281, there were no obvious changes in their use of the crop at the time of spraying. In contrast, the birds from nest box 361 tended to visit the crop more frequently after the insecticide spray, and those at nest box 275 did so less frequently. However, these results must be interpreted with caution because the departure of a young bird from this nest box to an unknown location on the day of spraying might have influenced the behaviour of its parents.

Overall, there was no evidence that the foraging behaviour of the birds studied was influenced by any effect of pesticide application. It is clear, however, that some breeding pairs visited the crop more frequently than others, suggesting that some birds were more likely to be exposed to pesticides than others.

Figure 15.3   Foraging destinations of pairs of tree sparrows rearing young in the Full Insurance area in July 1988. The frequency with which birds flew to the crop is shown as the proportion of trips to all known destinations in 30-minute observation periods. Symbols indicate periods containing more than five foraging trips. Vertical broken lines mark the days on which adjacent fields were sprayed with demeton-S-methyl. Data are from Hart (1990).

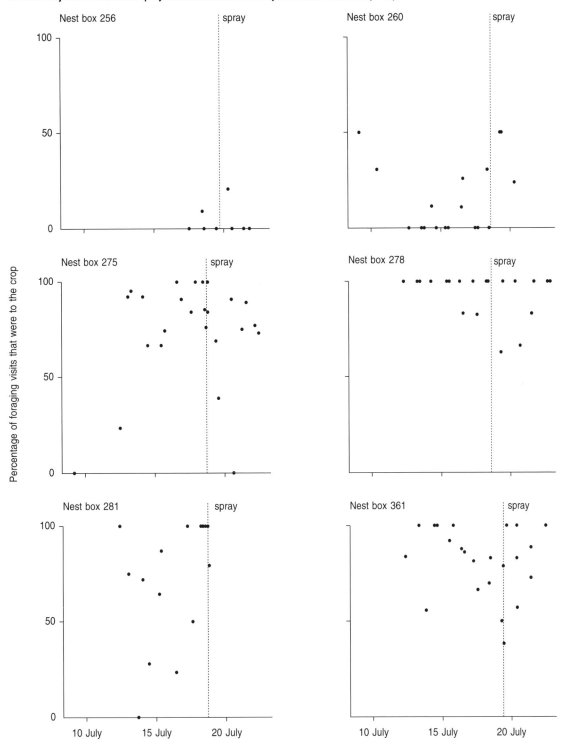

Of the ten nest boxes at which observations were made, four are not represented in Figure 15.3. Two are omitted because observations were discontinued before spraying occurred, in one case when the eggs were abandoned by the parents and in the other because the young fledged. Two breeding pairs which hatched young close to the time of spraying are excluded from Figure 15.3 because they made few foraging flights. For none of these breeding pairs was there evidence of any change in behaviour attributable to effects of the insecticide.

Conclusion

In summary, use of the winter wheat crop by tree sparrows in July differed widely among breeding pairs, and changed from day to day. Apparent changes in behaviour at the time of spraying were recorded for only two breeding pairs, and these were in opposite directions. Both might have been caused by factors other than effects of the insecticide. However, the great variability of behaviour means that small effects would be difficult to detect by observations of this kind.

Nest attendance

As young birds grow, their demand for food, and hence the rate at which their parents visit the nest to provision them, increases to reach a maximum several days before they leave the nest. Weather and the availability of food may also influence nest-visit rates. Pesticides might have an effect in several ways. Toxic exposure may kill one or both parents, or may impair their ability to gather food. Direct effects on the nestlings may reduce their food demands, or the vigour with which they beg for food. Indirectly, the density and activity of prey animals may be altered, making them less or more easy to capture. If parent birds have to forage farther afield, their trips may become longer and less frequent.

Several previous studies of birds have addressed the effects of pesticides on nest attendance (eg Powell, 1984; Spray *et al.*, 1987; Meyers *et al.*, 1990). Their methods were adapted for use at Boxworth in order to search for any effects on the behaviour of parent tree sparrows at the time of aphicide spraying.

Methods

Two methods were used to monitor nest visit rates. First, direct observations of visits were recorded in 1988 during the investigation of foraging destinations described earlier. Second, an electronic monitoring system was used to obtain a continuous record of activity at the entrance holes of selected nest boxes when they were occupied by breeding tree sparrows. Infra-red emitters were placed in the entrance of each of these boxes, such that any bird entering or leaving the box had to pass through two beams of infra-red light. Each beam was monitored by an infra-red sensor, connected to a data-logging device which recorded the beginning and end of each period for which the beam was obstructed. The record of beam-breaks was subsequently analysed by computer to provide an index of activity for each nest. A unit of activity was defined as a sequence of beam-breaks during at least part of which both beams were broken concurrently, and separated from other such sequences by an interval of at least 5 seconds. When the activity index was compared to visual observations, each visit scored two units of activity: one when the bird entered the nest and another when it left. Older nestlings were sometimes fed at the entrance, however, scoring only a single unit of activity per visit.

Some records of activity in the nest-entrance were caused by older nestlings looking out, or occasionally by birds of other species looking in. It is difficult to distinguish these different types of activity, and therefore the results must be interpreted with caution.

Results

Of the ten nests at which visual observations were made, seven were active throughout the period 18–20 July 1988 when the aphicide demeton-S-methyl was being applied to fields in the Full Insurance area. Data from two 30-minute periods, one in the morning and one in the afternoon, were combined to provide an estimate of hourly visit rate for each nest on each day (Figure 15.4). Three nests showed declines in visit rate as the young birds fledged. Three others, including two where the eggs had not yet hatched, showed similar nest visit rates before and after spraying.

At only one nest (Box 275) did the visual observations of visit rates suggest a short-term decrease after spraying, which might be attributed to a pesticide effect (Figure 15.4). The parent tree sparrows at this nest made extensive use of the sprayed wheat crop, flying to it on an average of 71% of occasions when they were observed leaving the nest (Figure 15.3). The field adjacent to the nest was sprayed late on 18 July. The reduction in the visit rate on 19 and 20 July occurred primarily in the afternoon observation period, as might be expected if the parents gradually accumulated a sub-lethal exposure to pesticide while foraging

Figure 15.4   Frequency of visits to their nests by pairs of tree sparrows breeding in the Full Insurance area, recorded during visual observations. Each line represents data for a different nest. Triangles denote nests at which the young birds fledged before 22 July.

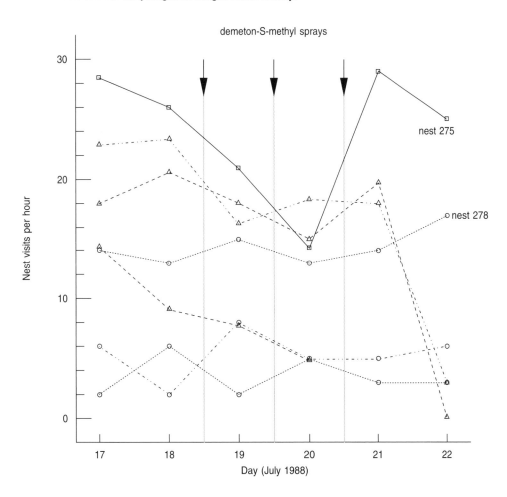

each morning, and recovered overnight. However, the decline could have been caused by other factors, and cannot be confidently attributed to the pesticide. In any case the effect was temporary, and caused no interruption to the growth of the five nestlings in this nest, which subsequently fledged successfully.

Daily totals of activity at nest entrances in July 1988, recorded by the automatic monitoring system, were generally consistent with the visual observations. The activity index for nest 275 showed a decline from 18 July to 19–20 July, as was found in the visual records for the same nest. For nest 278, the activity index showed a

Figure 15.5 Hourly totals of an index of nest visit rate for one pair of tree sparrows breeding in Nest 278 in the Full Insurance area in July 1988, recorded by automatic monitoring as described in the text. The field adjacent to the nest was sprayed with demeton-S-methyl on 18 July starting at 15:15. The growth of the young birds is indicated by their average bodyweight, measured daily.

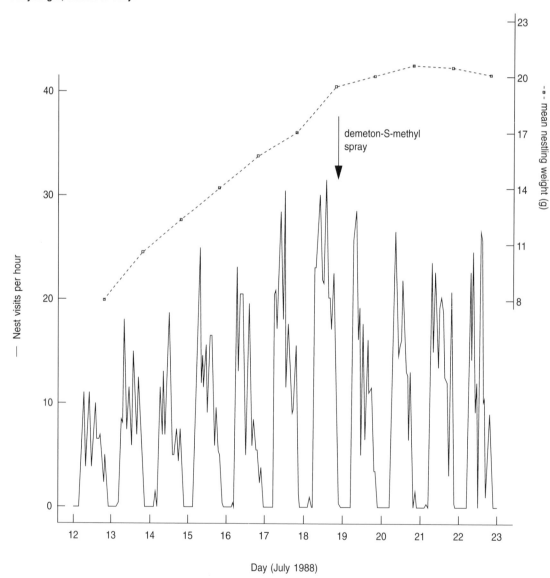

marked decrease from 18 to 19 July. This was not apparent from the visual observations for that nest (Figure 15.4). For this nest the activity index should provide a consistent measure of visit rate, as the parents always fully entered the nest before feeding their young. Hourly totals of the activity index showed a gradual increase from 12 to 18 July, consistent with the increasing size and food requirements of the nestlings (Figure 15.5). There was no indication of any marked change after spraying started in mid-afternoon on 18 July. On the day after spraying, however, activity dropped markedly after 08:00 (Figure 15.5). Subsequent days also showed an early morning peak followed by a decline, but activity never again reached the levels recorded on 18 July. Interpretation of these results is difficult because the initial decline on 19 July coincided with disturbance due to human activity nearby, and there were no comparable data from nests in unsprayed areas to indicate whether the pattern of activity at nest 278 was normal. The three nestlings in nest 278 showed no deviation from normal growth, and subsequently fledged successfully.

Conclusion

Neither visual observations nor automatic monitoring of nest visits showed any consistent major change attributable to the effects of pesticide applications. However, because relatively few nests were active at the times of spraying, few data were obtained, so that small changes in nest attendance, which might have biological consequences, could not reliably be detected.

Biochemical effects

Certain pesticides produce toxic symptoms by affecting the activity of enzymes, such as brain acetylcholinesterase (AChE), that are essential to normal body function. The principal summer aphicide used at Boxworth, demeton-S-methyl, is an organophosphorus chemical, which inhibits AChE and similar esterases. Measurement of this inhibition provides a means of simultaneously determining whether birds have been exposed to the pesticide, and assessing its adverse effects. Study of brain AChE would give a direct reflection of toxic action, but it is more convenient to carry out assays on blood esterases. Although inhibition of blood enzymes may not be directly damaging, it is likely to be accompanied by potentially damaging effects on the nervous system. The practical advantages of blood enzyme measurements are (a) that inhibition is a more sensitive indicator than for brain AChE (Zinkl et al., 1981), and (b) that a series of samples can be taken from the same individual birds, providing a picture of the time-course of their exposure to a pesticide.

Collection and analysis of samples

Biochemical studies of tree sparrow nestlings at Boxworth were undertaken in 1988, around the time of the two demeton-S-methyl applications in the Full Insurance fields, on 16 June and 18–20 July. Blood samples were collected from nestlings before and after each spray, and brain tissue was taken from a few birds following the first spray. Samples for control comparisons were obtained from both the Supervised + Integrated area and the Full Insurance area by selecting birds that would not have been exposed to any insecticide spray for at least two weeks.

Three types of esterase were investigated, using methods described by Thompson et al. (1988a): serum cholinesterase (ChE), serum carboxylesterase (CbE) and brain AChE.

Wing lengths were measured at the same time as blood samples were taken, in order to assess nestlings' growth stage. Serum ChE activity changes as nestling birds develop (Figure 15.6), following a two-stage increase that reflects their switch to homeothermic temperature regulation midway through their period in the nest. It is important to be able to judge the inhibition of esterases in relation to the level of activity expected under

Figure 15.6 Serum cholinesterase activity of nestling tree sparrows at various stages of growth. Filled squares represent normal values for nestlings not exposed to pesticides, for which the line is a fitted two-stage regression. Values for nestlings exposed to recent demeton-S-methyl spraying are shown by open triangles.

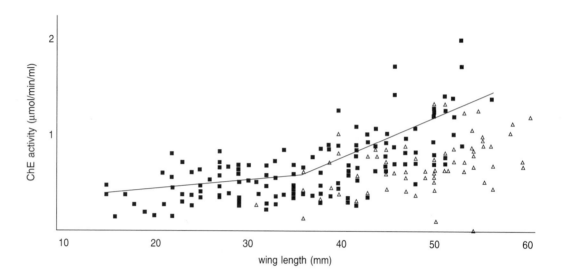

normal development, and an adjustment was made to take account of this increase.

Effects on esterase activity

Figure 15.6 includes data for blood samples collected after the second aphicide spray. Many of these fell below the levels of activity seen in birds from the untreated control area, during a period when there is normally a rapid increase in activity. This implies that many nestlings suffered inhibition of their blood enzymes. In order to investigate the effect further, serum ChE activity of birds in each nest was calculated as a percentage of that normally expected in a bird of the same wing length. An adjustment to take account of diurnal variation in esterase activity (Thompson *et al.*, 1988b) was necessary, because measurements around the first spray were taken in the morning, and those around the second spray were taken in the afternoon.

Table 15.4 shows the time-course of ChE inhibition in eleven nests as average levels of

Figure 15.7 Serum ChE activities of tree sparrow nestlings in one nest box. Samples were taken 3–4 hours after spraying (day 0) to 4 days post-spray. Activity is expressed as a percentage of the expected activity of a nestling of the same wing length (from Figure 15.6).

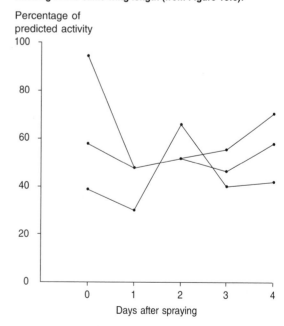

Table 15.4   Average inhibition of serum cholinesterase activity in nestling tree sparrows following demeton-S-methyl sprays in 1988. Values are percentages of the activity expected in normal development, with the number of nestlings examined given in brackets.

First Spray

Nest box	Days between hatching and spraying	% of predicted activity at three times after spraying		
		Day 1	Day 2	Day 3
268	12	41 (4)	37 (2)	65 (2)
272	7	110 (5)	89 (4)	–
278	13	23 (4)	23 (4)	–
279	14	30 (5)	37 (4)	–

Second Spray

Nest box	Days between hatching and spraying	% of predicted activity at five times after spraying				
		3–4 hours	Day 1	Day 2	Day 3	Day 4
256	11	43 (3)	60 (3)	50 (2)	–	–
260	15	80 (2)	114 (2)	–	–	–
275	12	114 (5)	88 (5)	75 (5)	71 (5)	71 (5)
278	8	68 (3)	44 (3)	59 (3)	50 (3)	58 (3)
281	13	5 (2)	–	–	–	–
361	16	81 (4)	65 (3)	43 (3)	84 (1)	–
384	14	109 (4)	76 (4)	–	–	–

activity for all nestlings. An example of variation between birds within one nest is given in Figure 15.7.

Following the first spray, the greatest effect observed was a reduction of ChE activity to 23% of the predicted normal value, whereas after the second spray it fell to 5% in one nest (Table 15.4), although most reductions were less extreme. Further samples on successive days, up to the time that birds left the nest, showed continuing ChE inhibition, although in some nests there was a suggestion of a return towards normal levels of ChE activity. Nevertheless, it appears that many of the nestlings that were exposed to demeton-S-methyl fledged with lower activity of serum ChE than they would otherwise have.

Similar assessments of serum CbE activity revealed a developmental increase in activity, though less predictable than for serum ChE. Inhibi-

tion of CbE occurred in nestlings in some nest boxes, but was not as severe as for ChE.

Brain AChE activity in four birds sampled two days after the first demeton-S-methyl spray ranged from 76% to 120% of that in 'control' birds at a similar stage of development. Depression of AChE activity is generally considered to indicate exposure to an organophosphorus pesticide only if it falls below 80% of control levels (Ludke et al., 1975). However, the two nests in which there was greatest AChE inhibition also revealed strong inhibition of serum ChE, suggesting that both effects were the result of exposure to the pesticide.

Conclusion

Overall, the biochemical data demonstrate that nestlings in all the nests studied were exposed to demeton-S-methyl. It is not possible to determine whether this exposure was to residues on the food brought to the nest, or to drift of the pesticide spray or vapour. However, there was considerable variation between nests, which probably reflects differences in the parent birds' behaviour, and the proximity of nests to the treated field. This offers an opportunity to predict the proportion of the tree sparrow population affected by esterase inhibition (p.192).

Nestling Growth

The growth of nestling birds may be a sensitive indicator of the effects of pesticides. Deviations from normal growth may be caused by depletion of food supplies, by changes in the behaviour of parent birds, or as a direct result of poisoning through the consumption of contaminated food. Growth may also be disrupted by disturbance at the nest. The growth of tree sparrow nestlings was investigated at Boxworth in 1987 and 1988 in conjunction with other studies.

Methods

The nests of breeding tree sparrows were inspected daily, from 2–3 days after the nestlings hatched until they attained wing lengths of over 50 mm. The nestlings were marked on their legs with individual codes, using non-toxic pens, until they were large enough to be fitted with numbered leg rings, and were weighed individually each day. During 1987 and after 17 June in 1988, wing lengths were measured daily, starting at the time of feather emergence. To avoid excessive cooling of the nestlings, no measurements were taken on very wet or windy days.

The data on nestling growth were analysed by two methods. First, the mean weight and wing length of nestlings in each brood on the tenth day after hatching were used as rough measures of growth performance. Second, daily changes in weight and wing length were examined for evidence of short-term deviations immediately after pesticide applications.

Results

Growth data were obtained for a total of 53 broods, for 22 of which the hatching dates were known precisely. The hatching dates of a further 21 broods were estimated from laying dates, assuming an average incubation period of 13 days, and in three further cases it was possible to estimate the age of the nestlings when first observed.

Table 15.5 Weights and wing lengths of 10-day old tree sparrow nestlings, showing differences between the treatment areas in 1987 and 1988. Values are means of brood means, ± the standard error, with the number of broods shown in brackets.

	Weight (g)		Wing length (mm)	
	1987	1988	1987	1988
Full Insurance area	16.0 ± 1.0 (10)	16.8 ± 1.1 (14)	39.9 ± 2.5 (10)	40.3 ± 2.6 (11)
Supervised + Integrated area	19.2 ± 0.5 (7)	18.6 ± 1.5 (3)	48.5 ± 1.3 (7)	42.7 ± 1.5 (3)

In both 1987 and 1988 the weights and wing lengths of 10-day old nestlings were lower in the Full Insurance area than in the Supervised + Integrated area (Table 15.5). In 1987 the difference was statistically significant for both measures of nestling growth (t-tests). These differences are not necessarily attributable to effects of the pesticide regimes, because other differences between the treatment areas may have been responsible. Regression analysis showed that most of the variation in growth between treatment areas could equally be explained by other factors, including differences in the timing or size of broods, the time at which measurements were taken, and the frequency of nest inspections. It is not possible to test whether growth rates differed between areas during the baseline years, because growth data were not collected until 1987.

Pesticide applications could reduce growth rates directly through the physiological effects of poisoning, or indirectly by a reduction in the availability of food. It should be possible to identify direct effects by comparing broods being reared at the time of a pesticide application with broods in the same treatment area at other times. No such differences were observed (Table 15.6), although only quite large differences would be detected with the small samples of data available.

Table 15.6 **Growth (weight and wing length at age 10 days) of broods of tree sparrow nestlings in the Full Insurance area that were exposed to demeton-S-methyl in the first ten days after hatching, compared to broods in the same area which were not exposed by that age.** The values shown are the means of brood means ± the standard error, with the number of broods given in brackets. The differences between exposed and non-exposed broods are not statistically significant (t-tests).

	Weight (g)	Wing length (mm)
Exposed	15.4 ± 1.8 (5)	38 ± 5 (4)
Not exposed	16.7 ± 0.9 (19)	41 ± 2 (17)

Biochemical effects were detected in nestlings within a few hours of aphicide applications (see pp. 187). The toxic action of organophosphorus pesticides is such that if the level of exposure was sufficient to influence growth then deviations from normal development would probably appear within a few days. No consistent effects of this nature are suggested by the growth curves obtained for broods that were being monitored at the times of aphicide applications in 1987 and 1988.

Growth in wing length appears to have continued normally after spraying in every case, as illustrated in Figure 15.8(a). Growth in weight was more variable, making it difficult to identify deviations after spraying. A number of broods stopped increasing in weight at around the time of aphicide applications (Figure 15.8 (b)-(d)), but these birds were close to the point when growth normally reaches a plateau. Only in July 1987 were concurrent data available from sprayed and unsprayed

areas, and on that occasion three of the four broods in the sprayed area had already shown slower growth before the date of spraying (Figure 15.8 (d)). Only two broods were exposed to spraying before reaching the normal range of fledging weights, and in both of these growth continued apparently normally after spraying (shown by the bottom lines in Figures 15.8 (b) and (d)).

Conclusion

In 1987 and 1988, tree sparrow nestlings in the Full Insurance area grew more slowly than those monitored in the Supervised + Integrated area. A difference of the size observed might jeopardise the survival of nestlings in the Full Insurance area after fledging (Garnett, 1981). Because of the lack of data from the pre-treatment years of the Project, it is not possible to determine whether the difference was attributable to effects of pesticides, such as a reduction in the availability of invertebrate prey in the Full Insurance fields. The data available suggest that there were not any major changes in growth coinciding with the exposure of chicks to the summer aphicide sprays, as would be expected if differences in growth were caused by poisoning.

Effects on breeding performance

Data described in Chapter 14 demonstrate that there were no major effects of the high pesticide input regime on the population density or breeding success of tree sparrows. However, the possibility exists that summer aphicide applications might cause a short-term depression of breeding performance. In order to examine this possibility, nesting records for tree sparrows were analysed to see if there were unusual numbers of nest-failures, or deaths of individual nestlings, about the time of spraying, and to look for any evidence of a change in nesting effort (eg in clutch size) after spraying.

Nesting records from 1987 and 1988 were selected because the frequency with which nests were checked was much higher than in earlier

Figure 15.8 Growth in weight and wing length of tree sparrow nestlings in relation to aphicide applications. Values plotted are brood means. In (a) to (c), all broods were in the sprayed (Full Insurance) area. In (d), solid lines indicate broods in the Full Insurance area which were sprayed, while broken lines indicate broods in the unsprayed Supervised + Integrated area. Vertical dotted lines indicate the dates on which demeton-S-methyl was applied.

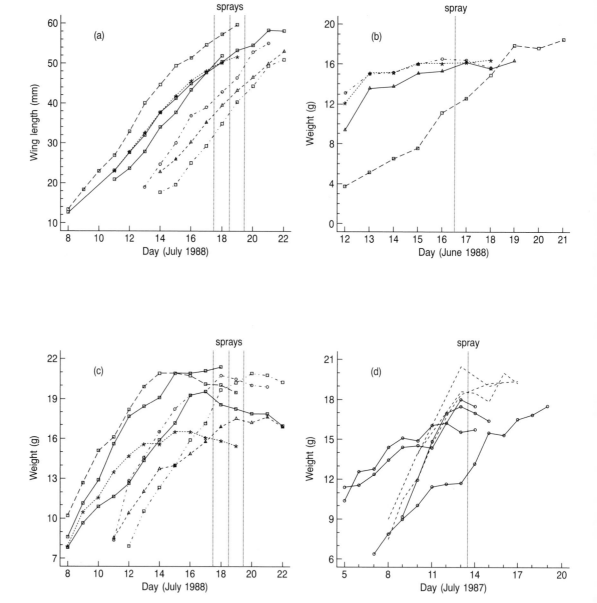

years. Within a period from 8 days before to 8 days after spraying of demeton-S-methyl in the Full Insurance area, there were 19 occasions on which mortality of nestlings (single chicks, or whole broods) was recorded. Only 4 of these were on the first or second days after spraying, and the pattern was similar for nests in the untreated area. Several of the deaths that occurred are thought likely to have been caused by disturbance during the intensive period of research around the spray dates, rather than by pesticide effects or 'natural' causes.

There were no detectable changes in clutch size, or in the number of occupied nests, coinciding with the time of spraying aphicides. It can be concluded that the exposure of tree sparrows to these chemicals did not lead to any major adverse effects on breeding performance in the years of this study.

Discussion

The studies of tree sparrows were concerned chiefly with possible direct effects of aphicide applications during the breeding season, and addressed three issues:

Were birds exposed to aphicides?

Adult tree sparrows might receive exposure to demeton-S-methyl by direct overspraying, by inhaling vapour, by contact with vegetation, or by eating food containing residues of the pesticide. It was not possible to catch adults in order to assess exposure directly, because of the disturbance this would have caused to nest attendance. However, examination of house sparrows (*Passer domesticus*) trapped just after spraying indicated that they had been exposed (see Chapter 16) and it seems likely that this was also the case for tree sparrows.

The major route by which the pesticide could have reached nestling tree sparrows is in their diet. Not only did the parent birds continue to forage in wheat fields during the first few days after spraying, but they switched their food-gathering to favour aphids, which are probably the most likely insects to harbour chemical residues. Biochemical changes in the activity of serum esterases confirmed the exposure of nestlings, although the data cannot exclude the possibility of a contribution from drift of spray or vapour at the time of application.

What were the effects of exposure?

The evidence suggested that exposure was very variable, since some broods received most of their food from wheat fields, while others did not. In part, this reflects the proximity of nests to fields, but there was some variability in exposure even among nests close together in the same hedge at the margin of Backside field.

In none of the nests studied was there a definite indication that exposure led to the death of nestlings, nor was there any consistent interference with their patterns of growth. The chief effect detected was a lowering of the activity of the blood enzymes ChE and CbE. For some birds, activity of ChE fell to very low levels within the first day after spraying, and it appeared that most of the broods exposed would have left their nests with lower ChE activity than normal. It is not clear what effects this may have had on their subsequent survival; there is little information available on the functional significance of reduced serum esterase activities, although it is known that many animals can survive with very low serum cholinesterase activity. The activity of AChE in the brain, for which a 50% inhibition is often associated with death, was not markedly reduced in the few tree sparrows examined.

Were these effects ecologically significant?

At the levels observed, the effects on nestlings evidently had no serious consequences for the local tree sparrow population. Also, direct monitoring of breeding density and nesting success, both at the time of aphicide application and over the whole season, failed to reveal any major differences between the treatment areas that might have been the result of the use of aphicides or any other pesticide. The question that remains is how close was the exposure of tree sparrows in the years studied to a level that *would* cause adverse changes to the population?

If it is correct that nestlings were affected chiefly because of contamination of their food, of which an average of 50% came from wheat fields, it is useful to speculate on the consequences of obtaining 100% of the food from a treated field. The data collected were insufficient to demonstrate a clear relationship between the percentage of foraging trips to the crop (which ranged from 1% to 71% in the six nests for which foraging observations and enzyme data were collected) and the measured depression of serum ChE. Therefore it is not possible to determine whether foraging exclusively in the crop would have greatly increased the largest inhibition of ChE activity observed in 1988.

Most of the measurements were taken at nests where the nestlings were already well developed. There is evidence from other species of birds that the youngest nestlings may be much more sensitive than adults to organophosphorus pesticides (Grue *et al.,* 1981; Hooper *et al.,* 1990). Therefore, the consequence of spraying could be more serious if it occurred soon after chicks had hatched.

It was observed that most of the nests in which there was significant inhibition of ChE activity on the day of spraying were downwind of the sprayed field. This highlights the possibility of exposure by spray or vapour drift. Theoretical calculations indicated that the maximum likely pesticide intake by vapour inhalation would be far below a lethal dose. Nevertheless, nests placed low in hedges on the downwind side of fields might be expected to receive some of their exposure through drift, despite being enclosed in nest boxes.

It appears possible that even the worst likely exposure of nestling tree sparrows to demeton-S-methyl may not seriously threaten their survival. Only a part of the population would be exposed at that high level. It is useful to estimate that proportion, to provide an indication of the possible consequences of using a more toxic aphicide.

Nestlings in nest boxes at the edge of fields are both more likely to be affected by spray drift and more likely to receive a large part of their diet from a treated field. On average, such nests formed 78% of the population at Boxworth, although this could be very different at other sites. The remainder were in woods or other habitats away from the fields. Demeton-S-methyl residues are rapidly lost from the crop after spraying (see Chapter 7), although it is not known how long they persist in insects killed by the spray. However, residues would probably be greatly reduced within a week after spraying. Only 61% of the nests at Boxworth contained young birds in any one-week period in late June/early July, when spraying occurred. Accordingly it may be concluded that about half of the nests (78% $\times$ 61%) would be at risk of exposure equal to or greater than the worst observed in this study. If adverse effects of pesticides are likely to occur only in very young, susceptible nestlings, this percentage would be lower.

Thus, although the data collected cannot exclude the possibility of some direct pesticide effects on nestling tree sparrows, it seems that any adverse effects were more likely to have been caused indirectly, by prey depletion. There was no firm evidence of major damage to the size or breeding performance of the local tree sparrow population. Further study would be required to fully understand the complexities of birds' exposure to pesticides sprayed on farmland.

Summary

The possible direct effects of summer aphicide applications on tree sparrows were investigated by a combination of behavioural, biochemical and dietary studies. Nestling tree sparrows were fed by their parents on invertebrates, many of which were obtained from the wheat crops, although the amount of foraging time in fields varied greatly among pairs of birds. After the fields were sprayed with demeton-S-methyl, the nestling diet changed to include more aphids and fewer beetles, and birds continued to forage in the treated fields. The implication that they might carry residues of the pesticide to their nestlings was confirmed by demonstration of lowered blood enzyme activity, indicating exposure to the pesticide, during the first few days after spraying. However, this inhibition of enzyme activity caused no obvious interruption of the nestlings' normal growth, nor did spraying consistently disrupt the behaviour of parent birds visiting their nests. Overall, the study demonstrates that tree sparrow nestlings were exposed to summer aphicides, but the consequences did not appear to be serious.

References

Davies, N B (1976). 'Food, flocking and territorial behaviour of the Pied wagtail (*Motacilla alba yarellii* Gould) in winter'. *Journal of Animal Ecology* **45**, 235–252.

Garnett, M C (1981). 'Body size, its heritability and influence on juvenile survival among Great Tits *Parus major*'. *Ibis* **123**, 31–41.

Grue, C E, Powell, G V N & Gladson, N L (1981). 'Brain cholinesterase activity in nestling starlings: implications for monitoring exposure of nestling songbirds to ChE inhibitors'. *Bulletin of Environmental Contamination and Toxicology* **26**, 544–547.

Hart, A D M (1990). 'The assessment of pesticide hazards to birds: The problem of variable effects'. *Ibis* **132**, 192–204.

Hooper, M J, Brewer, L W, Cobb, G P & Kendall, R J (1990). 'An integrated laboratory and field approach for assessing hazards of pesticide exposure to wildlife'. In: Somerville L and Walker C H (Eds.) *Pesticide Effects on Terrestrial Wildlife*. Taylor & Francis, London. pp. 271–284.

Ludke, J L, Hill, E F & Dieter, M P (1975). 'Cholinesterase response and related mortality among birds fed ChE inhibitors'. *Archives of Environmental Contamination and Toxicology* **3**, 1–21.

Meyers, S M, Cummings, J L & Bennett, R S (1990). 'Effects of methyl parathion on red-winged blackbird (*Agelaius phoeniceus*) incubation behaviour and nesting success'. *Environmental Toxicology and Chemistry* **9**, 807–813.

Moreby, S J (1988). 'An aid to the identification of arthropod fragments in the faeces of gamebird chicks (Galliformes)'. *Ibis* **130**, 519–526.

Powell, G V N (1984). 'Reproduction by an altricial songbird, the red-winged blackbird, in fields treated with the organophosphate insecticide fenthion'. *Journal of Applied Ecology* **21**, 83–95.

Spray, C J, Crick, H Q P & Hart, A D M (1987). 'Effects of aerial applications of fenitrothion on bird populations of a Scottish pine plantation'. *Journal of Applied Ecology* **24**, 29–47.

Thompson, H M, Walker, C H & Hardy, A R (1988a). 'Esterases as indicators of avian exposure to insecticides'. In: Greaves, M P, Greig-Smith, P W & Smith, B D (Eds.) *Field Methods for the Study of Environmental Effects of Pesticides*. British Crop Protection Council, Monograph No. 40, pp. 39–45.

Thompson, H M, Walker, C H & Hardy, A R (1988b). 'Avian esterases as indicators of exposure to insecticides—the factor of diurnal variation'. *Bulletin of Environmental Contamination and Toxicology* **41**, 4–11.

Zinkl, J G, Jessup, D A, Bischoff, A I, Lew, T E & Wheeldon, E B (1981). 'Fenthion poisoning of wading birds'. *Journal of Wildlife Diseases* **17**, 117–119.

Exposure of starlings, house sparrows and skylarks to pesticides

16

Helen M Thompson*, **Kenneth A Tarrant*** and **Andrew D M Hart** # (MAFF Central Science Laboratory, *Tolworth and # Worplesdon)

The tree sparrow *Passer montanus* was deemed to be one of the most likely bird species at Boxworth to be affected directly by pesticide applications. However, other birds also fed or nested in and around the Project fields and might have been exposed to certain pesticides. Although their populations were not seriously reduced by the Full Insurance programme (Chapter 14), it was thought possible that birds of several species might experience short-term sub-lethal effects from exposure to insecticides. Accordingly, studies of house sparrows (*Passer domesticus*) starlings (*Sturnus vulgaris*) and skylarks (*Alauda arvensis*) were carried out during the last two years of the Project, to evaluate their exposure to two organophosphorus pesticides.

House sparrows and starlings were examined biochemically, to identify changes in the levels of blood and brain enzymes that might be inhibited or induced through exposure to triazophos (starlings) or demeton-S-methyl (house sparrows). Samples of liver tissue were also taken, to search for associated damage to liver histology. Both these species have been investigated in this way in previous studies, in the field and in captivity (Hardy *et al.*, in prep.; Thompson *et al.* 1991).

The conspicuous song-flights made by skylarks, which nested in the wheat fields, prompted an observational study to determine whether their behaviour was affected by the application of demeton-S-methyl.

House sparrows

Nest boxes were occasionally occupied by house sparrows, which regularly fed on the wheat crop. It was possible to catch adult house sparrows using mist nets placed alongside the neighbouring hedgerows. A total of 46 birds was trapped in the Full Insurance area at about the time of the second of the two demeton-S-methyl aphicide applications in 1987 and 1988. 'Control' birds that would not have been recently exposed to demeton-S-methyl, either before the first application (1987) or well after the second application (1988), were also trapped.

Samples of blood were taken from all the birds captured, and the brain and liver were taken from some. The activity of serum cholinesterase (ChE), serum carboxylesterase (CbE) and brain acetylcholinesterase (AChE) were assayed using methods described by Thompson *et al.* (1988). The activity of glutamate oxaloacetate transaminase (GOT) in serum, which is a good indicator of cellular damage, was measured by the method of Bergmeyer & Bernt (1963).

Liver samples for histological evaluation were initially fixed in neutral buffered 10% formalin, paraffin-sectioned at 6 microns, and stained using a standard haematoxylin and eosin method. An image analysis system linked to a light microscope was used to measure liver cell size and the incidence of binucleation. These two parameters are sensitive indicators of the physiological response to poisoning (Tarrant, 1988). The numbers and sizes of inflammatory foci (which may indicate the extent of cell damage) were evaluated for each whole liver section.

The results of the enzyme studies are shown in Figure 16.1, which expresses the activity of the enzymes as a percentage of that recorded in the 'control' birds. In 1987, there was inhibition of serum ChE and brain AChE following the demeton-S-methyl spray, and an increase in the activity of serum CbE (Tarrant *et al.*, 1992). A similar increase of CbE was observed in laboratory studies when birds were dosed with organophosphorus pesticides (Thompson *et al.*, 1991). In 1988, serum

Figure 16.1   Levels of activity of enzymes (Brain AChE, serum ChE and CbE, and GOT) in house sparrows caught 1–3 days after demeton-S-methyl applications in 1987 and 1988, expressed as percentages of average enzyme activity in 'control' birds.

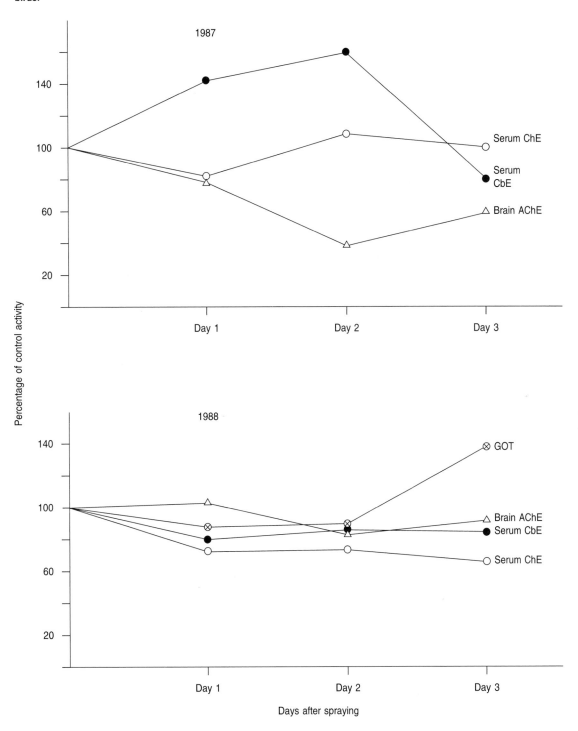

ChE inhibition was observed on the first three days after spraying, and was accompanied by a rise in GOT levels on the third day, which indicates that exposure may have resulted in liver damage after about two days.

This biochemical evidence of exposure to demeton-S-methyl was supported by histological evaluation of liver sections, which revealed increases in the numbers of cells and in the incidence of binucleation, relative to 'control' data (Figure 16.2). Such changes reflect increased cell replication, producing more liver cells to overcome the toxic challenge of the pesticide. There were also signs of liver cell damage in both 1987 and 1988, seen in the occurrence of inflammatory foci spread diffusely through the livers of birds exposed to the pesticide. The numbers and size of these foci, in 1988 particularly, were significantly greater than in the 'control' samples. Despite the presence of foci indicating local cell damage, the liver architecture was not generally abnormal and showed no signs of fibrous scarring or large areas of cell death.

The changes observed represent a low-key response to the presence of a foreign compound, similar to that observed in a previous pesticide field trial (Hardy et al., in prep.).

Starlings

To broaden the investigations beyond the study of summer aphicides, the exposure of starlings to winter applications of triazophos, which was used against yellow cereal fly in March, was examined. Before and after the applications of triazophos in 1987 and 1988, adult starlings were captured while they roosted overnight in nest boxes in Grange Wood (Full Insurance area) and Thorofare Spinney (in the unsprayed Supervised area).

Blood, brain and liver samples were collected and examined in the same way as for house sparrows, except that in 1987 blood samples were stored for 24 hours at -20° C before assaying for enzyme activity, which is liable to have caused

Figure 16.2   Histological evaluation of livers from house sparrows obtained after the application of demeton-S-methyl in 1988.

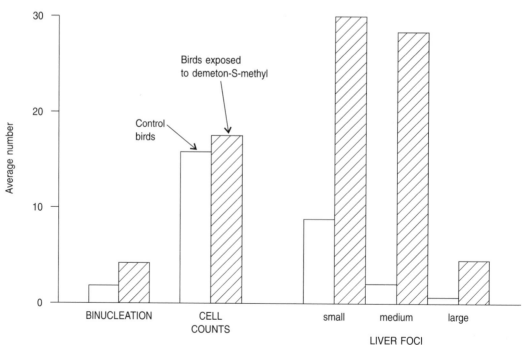

Figure 16.3 Average activities of brain and blood esterases of starlings caught one day after application of triazophos in 1987, expressed as percentages of the average enzyme activity in eight 'control' birds caught before spraying.

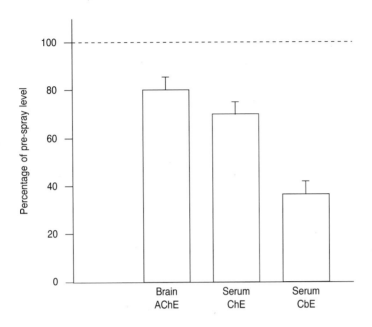

some spontaneous reactivation of inhibited cholinesterase.

Figure 16.3 shows the results of biochemical studies. In 1987 there was inhibition of serum ChE and CbE, and of brain AChE (which was probably underestimated because of reactivation during storage), on the first day after application. Only three birds were trapped in 1988, of which one showed inhibition of serum ChE and CbE.

Histological examination of the small number of liver samples showed no pronounced changes in liver morphology or structure after triazophos spraying. However, there was an indication of an increase in the number of inflammatory foci similar to that observed in house sparrows after demeton-S-methyl application. Again there was a predominance of small foci, which may have arisen from minor cell damage stimulating the immunological defence process.

Skylarks

The skylark was chosen for a study of possible effects of pesticides on behaviour for two reasons. The birds nest and feed in cereal fields at the time of summer aphicide applications, and the males make very conspicuous song-flights, which should permit any changes in their behaviour to be detected. Observations were carried out in the summers of 1987 and 1988, to document the singing activity of skylarks before and after applications of demeton-S-methyl.

Data were collected in 30-minute observation periods at two sites: the southern ends of Grange Piece and Shackles Aden (Full Insurance area, see Figure 2.1) and the eastern parts of Top Pavements and Thorofare (Supervised area). Each of these areas contained about 5–10 territorial pairs of skylarks. The birds rise steeply out of the crop to a height of 30 metres or more, where they 'hover' while singing continuously, sometimes for several minutes. The number of these flights made in 30 minutes was counted by an observer stationed at the edge of the fields. On each occasion, a session at one site was immediately followed by observations at the other, in order to provide paired records for which variations due to factors such as weather and time of day would be minimized.

The data collected in 1988 suggest that skylarks exhibit a daily pattern of singing activity, with the greatest number of song-flights in the morning followed by a decline during the afternoon (Figure 16.4). The biggest deviation from this pattern occurred following an application of demeton-S-methyl and a fungicide in the Supervised Area on 13 June. Singing activity in that area fell to nothing within half an hour of spraying, then rose to an unusually high level an hour later. The frequency of song-flights remained low throughout the next day, though it was noted that many skylarks were singing from positions on the ground. A more normal

pattern of song-flights was resumed on the second day after spraying. The Full Insurance area fields were sprayed with fungicides on 14 and 15 June, and with demeton-S-methyl on 16 June. Only for the latter was there any suggestion of an effect on skylark song-flights, which were less frequent than usual during the morning of the following day.

The results obtained in 1987 were more variable. The Full Insurance area fields were sprayed twice with demeton-S-methyl and once with fungicide during the period of observations, but the data showed no evidence of any effects on the frequency of song-flights.

Thus a marked change in skylark behaviour was observed following one out of four applications of demeton-S-methyl. Exposure to the pesticide on that occasion may have been particularly high, as it was the only time when the pesticide was applied to a crop with a large number of aphids, on which the skylarks may have been feeding. In 1987, skylarks

Figure 16.4   Changes in the frequency of song-flights made by skylarks during a period when insecticides and fungicides were applied in both treatment areas. A regular daily pattern was apparent over most of the period, but song-flight frequency seemed to be temporarily depressed following applications of demeton-S-methyl.

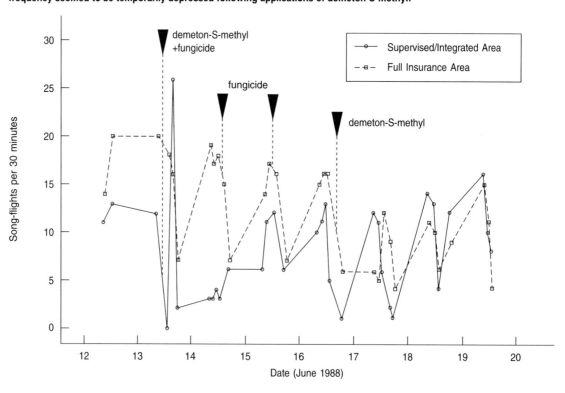

from both observation areas frequently visited adjacent crops of oilseed rape, perhaps reducing their exposure to the aphicide.

Even if the change in skylarks' behaviour was due to effects of demeton-S-methyl, it seems unlikely that the consequences would have been serious. The continuation of song by birds on the ground, and the resumption of song-flights within 48 hours, suggests that there was no mortality of adult birds. Nestling skylarks might be more susceptible than their parents to the effects of spraying. It is not known whether there was any mortality of nestlings associated with demeton-S-methyl applications in 1987 and 1988. On one occasion in 1984 two skylark nests were found in wheat fields in the Full Insurance area before a demeton-S-methyl application, and in both cases all the nestlings survived after spraying.

Conclusions

These studies demonstrate that both starlings and house sparrows were exposed to pesticide sprays used at Boxworth. Examination of house sparrows' crop contents confirmed that they were feeding on developing wheat grains from the fields, which was probably the principal route of exposure.

The differences in biochemistry and histology between birds sampled before and after spraying suggest that exposure resulted in effects at the cellular level in the liver, and may have affected the immunological status of the birds probably without any serious long-term consequences. However, in the short term birds could become more susceptible to endemic infections and more vulnerable to predation, and their response to additional physiological stresses such as moulting or egg production might be affected.

From observations of skylarks, it is clear that measurements of behaviour in the field are highly variable, and are likely to reveal only major changes in the activities of birds after pesticide applications. Results from this study suggested that adult skylarks may sometimes be sufficiently exposed to demeton-S-methyl to cause changes in

singing behaviour, but further work would be required to determine whether their breeding performance could be affected.

Summary

House sparrows trapped around the time of summer aphicide spraying, and starlings trapped before and after spring applications of triazophos, were examined biochemically and histologically to identify their exposure to these pesticides, and to reveal any sub-lethal effects. Birds of both species showed effects that were potentially debilitating. On one occasion following an aphicide application, the singing behaviour of skylarks was disrupted, although there was no indication of adult mortality due to this exposure.

References

Bergmeyer, H U & Bernt, E (1963). 'Glutamate oxaloacetate transaminase.' In: Bergmeyer H U (Ed.) *Methods of Enzymatic Analysis* 1st Edition. Verlag Chemie, Weinheim. pp. 837–842.

Hardy, A R, Greig-Smith, P W, Brown, P M, Fletcher, M R, Stanley, P I, Tarrant, K A, Westlake, G E & Lloyd, G A (in prep.) 'An intensive field trial to assess hazards to birds and mammals from the use of methiocarb as a bird repellent on ripening cherries.'

Tarrant, K A (1988). 'Histological identification of the effects of pesticides on non-target species.' In: Greaves M P, Greig-Smith P W and Smith B D (Eds) *Field Methods for the Study of Environmental Effects of Pesticides*. British Crop Protection Council, Monograph No. 40. pp. 313–317.

Tarrant, K A, Thompson, H M & Hardy, A R (1992) 'Biochemical and histological effects of the aphicide demeton-S-methyl on house sparrows under field conditions.' Bulletin of Environmental Contamination & Toxicology, 48, 360–366.

Thompson, H M, Walker, C H & Hardy, A R (1988) 'Esterases as indicators of avian exposure to insecticides'. In: Greaves M P, Greig-Smith P W & Smith B D (Eds) *Field Methods for the Study of Environmental Effects of Pesticides*. British Crop Protection Council, Monograph No. 40. pp. 39–45.

Thompson, H M, Walker, C H & Hardy, A R (1991) 'Changes in activity of avian serum esterases following exposure to organo-phosphorus insecticides.' *Archives of Environmental Contamination and Toxicology* 20, 514–518.

Summary and recommendations

P W Greig-Smith and **M J Griffin** (MAFF Central Science Laboratory, Worplesdon, and ADAS Cambridge)

Introduction

The Boxworth Project was set up with specific aims, to compare the environmental side-effects of three particular crop protection regimes used on cereals. Its large area and long duration were a major departure from conventional agricultural experiments. Although there was originally some scepticism about the need for such an investigation, this area of research rapidly attracted interest from many different points of view. As a result, the Boxworth Project was soon accorded wider significance, and its results have been used as the basis for general statements about the pros and cons of 'low input' systems and of 'whole farm' research studies. This is not surprising, in the absence of other comparable research.

In order that the implications of the Project may be fully exploited, it is important to recognise its limitations, so that interpretation of its results is not taken too far. This chapter examines the original objectives of the work, in order to identify what can, and what should not, be inferred from the results. The major implications are then discussed in relation to: (1) the ecological effects of intensive pesticide use on wild animals and plants, (2) the economics and operation of reduced-input crop protection systems, and (3) the design of research studies to investigate environmental consequences of farming practice.

Some developments that have taken place as a result of the Boxworth Project, including new research initiatives and proposals for changes to advisory policy, are also discussed.

Objectives of the project

The Boxworth Project had its origins during the 1970's when increased cereal production, with consequent intensification, was still encouraged (Chapter 1). This led to the widespread adoption of high-input prophylactic approaches to crop protection, involving heavy annual use of insecticides, herbicides and fungicides. Concern arose that the system of approval for the chemicals used in these programmes, although adequate to ensure the environmental safety of each product on its own, did not provide positive reassurance about the safety of the total 'package' of pesticides being used by many farmers. The Boxworth Project was established to provide information on the combined, cumulative effects of high pesticide inputs, through a large-scale experimental approach.

The essence of the Project was a comparison between high pesticide use (a 'Full Insurance' regime), following the kind of crop protection approach that was the result of pressures to maximise production, and a more cautious ('Supervised') philosophy then being developed, which involved monitoring of pests, weeds and diseases in order to determine whether pesticide applications were necessary. A third ('Integrated') treatment was added to examine additional means of reducing pesticide use, but was largely incidental to the principal comparison between 'high' and 'reduced' inputs.

The three aims of the Project were:

1. A survey of the **ecological effects** of the prophylactic Full Insurance programme, relative to two forms of reduced input systems (Supervised and Integrated). To address this aspect, studies were carried out on all the major forms of wildlife considered likely to be affected by pesticides, and for which suitable study methods were available. This covered both the direct effects of exposure to pesticides, through their toxic action, and (to a lesser extent) indirect effects such as depletion of prey populations, removal of

predators and competitors, and changes in habitat. Such indirect effects of pesticides may be more serious for animal and plant populations than direct mortality.

2. An **economic evaluation** of the three approaches to pesticide use. This was always intended to be subsidiary to the ecological interests of the Project, but was considered necessary in order to gauge the realism of the three systems. It was not planned as a test of what might be designed as a practical compromise between environmental safety and farming efficiency.

3. An assessment of the problems involved in **management** of supervised approaches to crop protection. At a time when advice to farmers was beginning to turn towards a more cautious use of pesticides, based on forecasting and monitoring of problems, the Boxworth Project provided an opportunity to examine in detail some practical aspects of the operation and efficiency of monitoring.

Limits to the interpretation of results

Because the Project inevitably selected particular pesticide regimes from the many possible versions of 'high' and 'reduced' input approaches, the findings of the work must be regarded as specific to those regimes. Whether results can be extrapolated to other crop protection systems depends on many factors, some of which were controlled at Boxworth, and some which were not. These aspects are discussed later in this chapter. Also, the work must be interpreted as specific to Boxworth EHF. Although the farm is typical in many ways of arable farmland in the eastern part of England, the influence of local features, which could not be identified in the course of this project, may have affected the results obtained. The validity of extrapolation from Boxworth to other sites depends on a variety of factors and remains uncertain.

The comparison between high- and reduced-input approaches is able to address only the issue of whether the former is more environmentally damaging than the latter. It does not help to answer the question of whether the reduced input programme itself was more damaging than avoiding pesticide use altogether. That issue also needs attention, particularly in relation to the costs that complete abstention from using chemical pesticides might have in lower crop production.

The design of the Project also imposes some limitations on the interpretation of its results. Certain aspects of husbandry, particularly the near monoculture of winter wheat grown throughout the Project, were constrained by the need for experimental rigour, but gave rather untypical farming conditions. Similarly, the need to keep the same pesticide programme through the whole study meant that by the end of the Project, the Full Insurance regime was further from the declining average usage on arable farms than it had been at the outset. Nevertheless, high inputs do still occur. The two mild winters since the Project ended in 1988 have seen an increase in average insecticide use on winter wheat and winter barley, and ADAS surveys have demonstrated that some cereal crops received three insecticide applications plus a molluscicide in the 1988/89 and 1989/90 seasons. This is not far short of the inputs to the Full Insurance fields at Boxworth, and confirms the relevance of examining the consequences of such intensive crop protection, even in the 1990's.

The time scale of the Project, although considerably longer than most previous field studies of pesticide use, was nevertheless quite short, compared to the rates of some ecological changes. Some effects would be expected to occur rapidly after the imposition of a heavy pesticide system, but others might develop slowly over many years. Similarly, the rates of recovery of some populations after a period of intensive pesticide usage are likely to be much slower than others.

The spatial scale of the Project may also have caused some possible biases, which are important in the interpretation of many aspects of the results, and will be discussed further later. In general, the problems concern the possibility that effects on wide-ranging species might have been masked by opportunities for 'escape' from exposure to pesticides and by replacement of animals that suffered from poisoning.

However, it is the experimental layout of the Project that has been questioned most strongly. For reasons outlined earlier (Chapter 2), a compromise was necessary to balance the need for replication with an adequate size of treatment areas in order to provide realistic conditions and to detect effects that might occur only on a large scale. It was recognised that this entailed sacrificing the benefits of replication and the ability to apply conventional statistical tests to determine whether any differences observed were significant.

The experimental layout adopted for the Project has revealed some effects and patterns of population change that would not have become apparent if the work had been conducted with much smaller plots. On the other hand, the confidence with which results can be interpreted is correspondingly reduced. This is acceptable because the Project was conceived as a preliminary exploration of environmental effects, which could be addressed in more detail by subsequent, more specific studies, designed to establish cause and effect relationships and to measure the size of the changes.

In the absence of the conventional methods of statistical analysis that are usually applied to replicated agricultural trials, different ways of detecting effects were necessary for this Project. For example, data gathered in the baseline years were used to identify any inherent differences between fields. Changes caused by pesticides could then be sought in comparisons of the treatment areas using an index of change from these initial conditions, either for individual fields or for the whole treatment blocks.

Apart from the issues of replication and statistical testing, the scale of the Project has a profound effect on several aspects of interpretation, for other reasons. There was incomplete control and knowledge of pesticide applications in surrounding land, outside the Project fields. Such information could not feasibly be obtained, but it would have been a valuable aid to the interpretation of results for the very wide-ranging species of wildlife, such as birds and certain invertebrate species. This raises two areas of uncertainty. First, what would have been the consequences if chemicals had been more widely applied, to larger areas of farmland? Second, was there any influence of one treatment area on the other, such as exposure of birds from the Supervised and Integrated areas when foraging in the Full Insurance area at the other end of the farm? The Project design attempted to minimise such biases, but their possible influence could not be completely eliminated, nor measured. There is also a question of what constitutes a typical pattern of pesticide use in an area of farmland. To what extent are large areas treated simultaneously as blocks or are they more usually treated in a patchwork fashion giving a mosaic of treated and untreated fields? The Project has emphasised the importance of non-crop refuges for some species of wildlife. Since the variation between farms is much greater for non-crop habitats than for conditions within fields of a particular crop, it is essential to regard the results as specific to Boxworth.

Summary of results from the ecological studies

In assessing the consequences of pesticide use for the environment, it is important to recognise the variety of ways in which effects may occur. Thus, pesticides may cause direct effects, due to lethal or sublethal poisoning, and indirect effects, acting through changes to the habitat or to populations of prey, predators and competitors. The severity of initial impacts on populations must be compared with 'natural' fluctuations in population size, and the time necessary for recovery. There is a possibility of cumulative effects, either as a result of repeated uses of a single chemical in successive years, or combined effects of different pesticides applied within a season. Accordingly, the consequences may be short- or long-term.

Bird studies

Birds are among the most valued forms of wildlife, because of their conspicuousness and aesthetic appeal. Their mobility suggests that free-flying adult birds may be unlikely to be directly exposed to pesticides at the time of application (by overspraying, for example). However, that is not the case for nests, and adults may also be exposed through contact with contaminated vegetation or prey, when they return to fields which have been treated. Indirect effects, too, may affect bird populations, by reducing their food supplies, for example. The bird studies at Boxworth were designed to detect any major consequences of these potential effects in combination, by monitoring the densities and breeding performance of common territorial songbirds. In addition, specific attention was paid to the consequences of one particular pesticide use—summer aphicides—for birds of a species that seemed likely to be heavily exposed to these chemicals.

Monitoring of bird populations by regular mapping of their territories revealed no obvious overall change in the density of the common farmland songbirds (Chapter 14). However, caution is needed in interpreting this result. Changes in the density of territories should reveal major changes in suitability for breeding, caused by lowered prey availability, or removal of vegetation cover. Nevertheless, it is possible that a steady number of territories over several years might conceal a large turnover of individual birds following pesticide-induced mortality each year. Too few birds were individually marked at Boxworth to examine this possibility. However, it seems unlikely that there was large-scale mortality, in view of the findings from breeding studies (Chapter 14) and detailed investigations of tree sparrows (*Passer montanus*) at the time of aphicide use (Chapter 15).

The only species which did show a change in density that is consistent with effects of the Full Insurance regime on prey populations was the starling (*Sturnus vulgaris*). Part of the observed reduction in numbers of starlings nesting in the Full Insurance area relative to the Supervised + Integrated area may have been due to effects of pesticides on leatherjackets (crane-fly larvae) (Chapter 14). However, the pattern could not be fully accounted for by this factor, and there were several other habitat changes during the course of the Project which are likely to have affected starlings. This illustrates the difficulties in fully explaining fluctuations in bird populations over a period of several years, which are influenced by many factors.

A further problem lies in the power of the comparisons between the high and reduced pesticide input areas to detect important effects. The density of most farmland birds is low, limiting the sample sizes that can be achieved. Their wide-ranging habits may have affected the comparisons by inclusion of some individuals that were little exposed to pesticides within the study-area. There is often considerable variability in birds' breeding performance. These sources of variation, together with the lack of conventional replication, mean that the study could be certain of detecting only large changes in density or breeding performance. Some effects with considerable biological significance could have been missed.

There is no straightforward solution to these problems. Sample sizes can be increased by expanding the study area, or by introducing large numbers of artificial nest sites, but such measures have costs, and often introduce further complications. At Boxworth, the compromise adopted was using nest boxes to encourage the development of a local population of tree sparrows, which provided a good model species to study the effects of those pesticides considered most likely to be damaging to birds in the breeding season.

Despite these limitations, the Project has demonstrated that there was not a sudden fall in the populations of many bird species when the Full

Insurance programme was initiated, as some had feared there might be. The resilience of bird populations is probably explained in part by their mobility, and the use of alternative, non-crop, habitats. The impact of a high-input pesticide programme may therefore depend more on the local distribution of habitats than on the pesticides themselves.

Enzyme inhibition and associated histological changes were demonstrated in tree sparrows, house sparrows and starlings, but these did not appear to have serious adverse effects, (Chapters 15 and 16). The bulk of the results indicate that any adverse effects are more likely to operate through indirect mechanisms involving birds' food or habitat. This conforms to the conclusions of other specific studies, such as on grey partridge (*Perdix perdix*) in cereal-growing areas (Potts, 1986), and reviews of trends in British bird populations, which attribute changes chiefly to altered land-use (O'Connor & Shrubb, 1986). Nevertheless, regular pesticide use is likely to place some additional pressure on bird populations. A recent study in Denmark (Braae, Nohr & Petersen, 1988) has shown substantial differences in songbird densities between land farmed organically and land subject to 'conventional' husbandry involving pesticides. It is likely that the reductions in many invertebrates caused by the Full Insurance regime at Boxworth (Chapter 9) would be reflected in the population densities of their bird predators in the long-term.

Studies of small mammals

Many of the limitations discussed above for birds also apply to the small mammal species studied at Boxworth. In addition, the inconspicuous habits of most species prevent detailed observations of behaviour. For this reason most emphasis was given to mark-release-recapture techniques to monitor population levels of the wood mouse (*Apodemus sylvaticus*), and to detect short-term changes caused by the use of particular pesticides.

Wood mouse populations vary greatly from year to year, but the results revealed no evidence for overall population changes that could be attributed to the Full Insurance regime (Chapter 12). As for birds, it is possible that this apparent stability hid some adverse short-term effects from which the population was able to recover rapidly. One such effect was revealed by studying the survival of individually-marked wood mice at the time of autumn applications of slug pellets containing methiocarb. The results (Chapter 12) demonstrated an immediate local impact, apparently through the death of predominantly adult wood mice that were resident in or visiting fields at the time. The population recovered quickly, by immigration of juvenile mice from nearby woods, fields or hedges.

The discovery of this effect of slug pellets has posed further questions, although its generality has not yet been established. Indeed, studies involving similar methods in another arable area have not shown the same impact (Tarrant et al., 1990). Further investigation is needed to identify the reasons for this difference, and to understand fully the nature of the hazard under a range of conditions. For example, it is not yet known whether wood mice were poisoned by feeding directly on the pellets or by eating contaminated slugs and earthworms. If the latter, the hazard could differ where there are more worms than at Boxworth, or in years of more serious slug problems. The availability of alternative food might affect the readiness of mice to eat pellets. The distribution of fields, hedges and other habitats is also likely to have an influence. If woods and thick hedges are scarce and remote from treated fields, there may be less chance of animals being exposed to molluscide poisoning, but the recovery of populations in the fields is also likely to be slower.

Although wood mice will eat invertebrates, their diet is not so heavily dependent on animal prey as is that of the shrews (*Sorex* spp.) at Boxworth. The data collected on the diet and distribution of common shrews (*S. araneus*) provided some indication of changes which were consistent with a reduction in prey availability under the Full Insurance regime.

Botanical studies

Much of the botanical work carried out at Boxworth was concerned primarily with monitoring and understanding the distribution of weeds. This entailed surveys within fields, where any plants other than the wheat itself can be regarded as 'weeds' (even though at their normal density they may have a negligible effect on the crop's performance). There was also study of plants growing in field boundaries and in the margins of the crop, where the possibility of undesirable weeds spreading into a field may be a major stimulus for herbicide use.

Patterns of weed densities in the fields reflected the efficiency of weed control measures, such that in the Full Insurance area, grass weeds were held at low density by herbicides. The supervised approach in the other areas was also effective, except for poor control of brome grasses. There was little difference between the areas in the density of broad-leaved weeds (Chapter 4).

Broad comparisons between the treatment areas are hampered by high variability from field to field, including the baseline years of the Project. Some fields were subject to heavy blackgrass infestations, and others to couch, for example. These differences can be attributed partly to the previous cropping history of fields. There were also differences in the nature of the field boundaries and their flora. The persistence of this variation throughout the Project demonstrates the considerable delays that may attend changes in flora following a change in husbandry. This reflects the existence of large numbers of viable seeds of annual species in the soil and slow colonization by perennials. It is probably for these reasons that no major differences were identified between the Project areas in field boundary or crop margin floras.

The data collected at Boxworth provided useful information to contribute to the debate about whether weeds spread into cereal fields from field boundary habitats (Way & Greig-Smith, 1987). Plant species could be assigned to four categories based on their field-edge distributions (Chapter 8). Very few were capable of spreading into fields, although those that do are annual species, such as cleavers (*Galium aparine*), which produce large amounts of seed and are considered to be important weeds. However, there is an implication that general control of potential weeds in field boundaries is neither necessary nor desirable, in view of the adverse side-effects on other flora.

Two other general conclusions can be drawn from the botanical work at Boxworth. First, although herbicides affected plant populations in the fields, more changes, both of seed in the soil and plants above ground, were attributable to physical disturbance such as ploughing than to the use of herbicides (Chapter 8). This has important consequences in the light of the growing pressure to farm without chemical inputs, and therefore with more reliance on tillage or other methods of weed control. Second, the substantial seed banks present in the soil of fields, and particularly at their margins, may provide a reservoir for the survival and reintroduction of rare arable weeds whose habitats are threatened nationally.

Invertebrates in the soil

The study of springtails collected in soil cores from the Project fields (Chapter 11) addressed an apparently contradictory ecological risk. Soil animals are likely to be protected from immediate direct exposure to pesticides by their subterranean habits, but are potentially vulnerable because of their lack of mobility.

Effects of the Full Insurance regime were mixed, resulting in reduced catches of some species (e.g. *Folsomia quadrioculata*) but others were apparently favoured by the high pesticide inputs, presumably through a reduction in competition or predator-prey interactions. Some changes were gradual, but others occurred within a single year, notably in 1986 when crop cover was sparse. These effects are important both for the activities of herbivorous and fungus-feeding species of springtails, but also because some species are prey for beneficial predatory arthropods, whose numbers might be indirectly reduced.

There have been relatively few field studies of the effects of pesticides on springtails in cereals, though a recent study by Frampton (1988) identified adverse effects of some chemicals (mainly fungicides) on surface-dwelling species. However, generalisation from Boxworth to other situations is unwise for several reasons, including the very great variability between fields (Chapter 11). Indeed, this caused problems for the interpretation of data within the Boxworth Project, as differences between fields during the baseline years greatly reduced the opportunities to detect later changes. In an attempt to overcome this problem, sampling was also carried out in the replicated plots sited in Shackles Aden field (Chapter 2). Although successful as a means of identifying the effects of pesticides on subterranean springtails, these unenclosed plots might not be as reliable for studying surface-dwelling species. This illustrates the practical difficulties of conducting an experimental field study of this type.

Crop invertebrates

The range of invertebrate species that may be present in cereal fields during the course of the year is very broad. There are pests, which affect the crop plants in various ways. Other species are predators or parasites of the pests, while some could have an indirect beneficial effect by competing with pests. Many, however, are apparently neutral, being neither damaging to the farmer's interests, nor beneficial. In terms of ecology, too, the range is wide, including species which are largely sedentary and some which are highly mobile, with a variety of dietary habits, from specialised predators to omnivores, herbivores and detritus-feeders.

The studies at Boxworth were designed to investigate three principal aspects of pesticide effects on crop invertebrates (Chapters 9 and 10). First, regular monitoring in all the fields provided information about changes in population densities of individual species or higher taxonomic levels in the three treatment areas. Second, results for different species were combined to assess changes in the composition of the invertebrate community. Third, effects of pesticides on the ability of 'natural enemies' to control pest populations were investigated by specific study of the interactions between aphids and their predators and parasites.

Many invertebrates are prey for birds and mammals, including the tree sparrows and shrews that were studied in other parts of the Project. Information about population changes of these species was valuable in the interpretation of the indirect consequences of insurance pesticide use for wildlife.

Chapter 9 discusses the patterns of changes in the densities of invertebrate populations, according to major ecological divisions (herbivores, carnivores and detritivores). This revealed a variety of responses to the imposition of the Full Insurance regime, from apparent resilience, to a sudden total absence in the Full Insurance fields almost as soon as the high input programme began (eg the lucerne-flea). There were also some species that increased in density under the high pesticide inputs. Examination of the behaviour of individual species suggested that these varied patterns were due to a complex of factors, including the degree of exposure experienced by the animals, their capacity for recolonisation of fields, and effects on their prey, competitors and natural enemies. Most differences between species in their response to the Full Insurance regime could be attributed to habits which affected exposure. In particular, these include the time of year that animals used cereal fields (those present in winter were exposed to insecticides, with little protective vegetation cover), their vertical distribution within the crop (animals occupying exposed positions on the upper leaves being more vulnerable than those under the canopy), and their dispersal ability (which determines how quickly populations recover within a season, if at all, and hence may explain long-term cumulative effects). These aspects are well illustrated by the results obtained for ground beetles (Burn, 1988 and Chapter 10).

The role of dispersal, affecting the annual cycle of animals' movements into and away from fields, and the recovery of populations after damaging pesticide applications, is of great importance. Although some species reach fields from long distances (eg aphids by flight, and spiders 'ballooning' in wind-currents), many are ground walkers, for which the nature of adjacent non-crop habitats may be crucial. As in so many parts of the Boxworth Project, the results indicate the major environmental significance of field margins as a source of fauna and flora. For example, many species of predatory beetles spread into fields from hedgerows, and the study of broad-leaved weeds indicated a variety of distribution patterns related to the field margin (Chapters 9 and 10). The management of field margins is likely to have as great an influence on populations as the husbandry of the fields themselves.

Although the changes in some invertebrate populations were clear and consistent, most showed considerable variation from year to year. This is not surprising in view of the effects that weather conditions are likely to have on growth, reproduction and migrations. However, some variations had particular adverse consequences for certain vulnerable species. For example, a number of sudden population changes occurred in 1986, when crop growth was sparse, probably because much of the protection afforded to invertebrates at the time of spring and summer sprays was lacking. Similarly, occasional late applications of insecticides may have affected species that would otherwise have escaped exposure because they migrate into fields after the normal time of application (eg *Agonum dorsale*; Chapter 10). These irregular but significant events are probably as important a cause of long-term population changes as are gradual cumulative effects occurring every year.

Overall, the density of all herbivorous invertebrates in the Full Insurance area declined by about 50% relative to that in the Supervised + Integrated area (Chapter 9). Changes in the total carnivores (predators and parasites) were of a similar size. In contrast, the density of detritivores was very little affected overall, although some individual species showed marked changes.

As well as these major reductions in herbivores and predators, there were changes in species composition, so that, for example, spiders became relatively less frequent in the Full Insurance area compared to the reduced-input area during the treatment years of the Project. Because of the variations in effects between species, some of which are much more important as 'natural enemies' than others, it is difficult to predict the consequences of population changes for the overall efficiency of pest control.

Studies were carried out to address this aspect directly, by measuring levels of predation and parasitism, and by experimentally manipulating the access of predators to aphid populations (Chapter 10). The results revealed that in some years, exclusion of predators allowed a more rapid build-up of aphids in the reduced-input area than the Full Insurance area. This implies that the high inputs had diminished a community of predatory species that otherwise would have exerted some control over pests. Such a pattern, if consistent, would suggest that in some situations chemical control of pests might have the reverse of its desired effect, depending on the relative vulnerability of pest and predator. However, at Boxworth the pattern varied greatly from year to year, and in years when aphids were abundant, predators had little restraining effect on their populations. Although the results of the Boxworth Project have made a valuable contribution, there is still much to understand before practical strategies for aphid control based on a combination of 'natural enemies' and chemicals can be made available (Jepson, 1989).

Environmental fate of pesticides

Studies were made of the distribution of selected pesticides as they were applied and their persistence in vegetation, soil and water. Knowledge of environmental contamination is important, both to help understand the exposure of animals, and because of increasing concern about the potentially far-reaching consequences of chemicals in groundwater, for example. Monitoring of spray drift (Chapter 7) revealed some effects of summer aphicide applications on hedgerow arthropods, although there was rapid recolonisation. The lack of major changes in field boundary floras (Chapter 8) suggests that drift was not substantial. Residues of pesticides in drainage water from fields where they were applied were absent or present only at low levels.

Features of the cereal ecosystem

For all the various animal and plant groups investigated during the Boxworth Project, the ecological effects of the Full Insurance regime depend on ability to escape exposure, on susceptibility to poisoning or other effects, and on the capacity to recover (Greig-Smith, 1991). There is likely to be great variability in any or all of these features, within each group.

Ways of 'escaping' exposure range from simply moving away when pesticide applications are in progress, as birds are able to do, to the occupation of protected habitats within the crop or beneath the soil surface. The annual migratory habits and life-cycles of species may mean that they are absent from fields when certain pesticides are used. It is impossible to generalise about the importance of these mechanisms, because even within a group of otherwise similar species there may be critical differences in habits that determine the greater exposure of certain species. This is well illustrated by the predatory beetles (Chapters 9 and 10).

Differences in susceptibility to poisoning depend on biochemical or physiological attributes which are largely unknown for most of the species and chemicals involved at Boxworth. Even empirical data showing the relative toxicity of individual pesticides to different animals and plants are sparse (eg Marshall, 1989). The Boxworth Project did not attempt to investigate such patterns, being concerned more with the combined impact of both exposure and toxicity. Nevertheless, the results have identified a number of cases for which it would be worthwhile conducting further research to unravel the relative importance of exposure and toxicity (for example, the differences in effects among species of predatory ground beetles and spiders, and the age-related changes in susceptibility of songbirds).

Severe reductions of animal or plant populations may not have serious consequences if rapid recovery is possible. That can be due to recolonisation (from nearby habitats, seedbanks, or more distant sites), and in some cases may be almost immediate, so that population changes are minimal.

For both exposure and recovery the question of scale is crucial. For example, if the mobility of animals is limited, and allows only short-distance recolonisation, the large size of modern cereal fields may prevent recovery of affected populations within a season. On the other hand, the areas that are treated simultaneously may be too small to inhibit very wide-ranging animals from escaping to untreated fields or non-crop habitats. In most cases, however, the amount, distribution and management of such refuge habitats are likely to be very important, and may have more influence on the nature of the local wildlife than the crops and husbandry of fields.

Throughout the Project, ecological effects were investigated mainly by looking for changes in individual species. For groups represented by a large number of species, it is also possible to examine changes at the community level, which may involve more than just the sum of effects on component species. For birds and plants, there was no obvious change in species-composition in the Full Insurance area after 1984, but some groups of invertebrates showed major changes (Chapter 9). Among both herbivorous and predatory crop invertebrates, the relative densities of certain taxonomic groups increased or decreased, owing to the combined effects of direct pesticide toxicity and changes in prey, predators or competitors of individual species. Because of the potential complexity of the interactions behind such patterns, it is very difficult to predict what the consequences of a particular pesticide use are likely to be. At present, there is no firm basis for knowing how typical of cereal farmland the results at Boxworth might be.

It is also difficult to predict the likely effect of a pesticide regime on the predatory or parasitic activities of species that are 'natural enemies' of aphids or other pests. The importance of these animals lies in their combined predation or parasitism, and replacement of some species by others may have no overall effect on pest populations. It is therefore essential to examine functional aspects such as predation as well as population changes (Chapter 10).

The Boxworth Project aimed to examine the implications of a whole pesticide package, rather than individual components. However, it is likely that a few applications within the annual programme were particularly damaging. Although not designed to identify these, the results of the Project imply that for invertebrates, it was the winter insecticide sprays that were most damaging. Winter applications occur when there is minimum vegetation cover, so reducing the opportunities for invertebrates present in the fields to escape exposure. Some species were vulnerable to other treatments (for example, wood mice were affected by slug pellets), while adverse effects might have been greater if products with greater toxicity had been used for the control of some pests (eg dimethoate as a summer aphicide instead of demeton-S-methyl). The environmental safety or otherwise of a pesticide programme may depend more on the inclusion of a few particular chemicals than on the number of active ingredients applied annually.

During the Project there were several instances of pesticides being applied late because of bad weather. Such occurrences sometimes have a disproportionate effect on some groups of wildlife, for example if a spray is delayed until after spring-migrating beetles have moved into the fields. Despite the best standards of husbandry,

mis-timed applications will happen from time to time. In a long-term research study, and in normal commercial farming, allowance must be made for these 'rare events', although it may be impossible to predict when they will occur.

Economic aspects of the crop protection programmes

As expected, the Full Insurance programme provided a distinct advantage in yield over the other two programmes (Chapter 6). On average, yields of wheat in the Full Insurance fields were 0.92 tonnes per hectare higher than in the Supervised fields and 1.35 tonnes per hectare higher than in the Integrated area, although there was considerable variation from year to year. Measures of grain quality also generally indicated a benefit from the Full Insurance programme.

However, this apparent advantage entailed extra financial costs, in addition to the environmental damage discussed in the previous sections. Calculation of gross margins for the three regimes (Chapter 6) showed that the Supervised approach was equally, if not more, profitable than the Full Insurance system, although the Integrated approach did not perform as well. The important conclusion is that very high inputs of crop protection chemicals are unlikely to be required in a well-managed crop, and prophylactic use is not likely to bestow any additional economic benefits.

It is tempting to use the economic results from Boxworth more fully, by quantifying the differences between high and low input systems. However, this is not considered to be valid, for several reasons.

First, the results apply to Boxworth, with its particular soils, and to the specific pesticide programmes tested. Varying these programmes, and working on other sites, would probably alter the degree of difference between gross margins, through changes in yields and/or costs.

Second, the treatments in the Project continued for only five years, which is too short a period to cover all extremes of weather and crop conditions. The year-to-year variability observed even in 1984–88 is a warning that average results could be strongly influenced by one or more unusual seasons.

Third, the calculation of financial costs and benefits can be made in several ways, laying more or less emphasis on fixed costs. Although it is straightforward to account for variable costs such as purchase of chemicals and running machinery, the levels of capital investment in labour and machinery under the different systems are more flexible, and could greatly influence the viability of high or reduced-input farming.

Finally, the relatively poor returns from the Integrated regime should not be taken as a general indication of the worth of integrated farming. The particular regime used at Boxworth was a compromise, recognised to be unsatisfactory in many ways (Chapter 2), and not fully representative of how a low-input, truly integrated system would be designed.

Thus, it is important to regard this economic information only as a broad indication of the benefits of supervised approaches compared to a prophylactic high-input philosophy. The design of good reduced-input systems is a different issue, which is better addressed by making changes to cropping and husbandry practices to help avoid pest, weed and disease problems, in addition to efficient forecasting of the need for crop protection measures. There is also scope to improve the basis of financial assessments of entire farming systems, rather than making comparisons on particular crops.

Management of reduced-input systems

Because of the experimental requirements of the Project, operation of the Supervised approach to crop protection was not exactly the same as it would be under normal farming practice. Despite this, the experience at Boxworth provides some useful guidance to the feasibility of a farmer operating such a system as an alternative to a prophylactic approach.

The key element of the supervised approach is the monitoring of pests, weeds and diseases that underlies decisions about spraying. If this is to be done by advisors or consultants, the only constraints are the cost and the ready availability of expertise at the right time. Monitoring costs must be built into the calculations of gross margins and can be justified if they are exceeded by average savings in the costs of chemicals and their application. Market forces would determine the acceptance of supervised crop protection, and the charges for monitoring, after a few years' experience.

However, many farmers would prefer to carry out their own monitoring or replace monitoring with a less structured judgement. In that case, success depends on whether monitoring procedures and the decision criteria that follow from them are easily understood, and on how closely informed local judgement approximates to more accurate assessments. Thus, 'do-it-yourself' monitoring can reduce costs substantially, but is likely to result in more reliance on the extra margin of safety afforded by insurance applications, particularly if confidence is damaged by occasional failures to anticipate problems.

Even detailed monitoring to a standard protocol is unlikely to identify all problems correctly. There is an inescapable contradiction between the advantages of early monitoring which allows action to be taken, and the unexpected changes that may occur after the assessment. With the need to minimise assessment visits for economy, it is inevitable that some developing crop damage problems will be missed, requiring later urgent remedial action. Moreover, it may not be possible to reduce monitoring and remain effective. Surveys of weeds indicated that accurate estimates of weed densities could be achieved only by even more intensive sampling than that used at Boxworth. It is much less likely that suspected problems will be falsely signalled, causing an unnecessary pesticide application. At Boxworth, the arrangement for detailed weed assessements on three occasions during the year led to several failures of prediction, because of the long interval between assessment of flowering grasses in July and the emergence of new plants the following spring (Chapter 4). The Project offered the first major opportunity to test the utility of Wilson's (1986) proposed 'crop equivalents' system for predicting the effects of broad-leaved weeds. Assessments of pests and diseases at Boxworth were closer in time than weed surveys to the periods when applications were made, and more closely resembled the kind of monitoring that would be done on farms, although visits were more frequent, to meet the needs of the research. Overall, the experience gained at Boxworth, together with continual improvements being made to forecasting and threshold criteria, has confirmed the feasibility of farming efficiently according to a supervised approach.

Some aspects of husbandry within the Project were not successful. In theory, the use of varietal blends of wheat should reduce the spread of disease within and between fields, and obviate the need for some fungicide applications. However, the use of blends in the Integrated area in 1986 did not achieve this (Chapter 5) and it was subsequently deemed preferable to return to a single variety with proven low resistance to the major disease, eyespot. Although it is not appropriate to generalise this conclusion freely to all conditions, results at Boxworth are a salutary example of the potential drawbacks in such strategies, and suggest that further work will be required before they can be used with confidence.

Research at Boxworth has also provided several pointers to apparent effects which might justify changes in the advice offered to farmers. For example, the studies of wood mice imply that slug pellets should be drilled with seed rather than broadcast, and effects on certain invertebrate predators suggest that the timing of insecticide use in winter and early spring may be critical. However, such results are specific to Boxworth, and are derived from a single study. Therefore, they need confirmatory work in other conditions before advisory changes are implemented.

The Project dealt only with pesticides applied over whole fields, and the results do not carry any direct implications for more selective management strategies such as part-field treatments. However, the evidence obtained for the importance of 'refuges' to some beneficial invertebrates suggests that management to encourage their overwinter survival and colonisation of fields would be advantageous. The creation of artificial habitat patches (Thomas & Wratten, 1988) or unsprayed 'conservation headlands' (Boatman & Sotherton, 1988), are examples of possible strategies.

Design of studies of pesticides and farming systems

There were specific reasons, discussed above, for setting up the Boxworth Project on a scale which approximated a whole farm. It was by default rather than design that this feature has been emphasised, in comparing the Boxworth Project to the few other studies conducted on a similar scale, but with rather different aims—the Game Conservancy's North Farm study in the south of England (Potts & Vickerman, 1974), the Lautenbach Project in southern Germany (El Titi, in press) and the Nagele experiment in the Netherlands (Vereijken, 1986).

There are two general reasons for undertaking research on whole farming systems, rather than on individual components (Greig-Smith, 1988). These are: first, to assess environmental consequences in a context as close as possible to the 'real world'; second, to test ideas about modifying agricultural systems for greater efficiency and environmental benefit. The Boxworth Project was directed mainly at the first of these, but also tried to address the second. This led to some problems of incompatibility, because the demands of the environmental studies restricted the options for economically-desirable changes such as the choice of crop rotations in the Integrated area. The environmental conclusions are not compromised, because the needs of the research were always given precedence. However, the constraints imposed on the kinds of low-input husbandry adopted in the Project mean that the economic interpretation cannot be pursued too deeply. In addition, the context in which the ecological results were obtained was not as typical of current farming practice as would have been ideal.

Table 17.1 Features of the Boxworth Project relevant to the design of 'whole-farm' research studies.

Aspect of design	Advantages	Disadvantages
Large spatial scale	Reduced edge-effects and similarity to the fields used in commercial practice.	Difficulties of replication and statistical testing.
Long duration	Ability to detect effects that were slow to emerge.	
Close control of husbandry	Reduced confounding variation due to unplanned changes.	Restricted the use of rotations and other integrated practices.
Comparison of fixed pesticide programmes	Allowed firmer interpretation of differences than possible by monitoring changes in pesticide use.	High input regime became unrealistic by the end of the study.
Baseline monitoring	Identified differences present before pesticide regimes were imposed.	
Study of many environmental aspects	Allowed interactions between species and indirect effects to be detected.	Size and layout of treatment areas were not equally suitable for all species. Also caused logistic difficulties and compromises.
Records of crop performance and farming costs	Economic appraisal of the contrasting systems.	
Focus on pesticides	Reduced confounding variation.	Excluded possible effects of fertiliser use and other farming practices.

The lessons of the Project highlight a number of advantages and drawbacks in its design (Table 17.1). Experience at Boxworth has supported the view that in the future, studies in this area would be better divided into those driven primarily by environmental concerns and those dictated by economic priorities, rather than combining them. That does not imply that each aspect should be addressed without reference to the other. On the contrary, the Boxworth Project has demonstrated the value of monitoring crop performance and costs to put ecological studies in context, while the environmental consequences of low-input systems should be measured along with their economics.

Within the area of environmental studies themselves, there is also merit in dividing work to accommodate the fact so clearly demonstrated at Boxworth, that different scales of space and time are appropriate for the study of various kinds of animals and plants. A good philosophy to follow in researching farming systems is, therefore, to undertake separate but linked experiments which can each be designed specifically to suit its own purpose, minimising the need for compromises. Among other benefits, this kind of approach should ease the problems of achieving adequate replication in large-scale and long-term studies.

As studies of farming systems increase, there will be a need to compare their findings, in order to understand fully the ecological and economic implications. Because aspects of design may influence the results obtained, comparisons will only be valid if such studies are planned consistently. It would be valuable, therefore, to agree uniform principles for the design of large-scale farm studies. These should include common approaches to the choice of methodology, statistical analysis, economic assessments, size and layout of experimental areas, and selection of indicator species. Without imposing inflexible constraints on future studies, this would ensure that their results are used to the greatest effect.

Future studies

As the Boxworth Project neared the end of its planned seven years, there was widespread discussion about the merits of prolonging the study. Some argued that the knowledge already gained would best be consolidated by continuing to monitor the Project fields under the same pesticide regimes, in order to detect long-term effects. Others felt that the compromises inherent in the Project's design, discussed in the previous section, would reduce the value of such an extension. In particular, the Full Insurance programme was already seen as extreme in 1988, and would become more of an anachronism if the work went on without change.

After detailed debate, including a Special Topic Review involving representatives of many research organisations (MAFF, 1988), the decision was taken not to continue the Boxworth Project in the same form. The resources thus freed would be devoted to new studies, directed at following-up some implications of the Boxworth Project's results (see below).

However, a limited continuation study at Boxworth was agreed, in order to address the question of whether, and how quickly, animal populations affected by the Full Insurance programme recover once the fields in that area are returned to a lower-input regime. Accordingly, half of the Project fields were retained for a further three years' study, under management that is summarised in Table 17.2. Thus, one field continued to receive the Full Insurance programme, to act as a 'high input control', whereas the other two were switched to the Supervised regime, and were matched by their 'triplet' partners (see Chapter 2) in the original Supervised area.

The emphasis of the continuation study was on those invertebrate populations which suffered under the Full Insurance programme. Therefore most of the work involved regular sampling using a D-vac suction sampler and pitfall traps (as described in Chapters 9 and 10), plus an aerial suction sampler and emergence traps to investigate the routes of recolonisation of fields by various types of invertebrates. In addition, soil micro-organisms, which were not studied during the main Project, were surveyed in order to identify any major differences which might be pursued further in later research. Results of the work carried out in the continuation period will be described in a separate report.

Away from Boxworth, studies have been planned to pursue the two main possible adverse effects on birds and mammals revealed by the Project. These are the poisoning of wood mice (and perhaps other small mammals) by molluscicides, and the sublethal effects on birds of exposure to insecticides. Rather than pursue the latter by continuing to study tree sparrows, a project has been initiated to trace the consequences of pesticide use in orchards on members of the thrush family such as blackbirds (*Turdus merula*). This will permit a fuller study than was feasible at Boxworth.

The principal research to follow on from the Boxworth Project concerns invertebrates, and has been planned to take the best elements of the Boxworth approach, combined with modifications to overcome some of its drawbacks. One priority was to confirm the major ecological conclusions suggested by results at Boxworth, which were tentative because of the lack of field-scale replication. Another was the need to determine whether the moderate Supervised programme, while being ecologically preferable to higher inputs, was itself neutral, or whether it caused environmental dam-

Table 17.2 Management of fields during the continuation phase of the Boxworth Project, which aims to examine recovery of invertebrates after stopping the Full Insurance programme.

Field	Pesticide regime 1984–88	Pesticide regime 1989–91
Backside	Full Insurance	Full Insurance
Grange Piece*	Full Insurance	Supervised
Pamplins North #	Full Insurance	Supervised
Knapwell*	Supervised	Supervised
Top Pavements #	Supervised	Supervised

(*, # indicate pairs of matched fields)

age compared to a lower input. Finally, the rapidly-growing interest in low-input, 'integrated' approaches to crop protection required experimental work to develop viable systems that would be environmentally beneficial. For reasons discussed above, the investigations of ecology and economics were planned independently, each forming the core of a separate project, though designed to closely complement each other.

These needs have led to two new projects, named SCARAB (Seeking Confirmation About Results At Boxworth) and TALISMAN (Towards A Low Input System Minimising Agrochemicals & Nitrogen), which are outlined by Cooper (1990). SCARAB is concerned with the effects of moderate-input and low-input crop protection programmes on crop invertebrates and plants. The work will involve rotations of cereals and common break crops, and will be carried out at three farms, with very different soil conditions and farming practices. Rather than use whole fields as plots, with the consequent problems of inherent variation that were found at Boxworth, the SCARAB Project is based on large paired plots within single fields. A total of eight pairs of moderate and low input plots will be studied over the whole project. These contrasting pesticide regimes are based on (a) current farm practice, including monitoring of pests, weeds and diseases for managed control, and (b) reduced inputs, involving minimum necessary use of fungicides and herbicides, and no insecticides.

The TALISMAN Project is concerned with the economics of reduced-input farming, and involves small plots in a replicated design on four farms using rotations, including the three where SCARAB is sited. The pesticide and nitrogen inputs are to be varied in a complex series of programmes which range from full inputs in line with current practice, to reductions of 50% in pesticides and nitrogen, and modifications to rotations or other husbandry which would further enhance crop performance.

The results of the two projects will be linked through limited monitoring of invertebrates and plants on some TALISMAN plots, and recording of yields on the SCARAB plots. Together, they will contribute towards a general understanding of the consequences of adopting low-input crop protection strategies.

Conclusions

The Boxworth Project successfully addressed three aspects of crop protection in intensive cereal production, by means of experimental comparisons between high-input and reduced-pesticide programmes: ecological effects on wild plants and animals, an economic evaluation, and assessment of management problems.

The results have provided new information about the risks to the cereal-field environment arising from high levels of pesticide use as an insurance against crop damage. For some forms of wildlife, there was little evidence that the high inputs changed the density or performance of populations, whereas others were seriously affected. These findings have led to further studies and have influenced the interpretation of research being conducted elsewhere.

Economic appraisal of the three pesticide programmes suggested that although there was an advantage to the Full Insurance approach in terms of yield, it was fully balanced by the extra costs entailed. Crop protection was achieved as efficiently by a moderate, supervised approach based on use of pesticides only when shown to be necessary.

Experience gained in the course of monitoring pests, weeds and diseases for decision-making about pesticide applications has made a valuable contribution to the development of strategies for farmers to adopt in managed crop protection. It also identified areas of doubt, in which more work is needed to support advice, and several possible changes to current practice that should be encouraged.

The lessons of the Project include some advantages and disadvantages of the approach taken at Boxworth. These have helped to shape the design of further studies, including two major projects (SCARAB and TALISMAN) which will continue the exploration of ecological and economic aspects of cereal production that was developed at Boxworth.

Through these varied consequences, the Boxworth Project has amply fulfilled its role as an innovative way to address the short- and long-term consequences of high pesticide inputs on a realistic farm-scale.

Acknowledgements

We are grateful to Drs P J Bunyan, G K Frampton, and E J P Marshall for their comments.

References

Boatman, N D & Sotherton, N W (1988). 'The agronomic consequences and costs of managing field margins for game and wildlife conservation.' *Aspects of Applied Biology* **17**, 47–56.

Braae, L, Nohr, H & Petersen, B S (1988). 'Bird faunas of conventionally and organically farmed land.' *Miljoprojekt nr 102*. Miljostyrelsen, Copenhagen, Denmark (In Danish, English summary).

Burn, A J (1988). 'Effects of scale on the measurement of pesticide effects on invertebrate predators and predation.' In: Greaves M P, Greig-Smith P W & Smith B D (Eds.). '*Field Methods for the Study of Environmental Effects of Pesticides*. British Crop Protection Council, Monograph No. 40. pp. 109–117.

Cooper, D A (1990). 'Development of an experimental programme to pursue the results of the Boxworth Project.' *Proceedings of the 1990 Brighton Crop Protection Conference—Pests and Diseases*. Vol. I, pp. 153–162.

El Titi, A (1991). 'Twelve years experience of integrated wheat production at the commercial farm of Lautenbach, South West Germany.' In: Firbank L G, Carter N, Darbyshire J F & Potts G R (Eds.) *The Ecology of Temperate Cereal Fields*. Blackwell Scientific Publications, Oxford pp. 399–411.

Frampton, G K (1988). 'The effects of some commonly-used foliar fungicides on Collembola in winter barley: laboratory and field studies.' *Annals of Applied Biology* **113**, 1–14.

Greig-Smith, P W (1988). 'Special Topic review of future research on environmental effects of farming practices.' In: *The Boxworth Project and subsequent studies—Topic Review*. MAFF, London. pp. 2–9.

Greig-Smith, P W (1991). 'The Boxworth experience: effects of pesticides on the fauna and flora of cereal fields.' In: Firbank L G, Carter N, Darbyshire J F & Potts G R (Eds.) *The Ecology of Temperate Cereal Fields*. Blackwell Scientific Publications, Oxford pp. 333–371.

Jepson, P C (Ed.) (1989). *Pesticides and Non-Target Invertebrates*. Intercept Press, Wimborne, Dorset.

MAFF (1988). *The Boxworth Project and subsequent studies—Topic Review*. MAFF, London.

Marshall, E J P (1989). 'Susceptibility of four hedgerow shrubs to a range of herbicides and plant growth regulators.' *Annals of Applied Biology* **115**, 469–479.

O'Connor, R J & Shrubb, M (1986). *Farming and Birds*. Cambridge University Press, Cambridge.

Potts, G R (1986). *The Partridge: Pesticides, Predation and Conservation*. Collins, London.

Potts, G R & Vickerman, G P (1974). 'Studies on the cereal ecosystem.' *Advances in Ecological Research* **8**, 107–197.

Tarrant, K A, Johnson, I P, Flowerdew, J R & Greig-Smith, P W (1990). 'Effects of pesticide applications on small mammals in arable fields, and the recovery of their populations.' *Proceedings of the 1990 Brighton Crop Protection Conference—Pests and Diseases*, Vol. I, pp. 173–182.

Thomas, M B & Wratten, S D (1988). 'Manipulating the arable crop environment to enhance the activity of predatory insects.' *Aspects of Applied Biology* **17**, 57–66.

Vereijken, P (1986). 'From conventional to integrated agriculture.' *Netherlands Journal of Agricultural Science* **34**, 387–393.

Way, J M & Greig-Smith, P W (Eds.) (1987). *Field Margins*. British Crop Protection Council, Monograph No. 35.

Wilson, B J (1986). 'Yield responses of winter cereals to the control of broad-leaved weeds.' *Proceedings of the European Weed Research Society Symposium 1986—Economic Weed Control* pp. 75–82.

Choice of fungicides and expected timing of applications

Target diseases	Full Insurance area
In wheat:	
Seed-borne diseases	Triadimenol + fuberidazole seed treatment (a gives early season protection against leaf and st base diseases).
Eyespot	Prochloraz + carbendazim at stem extens (growth stages 30–36) (April—May).
Septoria	Recommendation for 1983–1986: chlorothaloni flag leaf emergence (growth stages 37–39) (Ma June) and captafol and triadimefon at ear em gence (growth stages 51–59) (June—July) (tri imefon also gives some control of rusts a triazole-sensitive strains of mildew). Recommen tion for 1987–1988: flutriafol at flag leaf emerger and chorothalonil at ear emergence.
Mildew	Recommendation for 1983–1986: fenpropimor at flag leaf emergence (growth stages 37–3 (May—June). Recommendation for 1987–198 fenpropimorph at ear emergence (growth stag 51–59) (June—July) (this also gives control of ru and triazole-sensitive strains of mildew).
Late *Septoria* and ripening diseases	Carbendazim + maneb, or carbendazim maneb + sulphur, during or after anthesis (grov stages 61–71) (June—July).
In oilseed rape:	
Alternaria **pod spot**	Iprodione at petal fall and three weeks later.
Light leaf spot	Benomyl or prochloraz at early stem extension or January—March.
Sclerotinia	Vinclozolin at early flowering.

upervised and Integrated areas

outine organomercury seed treatment (for protection against seed-borne diseases except loose smut).

arbendazim at stem extension (growth stages 30–36) (April—May) if eyespot lesions affect 20% of stems, ut use prochloraz + carbendazim, or prochloraz, if MBC-resistant strains of eyespot are present.

ropiconazole, or prochloraz if not applied earlier, at flag leaf emergence (growth stages 37–39) (May— une) if *Septoria* is present or if the weather has favoured infection (prochloraz also gives protection against ate infections of eyespot where crops were not sprayed earlier). Propiconazole or triadimenol at ear mergence (growth stages 51–59) (June—July) if *Septoria* is present and no spray was applied at flag leaf mergence, or if the weather has favoured infection).

enpropimorph or tridemorph between flag leaf emergence and ear emergence (growth stages 37–59) May—July) if mildew is present (this also gives control of rusts and triazole-sensitive strains of mildew).

arbendazim + maneb during or after anthesis (growth stages 61–71) (June—July) if any of these diseases re present or if the weather is wet during anthesis.

ecommendation for 1983–1986: iprodione at the first sign of disease on the pods. Recommendation for 987–1988: up to two sprays during flowering if *Alternaria* reaches the upper pods or leaves.

enomyl at early stem extension if 50% of plants are infected.

o treatment recommended as *Sclerotinia* is unlikely to be a problem at Boxworth (based on previous xperience).

APPENDIX I (continued) THE PROPOSED PROGRAMMES OF INSECTICIDE
APPLICATIONS AT BOXWORTH EHF, 1983–1988

Choice of insecticides and expected timing of applications

Target pests

Full Insurance area

In wheat:

Aphids (vectors of BYDV)

Cypermethrin in November.

Rose-Grain Aphid

Demeton-S-methyl between mid-flowering a‹
grain watery-ripe (growth stages 65–71).

Grain Aphid

Demeton-S-methyl at complete ear emergen‹
(growth stages 59–60).

Frit Fly

Chlorpyrifos at 100% plant emergence (grow‹
stages 10–12) (October—November).

Yellow Cereal Fly

Triazophos at egg hatch (January—March).

Wheat Bulb Fly

Chlorfenvinphos at egg hatch (January—March)
the threshold number of eggs per ha
September—October is reached in the Supervise‹
or Integrated areas. Omethoate if and when th‹
threshold number of 'deadhearts' (damaged your‹
leaves) is reached in the Full Insurance area.

Slugs

Methiocarb in September—November.

pervised area	Integrated area
permethrin in November if the need for control is dicated by the aphid infectivity index and crop ndition (sowing date and the presence or sence of volunteer plants).	As in the Supervised area.
emeton-S-methyl if the threshold number per flag af is reached (usually June—July).	As in the Supervised area but use pirimicarb instead of demeton-S-methyl.
commendation for 1983–1988: pirimicarb if the reshold number per ear is reached at flowering rowth stage 60) (June—July). Recommendation r 1987–1988: pirimicarb if the threshold of infec- d tillers is reached at ear emergence (50% of lers at growth stage 50) or at flowering (66% of lers at growth stage 60) (both equivalent to 5 ohids per ear at flowering).	As in the Supervised area.
hlorpyrifos at 100% plant emergence (growth age 10–12) (October—November) if the need for ontrol is indicated by the area forecast and crop ondition (date of emergence and the presence or osence of volunteer plants).	As in the Supervised area.
iazophos at egg hatch (January—March) if the umber of eggs per soil sample is large in ecember, but omit this treatment if an autumn ohicide has been applied.	
hlorfenvinphos at egg hatch (January—March) if e threshold number of eggs per ha in eptember—October is reached. Omethoate if and hen the threshold number of 'deadhearts' (dam- ged young leaves) is reached.	As in the Supervised area.
ethiocarb before drilling if bait trap catches reach e threshold.	As in the Supervised area.

APPENDIX I (continued) THE PROPOSED PROGRAMMES OF INSECTICIDE APPLICATIONS AT BOXWORTH EHF, 1983–1988

Choice of insecticides and expected timing of applications

Target pests	**Full Insurance area**
In oilseed rape:	
Cabbage Stem Flea Beetle	Deltamethrin when the seedlings are fully established and in November. Also in spring if necesary.
Cabbage Seed Weevil & Brassica Pod Midge	Triazophos at petal fall if the threshold number adult weevils is reached during flowering (no damage threshold is currently available for pod midge
Slugs	Methiocarb after drilling.
Cabbage Root Fly	Unlikely to be a problem at Boxworth with t normally late drilling date but treatments (probab prophylactic applications of carbofuran or phorat will be used if necessary (currently no damag thresholds are available for cabbage root fly).
Stem Weevil & Cabbage Aphid	Unlikely to be a problem at Boxworth (based previous experience) but treatments (probably mathion for stem weevil and pirimicarb for cabba aphid) will be used if necessary (currently no damage thresholds are available for stem weevil a cabbage aphid).

pervised area	Integrated area

ltamethrin when the seedlings are fully estab-
ned if plant damage by adults is severe and in
tumn-winter if the threshold number of larvae per
int is reached. Also in spring if the number of
vae per plant is high.

As in the Supervised area but use pirimiphos-
methyl instead of deltamethrin.

in the Full Insurance area.

As in the Full Insurance area but use phosalone
instead of triazophos.

ethiocarb at the 3–4 rough leaf stage if severe
mage has occurred.

As in the Supervised area.

Choice of herbicides and expected timing of applications

Herbicides in wheat:		**Full Insurance area**
Autumn:	Pre-sowing for general weed control.	Paraquat as routine.
	Pre-emergence for brome grass control.	Tri-allate granules followed by metoxuron, necessary for brome grass control.
	Post-emergence for grass weed control.	Isoproturon (blackgrass) or metoxuron (brom depending upon predominant species.
	Post-emergence for broad-leaved weed control.	Broad-spectrum herbicide, probably mecopro bromoxynil + ioxynil, or fluoroxypyr.
Spring:	Broad-leaved weed control.	Broad-spectrum herbicide, probably mecopro bromoxynil + ioxynil, or fluoroxypyr.
	Wild-oat control.	Difenzoquat as routine.
Summer:	Pre-harvest for perennial weed control.	Glyphosate at the first sign of couch grass or oth perennial weeds.

Herbicides in oilseed rape:

Autumn:	Grass weed control.	Fluazifop-P-butyl as routine between the 1 true le growth stage and mid-October.
	Broad-spectrum control of grasses and broad-leaved weeds.	Clopyralid + propyzamide as routine from the 3-le growth stage.
Spring:	Broad-spectrum control of broad-leaved weeds.	Pyridate or a combination of clopyralid + benazo as routine at the 6-leaf growth stage.

Supervised area	Integrated area
Paraquat if wheat is to be direct drilled, or if necessary after shallow cultivation.	Paraquat, but only if wet weather prevents control of weeds and volunteer plants by cultivation.
As in the Full Insurance area.	No pre-emergence herbicide (as ploughing usually precludes the need for brome control).
As in the Full Insurance area but only if the need for control is indicated by weed counts.	As in the Supervised area.
As in the Full Insurance area but only if the need for control is indicated by weed counts and the relative crop/weed vigour.	As in the Supervised area.
As in the Full Insurance area but only if the need for control is indicated by weed counts and the relative crop/weed vigour.	As in the Supervised area.
Difenzoquat or flamprop-M-isopropyl depending upon the crop variety and weather but only if wild-oat counts indicate the need for control.	As in the Supervised area.
Glyphosate but only if perennial weeds reach significant levels.	As in the Supervised area.
As in the Full Insurance area.	As in the Full Insurance area but only if the need for control is indicated by grass weed counts.
Propyzamide or a combination of clopyralid + propyzamide from the 3-leaf growth stage.	As in the Full Insurance area.
No spring broad-spectrum herbicide application unless weed counts exceed the threshold.	As in the Supervised area.

Husbandry activities:	Full Insurance area
Primary cultivation	Nil or minimum.
Drilling	Mid-late September. Average seed rate 150 kg/ha.
Spring cultivations	Probably none.
Control of summer weed spread (from field boundaries)	Flail when necessary.

pervised area	Integrated area
in the Full Insurance area.	At least 10 cm deep, normally with inversion.
in the Full Insurance area.	From early October (later if blackgrass or brome are present). Average seed rate 175 kg/ha.
in the Full Insurance Area.	Perhaps harrow and/or roll, depending upon soil conditions.
tovated strip.	As in the Supervised area.

Pesticide applications and husbandry

These tables summarise the cultivations and agrochemicals used during the Boxworth Project. The list of agrochemicals shows the products used, their active ingredients and type of formulation and the rates at which they were used. Full product names are listed but for clarity some names have been shortened in the tables. The codes for formulation types are those used in the UK Pesticide Guide:

DS powder for seed treatment
EC emulsifiable concentrate
GR granules
LI liquid
MG microgranules
PT pellets
SC suspension concentrate (= flowable)
SG water soluble granules
SL soluble concentrate
SP water soluble powder
WP wettable powder

For products which were applied at more than one rate during the Project, letter codes (a–e) are given after the product name in the tables. These cross-refer to the application rates indicated in the list of agrochemicals. Additional letter codes are used to indicate the following:

(h) applied to the headland only
(i) molluscicide incorporated into the soil during drilling
(p) pre-harvest application, various dates

Unless indicated otherwise, molluscicide pellets were surface-broadcast and straw was incorporated during the cultivations following harvest.

The cultivation methods used in the Project were:

Tine cultivations (Flexitine, Paratine & Vi-Till)
Disc/tine combinations (Opico cultivator)
Rotary cultivation (Roterra & Tillerator)
Rotary spading (Rotadig)

PRODUCT	FORMULATION TYPE	ACTIVE INGREDIENT(S)	APPLICATION RATES
Herbicides			
Arelon Liquid	SC	isoproturon (533 g/l)	a 3.75 b 4.0 c 4.2 d 4.5 l/ha
Avadex BW Granular	GR	tri-allate (100 g/kg)	22.5 kg/ha
Avenge 2	SL	difenzoquat (150 g/l)	5.0 l/ha
Avenge 630	SP	difenzoquat (630 g/kg)	a 0.79 b 1.58 kg/ha

Herbicides—*continued*

Benazalox	WP	benazolin (300 g/kg) & clopyralid (50 g/kg)	1.0 kg/ha
Brittox	EC	bromoxynil (75 g/l) & ioxynil (75 g/l) & mecoprop (345 g/l)	3.5 l/ha
Carbetamex	WP	carbetamide (700 g/kg)	3.0 kg/ha
CMPP	SL	mecoprop (570 g/l)	*a* 1.5 *b* 3.0 *c* 3.5 *d* 4.2 l/ha
Commando	EC	flamprop (200 g/l)	3.0 l/ha
Compitox Extra	SL	mecoprop (570 g/l)	4.2 l/ha
Deloxil	EC	bromoxynil (190 g/l) & ioxynil (190 g/l)	*a* 1.5 *b* 2.0 l/ha
Dicurane 500 FW	SC	chlorotoluron (500 g/l)	7.0 l/ha
Dosaflo	SL	metoxuron (500 g/l)	*a* 8.48 *b* 8.75 l/ha
Fortrol	SL	cyanazine (500 g/l)	*a* 0.75 *b* 1.75 *c* 4.5 l/ha
Fusilade 5	EC	fluazifop-P-butyl (125 g/l)	*a* 1.0 *b* 1.5 l/ha
Gesatop 500L	SC	simazine (500 g/l)	2.3 l/ha
Glean C	WP	chlorsulfuron (6 g/kg) & methabenzthiazuron (700 g/kg)	*a* 2.9 *b* 3.5 kg/ha
Gramoxone	SC	paraquat (200 g/l)	3.0 l/ha
Gramoxone 100	SL	paraquat (200 g/l)	*a* 2.0–2.2 *b* 2.7 *c* 3.2 *d* 4.2 l/ha
Hobane	EC	bromoxynil (240 g/l) & ioxynil (160 g/l)	*a* 1.0 *b* 1.5 *c* 1.75 l/ha

Herbicides—*continued*

Hoegrass	EC	diclofop (380 g/l)	3.0 l/ha
Hytane 500	SC	isoproturon (500 g/l)	*a* 2.0 *b* 3.75 *c* 4.2 *d* 5.0 l/ha
Ivosit	LI	dinoseb acetate (523 g/l)	6.0 l/ha
Kerb 50W	WP	propyzamide (500 g/kg)	1.5 kg/ha
Matrikerb	WP	clopyralid (43 g/kg) & propyzamide (430 g/kg)	1.63 kg/ha
Musketeer	SC	ioxynil (50 g/l) & isoproturon (250 g/l) & mecoprop (180 g/l)	6.0 l/ha)
MCPA	SL	mecoprop (476 g/l)	3.0 l/ha
NaTA	SP	TCA-sodium (950 g/kg)	2.7 kg/ha
Oxytril CM	EC	bromoxynil (200 g/l) & ioxynil (200 g/l)	1.5 l/ha
Pradone Plus	WP	carbetamide (525 g/kg) & dimefuron (175 g/kg)	4.1 kg/ha
Roundup	SL	glyphosate (360 g/l)	*a* 1.3–1.6 *b* 2.0 *c* 3.0 *d* 4.0 l/ha
Runcatex CMPP 60	SL	mecoprop (570 g/l)	4.3 l/ha
Springclene 2	SL	benazolin (50 g/l) & ioxynil (50 g/l) & mecoprop (420 g/l)	*a* 4.0 *b* 5.0 l/ha
Starane 2	EC	fluoroxypyr (200 g/l)	*a* 0.75 *b* 1.0 l/ha
Stomp 330	EC	pendimethalin (330 g/l)	6.0 l/ha
Swipe 560 EC	EC	bromoxynil (56 g/l) & ioxynil (56 g/l) & mecoprop (448 g/l)	4.5 l/ha

Herbicides—*continued*

Tolkan Liquid	SC	isoproturon (500 g/l)	*a* 4.2 *b* 5.0 l/ha

Insecticides

Ambush	EC	permethrin (250 g/l)	0.25 l/ha
Ambush C	EC	cypermethrin (100 g/l)	0.25 l/ha
Aphox	SG	pirimicarb (500 g/kg)	*a* 0.28 *b* 3.36 kg/ha
Cypermethrin	EC	cypermethrin (100 g/l)	0.25 l/ha
Decis	EC	deltamethrin (25 g/l)	0.3 l/ha
Demeton-S-methyl	EC	demeton-S-methyl (500 g/l)	0.25 l/ha
Dimethoate 40	LI	dimethoate (400 g/l)	1.7 l/ha
Dursban 4	EC	chlorpyrifos (480 g/l)	1.5 l/ha
Dyfonate 10 G	MG	fonofos (100 g/kg)	16.7 kg/ha
Gamma-Col	SC	gamma-HCH (800 g/l)	0.75 l/ha
Hostathion	EC	triazophos (420 g/l)	*a* 1.0 *b* 1.5 l/ha
Metasystox	LI	demeton-S-methyl (570 g/l)	*a* 0.21 *b* 0.42 *c* 0.48 l/ha
Metasystox 55	EC	demeton-S-methyl (500 g/l)	0.21 l/ha
Spannit	EC	chlorpyrifos (480 g/l)	*a* 1.5 *b* 4.2 l/ha

APPENDIX II (continued)

Growth regulators

Arotex Extra	SL	chlormequat chloride (644 g/l) & choline chloride (32.2 g/l)	1.0 l/ha
CCC	SL	chlormequat chloride (*)	a 1.25 b 2.5 l/ha
Chlormequat	SL	chlormequat chloride (*)	1.75 l/ha
Chlormequat 5C	SL	chlormequat chloride (*)	a 1.5 b 2.5 l/ha
Cycocel	SL	chlormequat chloride (*)	a 0.75 b 1.0 c 2.5 l/ha
5C Cycocel	SL	chlormequat chloride (645 g/l)	a 1.25 b 1.5 c 1.75 l/ha

Molluscicide

Draza	PT	methiocarb (200 g/l)	a 2.2–2.7 b 5.5–5.6 kg/ha

Dessicant

Reglone 40	SL	diquat (200 g/l)	3.0 l/ha

Bird repellent

Guardsman	SC	aluminium ammonium sulphate (83 g/l)	25.0 l/ha

Wetters

Agral	LI	alkyl phenol ethylene oxide condensate	various
Frigate	LI	fatty amine ethoxylate	various
PBI Spreader	LI	nonyl phenol ethylene oxide condensate	various

* various concentrations in the range 400–725 g/l

Fungicides

Name	Formulation	Active ingredient	Rate
Bayleton	WP	triadimefon (250 g/kg)	*a* 0.25 *b* 0.5 kg/ha
Bayleton CF	WP	captafol (650 g/kg) & triadimefon (62.5 g/kg)	2.0 kg/ha
Baytan	DS	fuberidazole (30 g/kg) & triadimenol (250 g/kg)	seed dressing
Benlate	WP	benomyl (500 g/kg)	*a* 0.5 *b* 1.0 kg/ha
Bolda FL	SC	carbendazim (50 g/l) & maneb (320 g/l) & sulphur (100 g/l)	5.0 l/ha
Bravo 500	SC	chlorothalonil (500 g/l)	2.0 l/ha
Captafol	SC	captafol (480–500 g/l)	2.8 l/ha
Corbel	EC	fenpropimorph (750 g/l)	*a* 0.5 *b* 1.0 l/ha
Delsene M	WP	carbendazim (50 g/kg) & maneb (320 g/kg)	2.5 kg/ha
Delsene M Flowable	SC	carbendazim (50 g/l) & maneb (320 g/l)	5.0 l/ha
Derosal Liquid	SC	carbendazim (511 g/l)	0.5 l/ha
Dorin	EC	triadimenol (125 g/l) & tridemorph (375 g/l)	1.0 l/ha
Impact Excel	SC	chlorothalonil (246 g/l) & flutriafol (39 g/l)	2.0 l/ha
Kombat	WP	carbendazim (100 g/kg) & mancozeb (540 g/kg)	2.8 kg/ha
Mistral	EC	fenpropimorph (750 g/l)	*a* 0.5–0.66 *b* 0.75 *c* 1.0 l/ha

Fungicides—*continued*

Radar	EC	propiconazole (250 g/l)	0.5 l/ha
Ringer	EC	tridemorph (750 g/l)	0.23 l/ha
Rovral Flo	SC	iprodione (250 g/l)	2.0 l/ha
Sportak	EC	prochloraz (400 g/l)	1.0 l/ha
Sportak Alpha	EC	carbendazim (100 g/l) & prochloraz (266 g/l)	1.5 l/ha
Sprint	EC	fenpropimorph (375 g/l) & prochloraz (225 g/l) & xylene	1.5 l/ha
Tilt 250 EC	EC	propiconazole (250 g/l)	a 0.25 b 0.5 l/ha
Tilt Turbo 375 EC	EC	propiconazole (125 g/l) & tridemorph (250 g/l)	1.0 l/ha

Tables of cultivations and agrochemicals

1981–82	September	October	November	December	January
Grange Piece	**29** Gramoxone 100 *b*	**12** Gramoxone 100 *b* (h) **14** drill **28** Gesatop			**20–22** Carbetamex
Backside	**7–10** Paratine **23–28** disc **29** harrow (h)	**3** drill **14** Stomp	**10** Draza *a* (half field)		
Pamplins	**4** disc, roll **7–9** drill, harrow NaTA, harrow	**15** N fertiliser **15–17** Gamma-Col **26** Pradone Plus			
Shackles Aden		**23** drill			
Top Pavements	**23** drill, harrow	**13** Draza *a*	**2** Tolkan *a* CMPP *d*		
Thorofare	**25–30** plough **29–30** disc	**1** plough **1–12** disc **14–15** drill			
Knapwell (Field)	**24** Flexitine	**16** Gramoxone 100 *b* **16–17** drill			
(Pasture)	**18** Rotadig **22** disc	**15** Gramoxone 100 *b* **16** drill, harrow			
Bushes & Pits		**15–16** drill, harrow **29** N Fertiliser P_2O_5			
Extra Close	**17–20** plough, disc	**3** harrow **3–14** drill			
Eleven Acre Extra	**17** Flexitine **21–22** plough **23** disc	**9** drill, harrow	**9** Tolkan *a*		

February	March	April	May	June	July	August
		3 Ivosit		9 Benlate *b* 16 Tilt *a*		3–5 harvest 10 bale straw
	24 N fertiliser	5–6 Runcatex Oxytril CM 14 N fertiliser 15 Avenge, Derosal CCC *a*	25–26 Tilt *b*	15 Tilt *a*		(P) Roundup *d* (half field) 17 harvest 26 bale straw 24–30 plough
Carbetamex N fertiliser	25 Benlate *b* 26–27 N fertiliser			11 Hostathion *a* Rovral Flo Reglone	19–26 harvest 26–27 baled 29–30 mole drain	1 Draza *b* 6–14 Rotadig 26–27 disc, harrow
Hytane *d* Tolkan *a*	24 N fertiliser	5 Oxytril CM Runcatex 16 Avenge 630 *a* Derosal 20 N Fertiliser		2 Delsene M		(P) Roundup *d* 10–11 harvest 23 burn straw
	9 N fertiliser P_2O_5 20 Runcatex (h) Oxytril CM (h)	13 Runcatex Oxytril CM 15 N fertiliser 16 Avenge *a* Derosal, CCC *a*	25 Delsene M Bayleton *a*	14 Sportak		20–21 harvest 28 burnt
Tolkan *a* Runcatex Oxytril CM	9 N fertiliser P_2O_5	13 Runcatex Oxytril CM 14 N fertiliser 15 Avenge 630 *a* Derosal, CCC *b*	25 Delsene M	3 Delsene M Bayleton *a* (west) 14 Bayleton CF (east)		(P) Roundup *d* 19–20 harvest
Tolkan *a*	8 N fertiliser P_2O_5	16 Runcatex Oxytril CM 19 Avenge 630 *b* Derosal, CCC *a* 20 N fertiliser		1 Delsene M		6, 14 Roundup *d* 22–23 harvest
Tolkan *a*	8 N fertiliser	13 Runcatex Oxytril CM 17 Avenge 630 *a* Derosal, CCC *a* N fertiliser, P_2O_5		2 Delsene M		(P) Roundup *d* 23 harvest
Tolkan *a*	8 N fertiliser P_2O_5	5 Runcatex Oxytril CM 13 N fertiliser 14 Avenge 630 *a* Derosal, CCC *a*	7 N fertiliser 28 Tilt *b*	15 Tilt *a*		12–21 harvest 23 bale straw
Tolkan *a* (h) Brittox (h)	15 N fertiliser P_2O_5	15 N fertiliser 16 Avenge 630 *b* Derosal, CCC *a*	26 Delsene M	14 Bayleton CF		23–28 harvest
	8 N fertiliser P_2O_5	5 Runcatex Oxytril CM 13 N fertiliser 15 Avenge 630 *a* Derosal, CCC *b*	28 Delsene M Bayleton *a*	11 Bayleton CF		(P) Roundup *d* 17–19 harvest 24 bale straw

1982–83	September	October	November	December	January
Grange Piece	**16** Roundup *c*	**12** Draza *b* **19,25** Gramoxone 100*d* **27** drill **28** harrow			
Backside	**27** disc **28** harrow	**15** drill **18** harrow	**11** Draza *b*		
Pamplins		**15–16** drill **18** harrow	**13** Swipe		
Shackles Aden		**15** Gramoxone 100 *c* **28** drill **29** harrow			
Top Pavements		**1** Gramoxone 100 *a* **12–16** drill	**5** Hytane *a* **11** Ambush		
Thorofare		**1** Gramoxone 100 *a* **11** Draza *b* **20** drill	**11** Glean C *b*		
Knapwell (Pasture)	**17–20** Rotadig	**1** disc **15** Draza *b*	**30** Glean C *b*	**3,7,11** drill **12** Gramoxone 100 *a*	
(Field)	**24–25** Flexitine	**12** Draza *b* **19** drill	**1** drill		
Bushes & Pits	**28** Flexitine	**1** Flexitine **20** drill **20,28** harrow			**10** Roundup *c*
Extra Close	**9** drill (east) **11–12** lime **17** Dyfonate (trials area)	**10** Gramoxone 100 *c* **12** drill (west)	**5** Glean C *a* **11** Ambush		
Eleven Acre Extra	**13** Rotadig	**2** disc **20,27** drill, harrow	**11** Draza *b*		

February	March	April	May	June	July	August
	11 N fertiliser **12** Glean C *b* Dosaflo *b*	**29** CMPP *d* Oxytril CM Derosal 5C Cycocel *a* **30** N fertiliser	**25** Avenge 630 *a*	**18** Bayleton CF		**16–17** harvest
N Fertiliser	**9** N fertiliser Dosaflo *b*	**30** CMPP *d* Oxytril CM Derosal 5C Cycocel *c*	**5** N fertiliser **26** Avenge 630 *b*	**20** Kombat	**2** Aphox *a*	**9** harvest
	16 N fertiliser Hytane *c*	**15** CMPP *d* Oxytril CM Kombat **27** N fertiliser	**5** Avenge 630 *a* 5C Cycocel *c*	**17** Bayleton CF	**2** Aphox *a*	**23** harvest
	8 N fertiliser **12** Musketeer	**29** CMPP *d* Oxytril CM Derosal 5C Cycocel *a* **30** N fertiliser	**20** Avenge 630 *a*	**18** Bayleton CF		**15–16** harvest
	9 N fertiliser **18** Hytane *a*	**28** Swipe, Derosal 5C Cycocel *a* **29** N fertiliser	**25** Avenge 630 *a*	**10** Bayleton CF **28** Aphox *b*		**11–12** harvest
	8 Dosaflo *b* **9** N fertiliser	**28** Swipe, Derosal 5C Cycocel *c* **30** N fertiliser	**27** Avenge 630 *a*	**13** Bayleton CF **28** Aphox *a*		**10–19** harvest
	10 N fertiliser **11** Dosaflo *b* Dicurane	**30** CMPP *d*, Derosal	**5** N fertiliser **27** Avenge 630 *b* 5C Cycocel *a*	**21** Delsene M Bayleton CF		**24** harvest
	11 N fertiliser DIcurane	**30** CMPP *d*, Derosal 5C Cycocel *a*	**5** N fertiliser **26** Avenge 630 b	**21** Delsene M		**24** harvest
N fertiliser	**11** Hytane *c*	**28** N fertiliser Swipe, Derosal 5C Cycocel *a*		**20** Bayleton		**19–20** harvest
	9 N fertiliser	**16** CMPP *d*, Derosal 5C Cycocel *c* **28** N fertiliser	**20** Avenge 630 *b*	**13** Bayleton CF		**13–15** harvest
N fertiliser	**11** Tolkan *b* Hytane *c*	**26** Springclene 2 *a* Derosal **28** N fertiliser	**25** Avenge 630 *a*	**20** Kombat Bayleton CF **28** Aphox *a*		**19** harvest

1983–84	September	October	November	December	January
Grange Piece	**5–9** harrow mole drain Tillerator **9** P$_2$O$_5$	**4** Draza *b* **6** Gramoxone 100 *a* **8** drill (Baytan) **8–11** cross-harrow **24** Avadex	**15** Ambush C Dosaflo *b* **19** Compitox Extra Spannit *b*		
Backside	**8** P$_2$O$_5$ **12,20** Flexitine **26** Draza *b* **28** Gramoxone 100 *a*	**1** drill (Baytan) harrow **22** Avadex	**14** Ambush C Hytane *d* **19** Compitox Extra Spannit *a*		
Pamplins	**7** Flexitine **8** P$_2$O$_5$ **9** Flexitine **24** Draza *b* Gramoxone 100 *a* **26** drill (Baytan) cross-harrow	**18** Avadex	**11** Ambush C Hytane *d* **17** Compitox Extra Spannit *a*		
Shackles Aden	**8** harrow drill (Baytan) **10** drill (Baytan) P$_2$O$_5$ **12** cross-harrow **28** Flexitine (h)	**4** Draza *b* **8** Gramoxone 100 *a* **24** Avadex	**15** Ambush c Arelon *b* **30** Compitox Extra Spannit *a*		
Top Pavements	**7–8** Flexitine **23** Gramoxone 100 *a* **24** Draza *b* **26** drill	**18** Avadex	**12** Dosaflo *b* Compitox Extra		
Thorofare	**10** mole drain **10–12** Flexitine **20** mole drain **24,26** Flexitine **28** Draza *b* **30** drill cross-harrow	**20** Avadex	**12** Dosaflo *b*		
Knapwell	**30** Gramoxone 100 *a*	**4** Draza *b* **5** Vi-till, Flexitine **6** disc, Flexitine P$_2$O$_5$ **11** drill cross- harrow **21** Avadex			
Bushes & Pits	**9,25** Flexitine **30** disc (h)	**4** Vi-till, Draza *b* **6** drill **8** cross-harrow			

(continued overleaf)

February	March	April	May	June	July	August
24 Hostathion *b*	**8** N fertiliser	**9** 5C Cycocel *b* Swipe **19** N fertiliser **26** Avenge 630 *a* Sportak Alpha		**8** Bravo 500 Corbel *b* **18** Metasystox 55 **19** Bayleton *b* Captafol	**4** Delsene M Kombat Metasystox 55	**9** Roundup *a* **27–28** harvest **28** bale straw
	1 Hostathion *b* **15** N fertiliser	**9** 5C Cycocel *b* **13** Avenge 630 *a* Sportak Alpha **18** N fertiliser **19** Swipe		**8** Bravo 500 Corbel *b* **19** Bayleton *b* Captafol Metasystox 55	**6** Kombat Metasystox 55	**8** Roundup *a* **24–26** harvest **30** plough (h) burn straw, Roterra, roll
17 Hostathion *b*	**15** N fertiliser	**4** 5C Cycocel *b* Springclene 2 *b* **14** Avenge 630 *a* Sportak Alpha urea	**31** Bravo 500 Corbel *b*	**18** Bayleton *b* Captafol Metasystox 55	**5** Delsene M Metasystox 55	**8** Roundup *a* **17–18** harvest **18–21** Opico cultivator **21** disc
24 Hostathion *b*	**8,14** N fertiliser	**12** Arotex Extra Swipe **26** Avenge 630 *a* Sportak Alpha		**4** Bravo 500 Corbel *b* **19** Bayleton *b* Captafol Metasystox 55	**4** Kombat Metasystox 55	**8** Roundup *a* **28–29** harvest
	10 N fertiliser	**5** 5C Cycocel *b* **14** urea **24** Derosal		**6** Radar **24** Metasystox 55		**16–17** harvest **18** incorporate straw **20** disc
	14 N fertiliser	**5** 5C Cycocel *b* **18** N fertiliser **24** Derosal		**6** Radar, Tilt *b* **28** Metasystox 55		**12** P_2O_5 **22–24** harvest **25** burn straw **28** plough (h) **30** Roterra **31** roll
	7 Dosaflo *a* **8** N fertiliser	**13** Arotex Extra Swipe **19** N fertiliser **27** Avenge 630*a*	**3** Derosal	**9** Radar **23** Metasystox 55		**26–27** harvest **31** bale straw
	6 Arelon *a* **15** N fertiliser	**9** 5C Cycocel *b* **18** N fertiliser	**3** Derosal	**6** Tilt *b* **24** Aphox *a* **29** Corbel *a* Delsene M		**9** Roundup *a* **21** harvest **23** plough (h) **24** burn straw **25** disc (h) **31** Roterra, roll

1983–84 (continued)	September	October	November	December	January

Extra Close East	**23** Flexitine	**4** Vi-till
		5 Draza *b*
		8 drill
		   cross-harrow
		25 Avadex
			30 Compitox Extra
			   Dosaflo *b*		

Extra Close West	**7,26** Flexitine	**4** Vi-till **5** Draza *b* **8** drill cross-harrow		**1** Dosaflo *b*	
Eleven Acre Extra	**25** Flexitine **30** disc (h)	**5** Vi-till, Draza *b* **6** drill **7** cross-harrow		**1** Compitox Extra	

bruary	March	April	May	June	July	August
N fertiliser	**9** 5C Cycocel *b* **20** N fertiliser **25** Derosal		**8** Tilt *b* **24** Aphox *a* (part field)	**7** Aphox *a* (part field)	**26** harvest **28** bale straw	
N fertiliser	**5** 5C Cycocel *b* Springclene 2 *b* **14** urea **24** urea **25** Avenge 630 *b* Derosal		**6** Radar **24** Aphox *a*		**18** harvest **21** incorporate straw **22** disc	
N fertiliser	**4** 5C Cycocel *b* Springclene 2 *a* **20** N fertiliser **25** Derosal		**6** Tilt *b* **24** Aphox *a* **29** Corbel *a* Delsene M		**21** harvest **24** burn straw **25** plough (h) Roterra **31** roll	

1984–85	September	October	November	December	January
Grange Piece	**4** plough **6** disc, plough **14** roll, plough	**1** Draza *b* **8** roll (landpack) **9** Gramoxone **11** drill (Baytan) cross-harrow	**2** Avadex	**10** Ambush C Arelon *c*	
Backside	**6** drill **8** roll **28** Draza *b*	**13** Fusilade *a* + PBI Spreader			**2** Matrikerb
Pamplins	**28** Draza *b*	**1** Gramoxone **2** disc **4** cross-harrow **4–5** drill (Baytan) **9–11** cross-harrow	**2** Avadex	**8** Ambush C Dosaflo *b*	
Shackles Aden	**8,10** bale straw **11** bale straw plough (h) **12** Roterra (½ field) **19,26** plough (½ field) **27** disc	**1** disc **3** Gramoxone **4** disc **8–9** roll **10** direct drill (4.1 ha) **11** Draza *b* **12** drill (Baytan) **13** roll, harrow		**11** Avadex **13** Ambush C Arelon *c* Avadex	
Top Pavements		**1** Gramoxone **3** P$_2$O$_5$, disc **4** drill **5** cross-harrow **17** Avadex	**5** Dosaflo *b*		
Thorofare	**5** drill **7** roll	**12** Fusilade *a* + PBI Spreader			
Knapwell	**3,5** bale straw **6** plough **8** disc **12** plough **13–14** plough, disc **17** plough **19** disc **27–28** plough	**1,8** disc **5** P$_2$O$_5$ **9** cross-harrow **10** drill, cross- harrow **16** roll		**10** Dosaflo *b* **13** Avadex	
Bushes & Pits	**6** drill **7** roll	**2** P$_2$O$_5$ **14** Fusilade *a*			**3** Kerb

(continued overleaf)

February	March	April	May	June	July	August
	9 Hostathion *b* **15** N fertiliser	**17** Deloxil *b* Starane 2 *b* **18** N fertiliser **30** Avenge 630 *a* Chlormequat 5C *a* Sportak Alpha	**2** N fertiliser **31** Bravo 500 Corbel *b*	**18** Bayleton CF *b* Metasystox *a*	**5** Delsene M Flowable **10** Metasystox *a*	
	9 N fertiliser **10** Sportak	**4** Benazalox N fertiliser	**29** Rovral Flo	**17** Hostathion *b* Rovral Flo	**23** Reglone	**6** harvest **17,20** roll
	9 Hostathion *b* **13** N fertiliser	**17** N fertiliser **20** Deloxil *b* Starane 2 *b* **30** Avenge 630 *a* Chlormequat 5C *a* Sportak Alpha	**2** N fertiliser **3** Avenge 630 *a* Chlormequat 5C *b* Sportak Alpha	**1** Bravo 500 Corbel *b* **18** Bayleton *b* Captafol Metasystox *a*	**4** Delsene M Flowable **10** Metasystox *a*	**29–31** harvest
	9 Hostathion *b* **15** N fertiliser	**18** Deloxil *b* N fertiliser Starane 2 *b* **30** Avenge 630 *a* Chlormequat 5C *a* Sportak Alpha	**2** N fertiliser **30** Bravo 500 Corbel *b*	**18** Bayleton *b* Captafol Metasystox *a*	**5** Delsene M Flowable **10** Metasystox *a*	
	13 N fertiliser	**17** N fertiliser Starane 2 *b* (h)	**1** N fertiliser **6** Chlormequat 5C *b* Sportak Alpha	**1** Radar	**2** Bayleton *a* Delsene M Flowable	
27 Fortrol *a*, Kerb	**6** N fertiliser	**3,4** N fertiliser	**6** Benlate *b*		**23,31** Reglone	**3–7** harvest **21,23** roll **24** Rotadig (h)
	14 N fertiliser	**17** Deloxil *b* Starane 2 *b* **20** N fertiliser	**3** Avenge 630 *a* + PBI Spreader Chlormequat 5C *b* Sportak Alpha **30** Radar		**2,3** Bayleton *a* Delsene M Flowable	
	8 N fertiliser	**3** N fertiliser **17** Benazalox (h)				**5** harvest **20** Rotadig (h) **21** roll **27** roll

1984–85 (continued)	September	October	November	December	January
Extra Close East	**2–4** plough	**4** P$_2$O$_5$ **5** disc **8** harrow **11** drill **12** cross-harrow **16** roll		**9** Arelon *a* Starane 2 *b*	
Extra Close West		**4** Gramoxone P$_2$O$_5$ **5** disc **8** drill **9** cross-harrow		**8** Arelon *a* Starane 2*b*	
Eleven Acre Extra	**7** drill, roll **14** P$_2$O$_5$	**12** Fusilade *a* + PBI Spreader			**3** Kerb

bruary	March	April	May	June	July	August
	15 N fertiliser	**17** Sportak Starane 2 *b* (h) **18** Starane 2*b* **21** N fertiliser	**1** N fertiliser	1 Radar	**2** Bayleton *a* Delsene M Flowable	**23** rotovate (h) **29** Roundup *b*
	13 N fertiliser	**17** N fertiliser Starane 2 *b* (h) **18** Starane 2 *b* **20** Chlormequat 5C *b*, Sportak	**1** N fertiliser **4** Avenge 630 *a*	1 Radar	**2** Bayleton *a* Delsene M Flowable	**23** rotovate (h)
7 Fortrol *a* (h)	**8** N fertiliser **9** Benlate *a* + PBI Spreader	**3** N fertiliser **17** Benazalox (h)				**7** harvest **21** roll **23** Rotadig (h)

1985–86	September	October	November	December	January

Grange Piece	**7, 9** harvest
	11,12,17 bale
	  straw
	24 plough
	26 plough, disc
	27 disc
	28 roll
		8 rotovate (h)
		  Rotadig (h)
		11 disc
		12 drill
		15 harrow
		18 roll, Draza *b*
			29 Avadex	**17** Dosaflo *b* (plots
				  only)	

Backside		**2** disc **8** Roterra **9** Vi-till, drill (Baytan) **10–11** harrow, roll **12** roll **17** Draza *b*		**17** Avadex	
Pamplins	**1–2** harvest (continued) **29–30** roll	**5** disc **12** roll **14–15** disc **15** drill (Baytan) **16** harrow, roll **19** Draza *b*		**17** Avadex	
Shackles Aden	**9–10** harvest **12,15** bale straw **23–26** plough **26–30** disc	**1** disc, roll (landpack) **8** rotovate (h) **11** drill **12** harrow **16** roll **18** Draza *a*		**30** Avadex	
Top Pavements	**2,6** harvest **25** roll	**3** disc **4–5** roll **7** Rotadig (h) **13** disc **14** drill **15** harrow **16** roll		**20** Avadex	
Thorofare		**2** disc, roll **3** rotovate (h), disc, roll **9** disc, Vi-till **10** disc, Vi-till, drill **11–12** harrow **14–15** roll			
Knapwell	**10–11** harvest **16** bale straw **21–24** plough **27–28** disc, roll **28–29** plough (h) disc (h) **29–30** roll, disc	**11** drill **14** harrow **15** roll		**20** Avadex	

(continued overleaf)

February	March	April	May	June	July	August
	15 N fertiliser	**23** Hostathion *b* **29** Dosaflo *b* **30** N fertiliser	**1** Deloxil *b* Starane 2 *b* **13** N fertiliser **29** Avenge 630 *a* + PBI Spreader Sportak Alpha	**7** Bravo 500 Mistral *c* **27** Bayleton CF *b* Metasystox *b*	**14** Bolda Demeton- S-methyl	**9** Roundup *d*
	18 Arelon *a*	**2** N fertiliser	**1** Deloxil *b* Dimethoate Starane 2 *b* **2** N fertiliser **19** Chlormequat Sportak Alpha **30** Avenge 630 *a* + PBI Spreader	**7** Bravo 500 Mistral *c* **27** Bayleton CF *b* Metasystox *b* **30** Metasystox *b*	**14** Bolda Demeton- S-methyl	**25** Roundup *d*
N fertiliser		**29** N fertiliser **30** Dosaflo *b*	**2** Deloxil *b* Dimethoate Starane 2 *b* **19** Avenge 630 *a* + PBI Spreader Chlormequat Sportak Alpha	**7** Bravo 500 Mistral *c* **27** Bayleton CF *b* Metasystox *b*	**14** Bolda Demeton- S-methyl	**9** Roundup *d*
	3 N fertiliser **17** Ambush C (bottom end only)	**29** Dosaflo *b* **30** N fertiliser	**1** Deloxil *b* Dimethoate Starane 2 *b* **12** N fertiliser **30** Avenge 630 *a* + PBI Spreader Sportak Alpha	**7** Bravo 500 Mistral *c*	**1** Bayleton CF *b* Metasystox *b* **14** Bolda Demeton- S-methyl	**9** Roundup *d*
	17 N fertiliser	**29** Dosaflo *b*	**1** N fertiliser **3** CMPP *c* Deloxil *a* **13** N fertiliser **19** Chlormequat	**7** Sportak **25** Radar, Ringer		**8** Roundup *d*
		11 N fertiliser **30** Deloxil *b* Starane 2 *b*	**2** N fertiliser **16** Chlormequat	**7** Sportak **26** Radar, Ringer		**12** Roundup *d*
	18 N fertiliser	**29** Dosaflo *b* **30** MCPA Starane 2 *b*	**1** N fertiliser	**8** Mistral, *c* Sportak **26** Radar, Ringer		**9** Roundup *b*

1985–86 (continued)	September	October	November	December	January
Bushes & Pits		**2** roll **18** drill, harrow			
Extra Close East	**11** harvest **17** bale straw **19–20** plough **26** roll	**3–4** disc **5** roll **7** Rotadig (h) **19** drill, harrow, roll			
Extra Close West	**1** harvest **26** roll	**3–4** disc **5** roll **7** Rotadig (h) **19** drill, harrow, roll		**20** Avadex	
Eleven Acre Extra		**2** Roterra, roll **18** drill **19** harrow			

ebruary	March	April	May	June	July	August
	18 N fertiliser		**1** Deloxil *b* Starane 2 *b* **2** N fertiliser **19** Chlormequat	**7** Sportak	**7** Delsene M Flowable Mistral *a*	
	17 N fertiliser **18** Arelon *a*	**30** Deloxil *b* Starane 2 *b*	**2** N fertiliser **19** Chlormequat	**7** Sportak	**8** Delsene M Flowable Mistral *a*	
	17 N fertiliser **18** Arelon *a*	**30** CMPP *d* Deloxil *b*	**2** N fertiliser **19** Chlormequat	**7** Sportak	**8** Delsene M Flowable Mistral *a*	**9** Roundup *b*
	18 N fertiliser	**30** Deloxil *b* Starane 2 *b*	**2** N fertiliser **16** Chlormequat	**7** Sportak	**7** Delsene M Flowable Mistral *a*	

1986–87	September	October	November	December	January
Grange Piece	**6** harvest **18–20** plough **22** disc **23–24** roll **29** P_2O_5	**5** disc **6–7** Roterra **8** drill **9** harrow	**7** Draza *b* **12** Avadex	**24** Spannit *a*	
Backside	**11** harvest **17–18** Opico cultivator **23** Flexitine **27** P_2O_5	**1** disc **3** Roterra, drill **4** harrow **7** roll	**7** Draza *b*	**1** Avadex	**8** Spannit *a*
Pamplins	**1–2** harvest **4** burn straw **6–7** Roterra **8** roll **9–10** drill, roll **19** Draza *b* **30** P_2O_5		**28,30** Fusilade *b* + Agral	**23** Decis	
Shackles Aden	**7** harvest **19–22** plough **22–24** disc **27** roll **29** roll P_2O_5	**4–5** disc **8–9** Roterra **14** Roterra **29–30** drill, harrow	**11** Draza *b* **12** Avadex		**8** Spannit *a* (h)
Top Pavements	**4–5** harvest **7** burn straw Vi-till, Roterra (h), drill **8** harrow, drill **9** drill, roll		**28** Fusilade *b* + Agral		
Thorofare	**8–9** harvest **15–16** Opico cultivator **24** Flexitine **24–25** disc **29** roll	**2** rotovate (h) **3** Roterra **4** Roterra, drill harrow **5** drill, harrow **6** roll			
Knapwell	**9–10** harvest **12–13** plough, disc **15–16** plough **18–19** disc **24–25** disc **29** disc **30** roll	**5–6** Roterra **7–8** drill, harrow roll		**3** Avadex	

(continued overleaf)

February	March	April	May	June	July	August
	10 N fertiliser **11–12** Hostathion *b* **14** Arelon *a*	**15** Hobane *b* N fertiliser Starane 2 *a* **17** Sportak Alpha **24** Avenge 630 *b* **28** N fertiliser	**26** Impact Excel	**29** Bravo 500 Metasystox *b* Mistral *b*	**3** N fertiliser **7** Delsene M Flowable **13** Metasystox *b*	
	12 Hostathion *b* **13** N fertiliser **15–16** Arelon *a*	**14** N fertiliser **15** Chlormequat Hobane *b* Starane 2 *a* **17** Sportak Alpha **23** Avenge 2 Cycocel *a* **28** N fertiliser	**26** Impact Excel	**29** Bravo 500 Metasystox *b* Mistral *b*	**7** Delsene M Flowable **13** Metasystox *b*	**12** Roundup *d*
6 Guardsman **8** N fertiliser **1** Guardsman		**13** N fertiliser **14** Sportak Alpha	**8** Rovral Flo	**11** Hostathion *b*	**17** Reglone + Agral (top 8 ha) **25** harvest (top 8 ha)	**4** Reglone + Agral (bottom 5 ha) **14** harvest (bottom 5 ha) plough **15** plough, disc **16** plough **17** plough, disc **18–19** disc
	10 Hostathion *b* **11** N fertiliser **15** Arelon *a*	**15** Hobane *c* N fertiliser Starane 2 *a* **17** Cycocel *b* Sportak Alpha **24** Avenge 630 *b* **28** N fertiliser	**26** Impact Excel	**29** Bravo 500 Metasystox *b* Mistral *b*	**3** N fertiliser **7** Delsene M Flowable **13** Metasystox *b*	
21 Kerb **24** Kerb	**19** N fertiliser **31** N fertiliser		**25** Radar	**21** Radar		**1–2** harvest **5** plough **6–8** disc
	14 Arelon *a* **16** N fertiliser	**14** Chlormequat CMPP *c* **15** N fertiliser **16** Sportak **24** Cycocel *a* **29** N fertiliser				
	10 N fertiliser **16** Dosaflo *b*	**14** CMPP *d* Hobane *a* **15** N fertiliser **16** Cycocel *b* Sportak **29** N fertiliser	**25** Radar	**24** Radar	**3** N fertiliser	**19–20** harvest **22–24** plough

1986–87
(continued)

	September	October	November	December	January
Bushes & Pits	**8** harvest **15** Opico cultivator **23** Flexitine **24** disc **25–26** roll	**4** Roterra **6** drill, harrow **7** roll			
Extra Close East	**5–6** harvest **18–19** plough **20,23** disc	**6** disc **8** Roterra **9** drill **10** harrow **13** roll			
Extra Close West	**3–4** harvest **6** burn straw Vi-till **7** Roterra (h) **8** roll **10–11** drill, roll		**21,27** Fusilade *b* + Agral		
Eleven Acre Extra	**7** harvest **13** Opico cultivator **24** Flexitine **27** roll	**4** Roterra **6** drill, harrow **7** roll			

February	March	April	May	June	July	August
	14 Arelon *a* **16** N fertiliser	**14** N fertiliser **15** Chlormequat CMPP *d* Hobane *b* **23** Cycocel *a* **27** N fertiliser	**26** Radar	**26** Radar		
	13 N fertiliser **14** Arelon *a*	**14** Chlormequat CMPP *d* Hobane *b* N fertiliser **23** Cycocel *a* **28** N fertiliser	**25** Radar	**26** Radar	**3** N fertiliser	
	20 N fertiliser **31** N fertiliser			**11** Rovral Flo		**2–3** harvest **6–7** plough **11** disc
	14 Arelon *a* **16** N fertiliser	**14** N fertiliser **15** Chlormequat CMPP *d* Hobane *a* **23** Cycocel *a* **27** N fertiliser	**26** Radar	**21** Radar		

1987–88	September	October	November	December	January
Grange Piece	**3–4** harvest **6–9** plough **10–11** drill **28** Roterra **29** Roterra, drill harrow **30** roll	**14** Draza *b* **26** Avadex	**7** Cypermethrin Dosaflo *b*	**5** Dursban	
Backside	**14** harvest **17–19** plough **24** disc	**2–3** disc **5–6** Roterra, roll **6** drill, harrow	**5** Avadex **6** Draza *b* **7** Dosaflo *b* **10** Arelon *d* Cypermethrin	**5** Dursban	
Pamplins	**24–25** Vi-till **26** harrow **28** drill	**14** Draza *b* **26** Avadex	**6** Arelon *d* Cypermethrin	**5** Dursban	
Shackles Aden	**11–12** harvest **21–24** plough **25** plough, disc **28** disc **30** plough	**2** disc	**5** Roterra **6–7** drill Draza *b* (i) harrow		
Top Pavements	**23–24** Vi-till **26** harrow **28–29** drill	**1** roll **6** Draza *b*			
Thorofare	**10–11** harvest **12–13** plough **14** plough, disc **15** plough **16** plough, disc **21–24** disc **26–27** disc	**1** disc **2** disc, Roterra, roll **3** Roterra, roll **5** Roterra **7–8** drill **19–20** Draza *b* **29** Avadex		**4–5** Fortrol *c*	
Knapwell	**3,10,12,25** disc **26** disc, Roterra **27** Roterra **29–30** Roterra, drill, harrow	**2** roll	**4–5** Avadex **7** Dosaflo *b*		

(continued overleaf)

February	March	April	May	June	July	August
3 Hostathion b	8 N fertiliser	5 N fertiliser 13 CMPP d Oxytril CM 17 N fertiliser 19 Cycocel c Sportak 24 Mn sulphate + Agral	6 Commando 17 Impact Excel	15 Bravo 500 Mistral c 16 Metasystox b	8 N fertiliser 9 Delsene M Flowable 19 Metasystox c	16 harvest 25 burn straw
3 Hostathion b	9 N fertiliser	13 CMPP c Deloxil a 17 N fertiliser 19 Cycocel c Sportak	7 Commando 17 Impact Excel	14 Bravo 500 Mistral c 16 Metasystox b	9,12 Delsene M Flowable 18 Metasystox c	18 harvest 25 burn straw
9 Hostathion b	7 N fertiliser	11 CMPP d Oxytril CM 13 N fertiliser 17 Cycocel c Sportak 24 Mn sulphate + Agral	6 Commando 17 Impact Excel	14 Bravo 500 Mistral c 16 Metasystox b	9 Delsene M Flowable 20 Metasystox c	23 harvest 26 burn straw
	2 Hostathion b 11 N fertiliser	2 Arelon a 19 CMPP d Oxytril CM 21 Cycocel c Sportak 25 Mn sulphate + Agral 30 N fertiliser	18 Avenge 630 b 27 Impact Excel	15 Bravo 500 Mistral c 16 Metasystox b	8 N fertiliser 19 Metasystox c	(harvest in September)
7 P₂O₅ 9 Arelon a Hoegrass	7 N fertiliser	3 Sportak 12 CMPP d Oxytril CM 13 N fertiliser 17 Cycocel c	20 Radar	13 Dorin Metasystox b		10 Roundup b + Frigate 29 harvest
7 P₂O₅ 9 Fortrol b Hytane a	8 N fertiliser	2 Sportak 17 CMPP b Cycocel c 19 N fertiliser 24 Mn sulphate + Agral	21 Radar	13 Dorin Metasystox b		(harvest in September)
18 P₂O₅	9 N fertiliser	3 Sportak 6 N fertiliser 18 N fertiliser 25 CMPP d Cycocel c	21 Tilt Turbo	13 Bravo 500 Metasystox b Mistral b	8 N fertiliser	10 Roundup b + Frigate 17 harvest 23 burn straw

1987–88 (continued)	September	October	November	December	January
Bushes & Pits	**9** harvest **20–21** plough **26–27** plough **28–30** disc	**1** disc **29** drill, Draza *b* (i)			
Extra Close East	**4–6** harvest **7–8** plough **14** disc **17** plough **22** disc **26** harrow **28** disc		**5** drill, Draza *b* (i)		
Extra Close West	**21** Vi-till	**5–6** drill Draza *b* (i) harrow			
Eleven Acre Extra	**9** harvest **14** plough **17** disc **18,21** plough **22** disc	**13–14** Vi-till **14** drill, Draza *b* (i)		**7** Arelon *d* **8** Avadex (h)	

ebruary	March	April	May	June	July	August
3 P_2O_5	10 N fertiliser	2 Arelon *a* 21 N fertiliser 26 CMPP *c* Oxytril CM	6 CMPP *c* 13 Commando 20 Sprint	16 Dorin		(harvest in September)
	10 N fertiliser	21 N fertiliser 24 CMPP *c* Oxytril CM		23 Aphox *a*		10 Roundup *b* + Frigate (harvest in September)
3 P_2O_5 9 Hytane *b*	7 N fertiliser	11 CMPP *d* Oxytril CM 13 N fertiliser 17 Cycocel *c*	20 Sprint	16 Dorin 23 Aphox *a*		10 Roundup *b* + Frigate (harvest in September)
7 P_2O_5	10 N fertiliser	14 N fertiliser 19 Mn sulphate + Agral 24 CMPP *a* Starane 2 *a*	20 Sprint	16 Dorin 23 Aphox *a*		(harvest in September)

Weather records

A meteorological station is operated at Boxworth EHF. Daily records are taken by farm staff of rainfall, air and soil temperatures, wind speed and hours of sunshine. This information is relevant to many aspects of the Project, as weather conditions affect crop growth, farming operations, and most types of wildlife, either directly or indirectly.

Monthly averages of the major measurements taken during the years of the Project are included in the 1988 Annual Report. The two aspects most likely to have influenced events (rainfall and temperature) are presented in Figure A1. For each year, the following paragraphs by the Farm Director, Mr. R H Jarvis, provide a summary of the main features of the weather, and the effects it had on sowing dates, pesticide applications and crop development.

Weather in 1981–82

All fields were cropped with winter wheat except Grange Piece (winter beans) and Pamplins North (winter oilseed rape). Both September and October gave above-average rainfall which interrupted cultivation and drilling; sowing of wheat was not completed until 23 October. Temperatures were above average in September but they fell below the mean in October although the number of sunshine hours stayed high. A dry November with slightly above-average air temperatures and no frosts enabled good progress to be made in applying herbicides for grass weed control.

In December and January 1982 rainfall was below average but the main features of these months were the extremely cold spells between 8 and 28 December and again between 6 and 16 January. The lowest temperatures recorded were − 11.5°C in December and − 16.0°C in January.

There were appreciable snowfalls, with snow lying on 16 days in December and 9 days in January.

February was another month of low rainfall with below-average sunshine and temperatures. At the beginning of the month it became apparent that the severe winter weather had done little damage to autumn-sown crops. Arrears of herbicide application were made good early in the month.

March was unusually sunny, with near-average rainfall and temperatures. April too was sunnier than average and with normal temperatures but very little rain. Conditions were good for spring work in all crops. In May, rainfall was slightly above average, following a six month period in which five months gave below-average figures. Conditions for spraying were generally good. It was also a very sunny month, with near-average temperatures. In contrast, June was very wet and dull, although reasonably warm. Frequent rainfall delayed the application of late-season fungicides, in some cases well past the planned date. Both July and August were relatively dry, sunny and warm. Harvesting finished on 28 August.

Weather in 1982–83

All fields were cropped with winter wheat. September was a warm month with sunshine and temperatures slightly above average and below-average rainfall. In contrast, October was very wet and rather cool, with very little sunshine. Although the first wheat was sown in late September, progress was spasmodic and drilling was not completed until the end of October. November weather was similar but rather less wet, with a short spell of dry, cold weather and night frosts at the end of the month. In December, rainfall, sunshine and temperatures were all near average.

January was appreciably milder than average with less rain than usual but more sunshine and higher temperatures. February was very sunny but rather cold, with near-average rainfall. Air frosts were recorded each night from the 2nd to the 24th of the month with only one exception, but the frosts were not severe and did not persist by day. Rainfall was near average and in March both rainfall and sunshine were below average with temperatures near to normal. It was fortunate that the weather conditions in March enabled good progress to be made with early spring work, for April was the wettest recorded since weather data were first collected at Boxworth in 1950. Rain fell on 21 days but often in small amounts, which may explain why the total sunshine for the month was above average and some field work was possible. Temperatures were near normal.

May was very dull and cool, with the lowest number of sunshine hours ever recorded for the month at Boxworth. It was also appreciably wetter than April with rain falling on 24 days, the longest dry spell being three days. June brought a respite, with below-average rainfall, which fell in measurable quantities on only eight days. This improvement provided the opportunity to catch up with field work, particularly fungicide applications. Temperatures were near average for the month but sunshine amounts were below normal.

July was hot, with both temperatures and sunshine hours well above average. Measurable rain was recorded on only nine days, but on one of these 31.8 mm fell, to raise the total for the month to well above average. The weather in August was redolent of the halcyon weather which is supposed to accompany harvest but is so rarely recorded. Rainfall was only 18 per cent of average, making it the driest August ever recorded at Boxworth, and both temperatures and sunshine hours were well above the mean figures.

Weather in 1983–84

All fields were cropped with winter wheat. With above-average rainfall in September following a dry August, the land was in a reasonable state for cultivation but ploughing was avoided on most fields because of the risk of bringing up clods. September temperatures were close to average but there was less sunshine than usual. In October, rainfall and air temperatures were both near average but sunshine figures were up. In these conditions crops emerged quickly and evenly. November was drier than usual with near-average temperatures in spite of a low figure for sunshine hours, and provided ideal conditions for autumn herbicide application. December too had rather below-average rainfall, but in other respects was close to average.

January began with reasonably mild weather but turned colder in mid month with some snow. Precipitation was about average but the number of hours of bright sunshine was the highest recorded for this month at Boxworth. The first half of February was warm, but this was followed by much cooler weather. Rainfall and sunshine hours were both close to average. The below-average temperatures persisted throughout March and the hours of bright sunshine were the lowest recorded for the month at Boxworth. Rainfall was below average. April too was an unusually dry month. Temperatures were slightly above average and the sunshine figure was the highest recorded at Boxworth, a very marked contrast to the March figure.

For much of May, the dry, rather cool weather persisted but in the last 10 days of the month 70 mm of rain fell, more than 40 per cent above the average for the whole month. Sunshine figures were well below average and mean temperatures were rather low. In June, temperatures, sunshine hours and rainfall were all near average; the rain fell mainly in the first week and on the 20th of the month. In July rainfall was well below average, sunshine rather above average and temperatures near normal; conditions in May-August were ideal for the timely application of spray chemicals, the later stages of crop development and ripening, and for harvesting.

Weather in 1984–85

Backside, Thorofare, Eleven Acre Extra and Bushes & Pits were cropped with winter oilseed rape and the rest with winter wheat. Following a dry harvest period, September was a very wet month with rain falling on 17 days, to give a total of 96.5 mm. Rape sown early in the month grew away well. The wet weather did not continue into October and there were no serious delays in seedbed preparation or drilling of winter wheat. For much of November, wet and windy conditions prevailed. Temperatures were rather above average and sunshine hours below. In contrast, December was appreciably drier, sunnier and warmer than usual and provided an opportunity to catch up with autumn herbicide application.

January was a cold month with frost recorded on 20 of the first 21 nights, the lowest air minimum being – 9.6°C on the night of 17 January. A partial thaw with heavy rain set in four days later but night frosts continued to the end of the month. February too was cold, with below-average precipitation. Snow fell at the beginning of the second week of the month and strong, cold north-easterly winds prevailed. The lowest air temperature recorded was – 11.2°C on the night of 12 February. The snow drifted extensively and there was frost burn on both wheat and rape, and the latter also suffered some pigeon damage. (It became clear subsequently that no lasting damage had been done). Temperatures in early March were near normal but there was a colder spell with night frosts in mid month. Rainfall was just below average. In April rainfall and sunshine hours were below average. For most of the month temperatures were above the normal level but the last 10 days were cooler, with some night frost.

Apart from July, which saw near-normal weather conditions, the last spring and summer months were characterised by below-average temperatures and, in May and June, sunshine too was much less than average. However, only in June was above-average rainfall recorded, although in both May and August the number of rain days was somewhat above average. The June rain led to moderate lodging in some of the Project fields but did not prevent the timely applications of late fungicides. There was some delay in harvest due to the wet conditions.

Weather in 1985–86

All fields were cropped with winter wheat. September and October were both dry months and although the drought broke in early November, amounts of rainfall in the rest of that month were small. These conditions inevitably led to difficulties in seed-bed preparation and to slow, uneven emergence of the crop. However, December was a wet month and this, combined with above-average temperatures in the first three weeks of the month, resulted in an all-round improvement. Following a cold spell in late December, both temperature and rainfall in January were close to the averages for the month.

The weather in February and early March was much colder. Air frosts were recorded every night, and from 4 February to 4 March and on 15 days in February the air temperature failed to rise above freezing point. Average temperature was the lowest for the month since records were first kept at Boxworth in 1950. Snow lay on the ground from 6–24 February but the maximum depth was no more than 11 cm.

The rest of March was relatively mild, but measurable rain fell on 20 days, although the total for the month was only slightly above average. In spite of these relatively good conditions, most of the winter wheat was slow to grow away. It showed little or no response to early spring nitrogen and, not surprisingly, spring plant populations remained thin.

April was a wetter month than March; on only three days was there no measurable rainfall. The first half of the month was rather cold with frequent night frosts but the second half was substantially warmer. Weather conditions in May, June and July were close to normal except that June was drier than usual. All late spring fungicides were applied to winter wheat at or very close to the intended growth stages. August, in contrast, was appreciably wetter and colder than average. Some of the rain was heavy and led to lodging in several fields, and cereal harvest was not completed until 12 September.

Weather in 1986–87

Pamplins North, Top Pavements and Extra Close West were cropped with winter oilseed rape and the rest with winter wheat. September 1986 was a low rainfall month with above-average sunshine, although temperatures were rather low. Similar weather continued into the first half of October. In spite of the dry weather, autumn seedbeds were much better than in the previous year and oilseed rape, as well as wheat, established well.

The second half of October was appreciably wetter than average; so too were November and December, when temperatures were a little above average. January 1987 brought a change to much colder weather. Air frosts were recorded on 15 consecutive nights from 5 January and on the night of 12 January the temperature fell to − 11°C, not having risen above − 6.5°C on the previous day. There was very little snowfall and the total precipitation for the month, 10.2 mm, was the lowest recorded at Boxworth. Subsequently there were no signs of significant frost damage in either rape or wheat.

In contrast, weather conditions in February were close to average in all respects. March was rather cooler than usual and below-average rainfall was recorded. April brought much warmer weather, particularly in the second half of the month. Sunshine was above average and rainfall close to average. In May, rainfall was close to average but temperatures and sunshine hours were slightly less than usual.

June was a much wetter month than average (75 per cent more rain than average), with less bright sunshine (45 per cent below average) and the average temperature was rather lower than usual. There were only two days in the month with no measurable rain. In July, rainfall was 66 per cent above average, and sunshine was 17 per cent below average; temperatures were near normal. August too was wetter than usual and although rape harvest began on 25 July, it was not completed until 17 August. There was clearly some loss from shedding, but in general the yield levels were acceptable. However, for wheat harvested between 19 August and 15 September, the average yield was the lowest recorded at Boxworth since 1979. With very few exceptions, fungicides had been applied on time in spite of the difficult weather and there appeared to be little in the way of fungal disease of leaf or ear. However, in many cases the top 4–8 spikelets on each ear produced no grain, and lower down the ear there were no more than two grains per spikelet. It is not clear whether there was a failure to pollinate or a failure to fill. The generally small size of grain in 1987 suggests the latter.

Weather in 1987–88

All fields were cropped with winter wheat. September rainfall was below average and the dry spell continued into early October, enabling a good start to be made with autumn cultivations. All fields in the Full Insurance and Supervised treatment areas were drilled in good order in the period 28 September–8 October. However, the weather then broke, with frequent heavy rainfall through the rest of the month. In all 141.5 mm fell in October, the wettest at Boxworth since records began in 1950. It was not possible to complete drilling in the Integrated treatment area until 5 November. In contrast to October, November and December gave below-average rainfall; temperatures were slightly below average in November and above average in December.

January 1988 was reasonably mild, with only eight air frosts recorded, but there was 84.9 mm of rain, almost twice the average. Temperatures rather above normal continued in February and March. February provided the opportunity to make progress with much of the grass weed control spray programme which had been impossible in the autumn, but March was another wet month. April, May and June all produced near-average rainfall and provided the opportunity first to catch up with the remaining backlog of grass weed control and subsequently for spring and early summer applications of agrochemicals to go ahead as planned. The number of sunshine hours was below average in all three months; air temperatures were a little below average in April but near normal in May and June. July temperatures too were near normal but the rainfall was high and the sunshine figure low. In August, rainfall was a little above average but the sunshine figure was high and there was no serious interference with cereal harvesting, which was completed on 7 September.

Figure A1   Total monthly rainfall and mean air temperatures at Boxworth, 1981 to 1988.

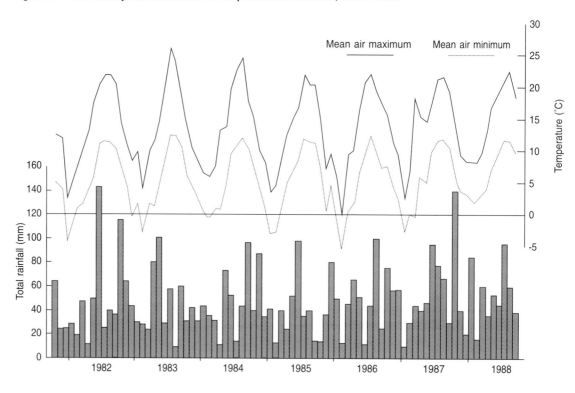

APPENDIX IV

Species recorded at Boxworth

(1) Birds

Ardea cinerea	Grey Heron
Anas platyrhynchos	Mallard*
Accipiter nisus	Sparrowhawk†
Falco tinnunculus	Kestrel*
Falco subbuteo	Hobby
Alectoris rufa	Red-legged Partridge*
Perdix perdix	Grey Partridge*
Phasianus colchicus	Pheasant*
Gallinula chloropus	Moorhen*
Pluvialis apricaria	Golden Plover
Vanellus vanellus	Lapwing
Gallinago gallinago	Snipe
Larus ridibundus	Black-headed Gull
Larus canus	Common Gull
Larus fuscus	Lesser Black-backed Gull
Larus argentatus	Herring Gull
Columba livia var.	Feral Pigeon*
Columba oenas	Stock Dove*
Columba palumbus	Woodpigeon*
Streptopelia decaocto	Collared Dove*
Streptopelia turtur	Turtle Dove*
Cuculus canorus	Cuckoo*
Tyto alba	Barn Owl
Athene noctua	Little Owl*
Strix aluco	Tawny Owl*
Apus apus	Swift
Picus viridis	Green Woodpecker*
Dendrocopos major	Greater Spotted Woodpecker*
Dendrocopos minor	Lesser Spotted Woodpecker*
Alauda arvensis	Skylark*
Riparia riparia	Sand Martin
Hirundo rustica	Swallow*
Delichon urbica	House Martin
Anthus pratensis	Meadow Pipit†
Motacilla flava	Yellow Wagtail†
Motacilla cinerea	Grey Wagtail†
Motacilla alba	Pied Wagtail*
Troglodytes troglodytes	Wren*
Prunella modularis	Dunnock*
Erithacus rubecula	Robin*

(1) Birds—*continued*

Luscinia megarhynchos	Nightingale*
Oenanthe oenanthe	Wheatear
Turdus merula	Blackbird*
Turdus pilaris	Fieldfare
Turdus philomelos	Song Thrush*
Turdus iliacus	Redwing
Turdus viscivorus	Mistle Thrush†
Locustella naevia	Grasshopper Warbler†
Acrocephalus schoenobaenus	Sedge Warbler*
Sylvia curruca	Lesser Whitethroat*
Sylvia communis	Whitethroat*
Sylvia borin	Garden Warbler*
Sylvia atricapilla	Blackcap*
Phylloscopus collybita	Chiffchaff*
Phylloscopus trochilus	Willow Warbler*
Regulus regulus	Goldcrest†
Muscicapa striata	Spotted Flycatcher*
Aegithalos caudatus	Long-tailed Tit*
Parus palustris	Marsh Tit†
Parus ater	Coal Tit†
Parus caeruleus	Blue Tit*
Parus major	Great Tit*
Sitta europaea	Nuthatch*
Certhia familiaris	Treecreeper*
Garrulus glandarius	Jay†
Pica pica	Magpie†
Corvus monedula	Jackdaw*
Corvus frugilegus	Rook
Corvus corone	Carrion Crow*
Sturnus vulgaris	Starling*
Passer domesticus	House sparrow*
Passer montanus	Tree sparrow*
Fringilla coelebs	Chaffinch*
Carduelis chloris	Greenfinch*
Carduelis carduelis	Goldfinch*
Carduelis cannabina	Linnet*
Carduelis flammea	Redpoll†
Pyrrhula pyrrhula	Bullfinch*
Emberiza citrinella	Yellowhammer*
Emberiza schoeniclus	Reed Bunting*
Miliaria calandra	Corn Bunting*

* definite breeder (52)
†probable breeder (12)

(2) Mammals

Nomenclature follows Corbet, G B, and Southern H N (1977). *The Handbook of British Mammals*, Blackwell Scientific Publications, Oxford.

Insectivora

Sorex araneus	Common shrew
Sorex minutus	Pygmy shrew

Lagomorpha

Oryctolagus cuniculus	Rabbit
Lepus capensis	Brown Hare

Rodentia

Clethrionomys glareolus	Bank Vole
Microtus agrestis	Field Vole
Apodemus sylvaticus	Wood Mouse
Micromys minutus	Harvest Mouse
Mus musculus	House Mouse

Carnivora

Mustela nivalis	Weasel

(3) Higher Plants

Nomenclature according to Clapham, A R, Tutin, T G & Warburg, E F (1981). *Excursion Flora of the British Isles*. Cambridge University Press.

English names according to Dony, J G, Jury, S L & Perring, F H (1986). *English Names of Wild Flowers*. 2nd. Edn. Botanical Society of the British Isles.

Pteridophyta
Equisetaceae

Equisetum arvense	Field horsetail

Angiospermae
Dicotyledones
Ranunculaceae

Clematis vitalba	Traveller's-joy
Ranunculus acris	Meadow buttercup
Ranunculus repens	Creeping buttercup
Ranunculus ficaria	Lesser celandine

Angiospermae—*continued*

Dicotyledones—*continued*

Papaveraceae	
Papaver rhoeas	Common poppy
Papaver dubium	Long-headed poppy
Papaver lecoquii	Yellow-juiced poppy
Cruciferae	
Sinapis arvensis	Charlock
Coronopus squamatus	Swine-cress
Capsella bursa-pastoris	Shepherd's-purse
Armoracia rusticana	Horse-radish
Cardamine sp.	Bittercress
Barbarea vulgaris	Winter-cress
Alliaria petiolata	Garlic mustard
Sisymbrium sp.	—
Violaceae	
Viola odorata	Sweet violet
Viola reichenbachiana	Early dog-violet
Viola arvensis	Field pansy
Hypericaceae	
Hypericum hirsutum	Hairy St. John's-wort
Caryophyllaceae	
Silene dioica	Red campion
Silene alba	White campion
Silene vulgaris	Bladder campion
Cerastium fontanum	Common mouse-ear
Stellaria media	Common chickweed
Chenopodiaceae	
Chenopodium album	Fat-hen
Atriplex patula	Common orache
Malvaceae	
Malva sylvestris	Common mallow
Geraniaceae	
Geranium pratense	Meadow crane's-bill
Geranium dissectum	Cut-leaved crane's-bill
Geranium molle	Dove's-foot crane's-bill
Geranium rotundifolium	Round-leaved crane's-bill
Geranium robertianum	Herb-Robert
Oxalidaceae	
Oxalis acetosella	Wood-sorrel
Aceraceae	
Acer campestre	Field maple
Acer pseudoplatanus	Sycamore

Angiospermae—*continued*

Dicotyledones—*continued*

Celastraceae
    Euonymus europaeus   Spindle
Buxaceae
    Rhamnus catharticus   Buckthorn
Leguminosae
    Trifolium repens   White clover
    Vicia sp.   Vetch
Roseceae
    Filipendula ulmaria   Meadowsweet
    Rubus fruticosus   Bramble
    Rubus caesius   Dewberry
    Potentilla anserina   Silverweed
    Potentilla reptans   Creeping cinquefoil
    Geum urbanum   Wood avens
    Agrimonia eupatoria   Agrimony
    Rosa arvensis   Field-rose
    Rosa canina   Dog-rose
    Prunus spinosa   Blackthorn
    Prunus domestica   Wild plum
    Crataegus monogyna   Hawthorn
    Crataegus laevigata   Midland hawthorn
    Malus sylvestris   Crab apple
Thymelaeaceae
    Daphne laureola   Spurge-laurel
Onagraceae
    Epilobium hirsutum   Great willowherb
    Epilobium montanum   Broad-leaved willowherb
    Epilobium roseum   Pale willowherb
    Chamerion angustifolium   Rosebay willowherb
Araliaceae
    Hedera helix   Ivy
Umbelliferae
    Chaerophyllum temulentum   Rough chervil
    Anthriscus sylvestris   Cow parsley
    Scandix pecten-veneris   Shepherd's-needle
    Conopodium majus   Pignut
    Aethusa cynapium   Fool's-parsley
    Conium maculatum   Hemlock
    Heracleum sphondylium   Hogweed
    Torilis japonica   Upright hedge-parsley
    Daucus carota   Wild carrot

Angiospermae—*continued*
Dicotyledones—*continued*

Cucurbitaceae
Bryonia dioica — White bryony
Euphorbiaceae
Euphorbia exigua — Dwarf spurge
Polygonaceae
Polygonum aviculare — Knotgrass
Polygonum persicaria — Redshank
Fallopia convolvulus — Black bindweed
Rumex crispus — Curled dock
Rumex obtusifolius — Broad-leaved dock
Rumex sanguineus — Wood dock
Rumex conglomeratus — Clustered dock
Urticaceae
Urtica dioica — Common nettle
Achillea millefolium — Yarrow
Tripleurospermum inodorum — Scentless mayweed
Arctium lappa — Greater burdock
Arctium minus — Lesser burdock
Carduus acanthoides — Welted thistle
Cirsium vulgare — Spear thistle
Cirsium arvense — Creeping thistle
Centaurea scabiosa — Greater knapweed
Centaurea nigra — Common knapweed
Lapsana communis — Nipplewort
Hypochoeris radicata — Cat's-ear
Picris echioides — Bristly oxtongue
Tragopogon pratensis — Goat's-beard
Lactuca serriola — Prickly lettuce
Sonchus arvensis — Perennial sowthistle
Sonchus oleraceus — Smooth sowthistle
Sonchus asper — Prickly sowthistle
Hieracium sp. — Hawksbeard
Crepis capillaris — Smooth hawksbeard
Taraxacum officinale — Common dandelion

(3) Higher Plants—*continued*

Angiospermae—*continued*

Dicotyledones—*continued*

Ulmaceae
    Ulmus spp.   Elm
    Ulmus glabra   Wych elm
Betulaceae
    Alnus glutinosa   Alder
Corylaceae
    Corylus avellana   Hazel
Fagaceae
    Quercus sp.   Oak
Salicaceae
    Salix caprea   Goat willow
    Salix fragilis   Crack willow
Primulaceae
    Primula veris   Cowslip
    Anagallis arvensis   Scarlet pimpernel
Oleaceae
    Fraxinus excelsior   Ash
    Ligustrum vulgare   Wild privet
Boraginaceae
    Myosotis arvensis   Field forget-me-not
    Lithospermum arvense Field gromwell
Convolvulaceae
    Convolvulus arvensis Field bindweed
    Calystegia sepium   Hedge bindweed
Solanaceae
    Solanum dulcamara   Bittersweet
Scrophulariaceae
    Kickxia spuria   Round-leaved fluellen
    Kickxia elatine   Sharp-leaved fluellen
    Scrophularia auriculata Water figwort
    Veronica montana   Wood speedwell
    Veronica chamaedrys Germander speedwell
    Veronica arvensis   Wall speedwell
    Veronica hederifolia Ivy-leaved speedwell
    Veronica persica   Common field-speedwell
    Odontites verna   Red bartsia
Labiatae
    Stachys sylvatica   Hedge woundwort
    Ballota nigra   Black horehound
    Lamium purpureum   Red dead-nettle
    Lamium album   White dead-nettle

(3) Higher Plants—*continued*

Angiospermae—*continued*

Dicotyledones—*continued*

Galeopsis tetrahit	Common hemp-nettle
Glechoma hederacea	Ground-Ivy

Plantaginaceae

Plantago major	Greater plantain
Plantago media	Hoary plantain
Plantago lanceolata	Ribwort plantain

Lobeliaceae

Galium cruciata	Crosswort
Galium verum	Lady's bedstraw
Galium aparine	Cleavers

Caprifoliaceae

Sambucus nigra	Elder
Viburnum lantana	Wayfaring-tree
Lonicera periclymenum	Honeysuckle

Compositae

Dipsacus fullonum	Teasel
Senecio vulgaris	Groundsel
Tussilago farfara	Colt's-foot

Monocotyledones

Liliaceae

Hyacinthoides non-scripta	Bluebell
Allium vineale	Wild onion

Juncaceae

Juncus inflexus	Hard rush

Dioscoreaceae

Tamus communis	Black bryony

Araceae

Arum maculatum	Lords-and-Ladies

Typhaceae

Typha latifolia	Bulrush

Gramineae

Festuca pratensis	Meadow fescue
Festuca gigantea	Giant fescue
Festuca rubra	Red fescue
Lolium perenne	Perennial ryegrass
Lolium multiflorum	Italian ryegrass
Poa annua	Annual meadow-grass
Poa pratensis	Smooth meadow-grass
Poa trivialis	Rough meadow-grass
Dactylis glomerata	Cock's-foot

(3) Higher Plants—*continued*

Angiospermae—*continued*

Monocotyledones—*continued*

Bromus sterilis	Barren brome
Bromus ramosus	Hairy brome
Bromus hordeaceus	Soft brome
Bromus commutatus	Meadow brome
Brachypodium sylvaticum	False brome
Elymus caninus	Bearded couch
Elymus repens	Common couch
Hordeum murinum	Wall barley
Avena fatua	Wild-oat
Arrhenatherum elatius	False oat-grass
Holcus lanatus	Yorkshire-fog
Agrostis stolonifera	Creeping bent
Phleum pratense	Timothy
Alopecurus myosuroides	Black-grass
Alopecurus pratensis	Meadow foxtail

(4) Common predatory beetles and money spiders

Nomenclature follows Kloet, G S & Hincks, W D (1976). 'A check list of British insects, Part 3: Coleoptera and Strepsiptera' and Roberts, M J (1987) 'The spiders of Great Britain and Ireland, Volume 2: Linyphiidae'.

Carabidae (ground beetles)
- *Carabus violaceus*
- *C. granulatus*
- *C. monilis*
- *Leistus spinibarbis*
- *L. ferrugineus*
- *Nebria brevicollis*
- *Notiophilus biguttatus*
- *Loricera pilicornis*
- *Clivina fossor*
- *Trechus quadristriatus*
- *T. obtusus*
- *Asaphidion flavipes*
- *Bembidion obtusum*
- *B. lampros*
- *B. guttula*
- *B. quadrimaculatum*
- *B. bipunctatum*
- *B. lunulatum*
- *B. aeneum*

Carabidae—*continued*
- *Stomis pumicatus*
- *Pterostichus madidus*
- *P. melanarius*
- *P. macer*
- *P. versicolor*
- *P. strenuus*
- *P. niger*
- *Abax parallelepipedus*
- *Calathus fuscipes*
- *C. melanocephalus*
- *Synuchus nivalis*
- *Agonum dorsale*
- *A. obscurum*
- *Amara plebeja*
- *A. familiaris*
- *Harpalus rufipes*
- *H. aeneus*
- *H. rufibarbis*
- *Acupalpus meridianus*

(4) Common predatory beetles and money spiders—*continued*

Carabidae—*continued*

    Badister bipustulatus
    B. sodalis
    Demetrias atricapillus
    Dromius linearis
    Metabletus obscuroguttatus

Staphylinidae (rove beetles)

    Tachyporus hypnorum
    T. obtusus
    T. chrysomelinus
    T. nitidulus
    Mycetoporus splendidus
    Tachinus signatus
    T. marginellus
    Staphylinus olens
    S. globuliferus
    Xantholinus laevigatus
    X. tricolor
    X. linearis
    X. glabratus
    Lathrobium fulvipenne
    Othius punctulatus
    Quedius nigrocaeruleus
    Q. fuliginosus
    Q. nemoralis
    Q. lateralis
    Philonthus fuscipennis
    Bolitobius analis
    Sepedophilus sp.
    Stenus sp.

Linyphiidae (money spiders)

    Erigone atra
    E. dentipalpis
    Walckenaeria clavicornis
    W. nudipalpis
    W. stylifrons
    W. antica
    Bathyphantes gracilis
    Lepthyphantes tenuis
    Pachygnatha degeeri
    Oedothorax retusus
    O. apicatus
    O. fuscus
    Savignya frontata
    Meioneta rurestris
    Diplocephalus latifrons
    Scotinotylus evansi
    Centromerus sylvaticus
    Dicymbium nigrum
    Gongylidium rufipes
    Diplostyla concolor
    Porhomma pygmaeum
    P. microphthalmum
    Panamomops sulcifrons
    Micrargus herbigradus
    Baryphyma pratense
    Saloca diceros

(Lists of other invertebrates recorded during the Boxworth Project
are given in Chapters 9 and 11)

(5) Plant diseases

Alternaria sp.	(associated with leaf spotting on wheat)
Alternaria brassicae	(dark leaf spot of oilseed rape)
Alternaria brassicicola	(dark leaf spot of oilseed rape)
Botrytis cinerea	(ear blight of wheat)

(5) Plant diseases—*continued*

Didymella exitialis   (associated with leaf spotting of wheat)

Erysiphe graminis var. *tritici*   (mildew of wheat)

Fusarium spp.   (brown foot rot and ear blight of wheat, also associated with leaf spotting)

Gaeumannomyces graminis   (take-all of wheat)

Leptosphaeria maculans
(= Phoma lingam)   (phoma leaf spot and canker of oilseed rape)

Peronospora parasitica   (downy mildew of oilseed rape)

Pseudocercosporella herpotrichoides   (eyespot of wheat)

Pyrenopeziza brassicae
  (= Cylindrosporium concentricum)   (light leaf spot of oilseed rape)

Puccinia recondita   (brown rust of wheat)

Puccinia striiformis   (yellow rust of wheat; not seen but action was taken against it)

Rhizoctonia cerealis   (sharp eyespot of wheat)

Septoria nodorum   (leaf spot and glume blotch of wheat)

Septoria tritici   (speckled leaf spot of wheat)

APPENDIX V

List of publications concerning the Boxworth Project

Burn, A J (1988). 'The effects of intensive pesticide use on arthropod predators in cereals.' In: Cavalloro, R & Sunderland, K D (Eds). *Integrated Crop Protection in Cereals.* A A Bolkema, Rotterdam. pp. 147–152.

Burn, A J (1988). 'Effects of scale on the measurement of pesticide effects on invertebrate predators and predation.' In: Greaves, M P, Greig-Smith, P W & Smith, B D (Eds). *Field Methods for the Study of Environmental Effects of Pesticides.* Monograph No. 40, British Crop Protection Council, Croydon. pp. 109–117.

Burn, A. J. (1988). 'Assessment of the impact of pesticides on invertebrate predation in cereal crops.' *Aspects of Applied Biology* **17**, 279–288.

Burn, A J (1989). 'Long-term effects of pesticides on natural enemies of cereal crop pests.' In: Jepson, P C (Ed.). *Pesticides and Non-target Invertebrates.* Intercept Press, Wimborne, Dorset. pp. 177–193.

Cooper, D A (1990). 'Development of an experimental programme to pursue the results of the Boxworth Project.' *Proceedings of the 1990 Brighton Crop Protection Conference—Pests and Diseases, Vol. 1.* pp. 153–162.

Cousens, R, Marshall, E J P and Arnold, G M (1988). 'Problems in the interpretation of effects of herbicides on plant communities.' In: Greaves, M P, Greig-Smith, P W & Smith, B D (Eds). *Field Methods for the Study of Environmental Effects of Pesticides.* Monograph No. 40, British Crop Protection Council, Croydon. pp. 275–282.

Doberski, J (1989). 'The Boxworth experiment and the pesticide tightrope.' *Ecos* **10**, 18–21.

Flowerdew, J R (1988). 'Methods for studying populations of wild mammals.' In: Greaves, M P, Greig-Smith, P W & Smith, B D (Eds). *Field Methods for the Study of Environmental Effects of Pesticides.* Monograph No. 40, British Crop Protection Council, Croydon. pp. 67–76.

Flowerdew, J R, Johnson, I P & Hare, R (1990). 'Rodent populations in cereal fields with high and low pesticide inputs.' *2nd International Conference on Pests in Agriculture*, ANPP, Versailles. pp. 209–216.

Greig-Smith, P W (Ed.) (1987). *The Boxworth Project: 1986 Annual Report.* ADAS, Tolworth. 116 pp.

Greig-Smith, P W (1987). 'Keeping a friendly eye on birds and small mammals.' *Farmers' Weekly* **23**, 5 June: 22.

Greig-Smith, P W (1987). 'A spray in the life . . . ' *Farmers' Weekly* **22**, 29 May: 15–16.

Greig-Smith, P W (1987). 'The Boxworth Project.' *ADAS Oxford Divisional Bulletin*, October 1987.

Greig-Smith, P W (Ed.) (1988). *The Boxworth Project: 1987 Annual Report*. ADAS, Tolworth. 137 pp.

Greig-Smith, P W (Ed.) (1989). *The Boxworth Project: 1988 Annual Report*. ADAS, Tolworth. 138 pp.

Greig-Smith, P W (1989). 'Effects of non-persistent insecticides on bird and mammal populations on farmland.' In: Walker, C H & Anderson, D (Eds). *Ecotoxicology*. Institute of Biology, London. pp. 12–18.

Greig-Smith, P W (1989). 'Intensive study versus extensive monitoring in field trials of pesticides.' In: Somerville, L & Walker, C H (Eds). *Effects of Pesticides on Terrestrial Wildlife*. Taylor & Francis, London. pp. 217–239.

Greig-Smith, P W (1989). 'The Boxworth Project—Environmental effects of cereal pesticides.' *Journal of the Royal Agricultural Society of England* **150**, 171–187.

Greig-Smith, P W (1990). 'The Boxworth Project.' *Pesticide Outlook* **3**, 16–19.

Greig-Smith, P W (1991). 'The Boxworth experience: effects of pesticides on the fauna and flora of cereal fields.' In: Firbank, L G, Carter, N, Darbyshire, J F & Potts, G R (Eds). *The Ecology of Temperate Cereal Fields*. Blackwell Scientific Publications, Oxford pp. 333–371.

Hardy, A R (Ed.) (1983). *The Boxworth Project: 1982 Annual Report*. ADAS, Tolworth, 78 pp.

Hardy, A R (Ed.) (1984). *The Boxworth Project: 1983 Annual Report*. ADAS, Tolworth, 108 pp.

Hardy, A R (Ed.) (1985). *The Boxworth Project: 1984 Annual Report*. ADAS, Tolworth. 101 pp.

Hardy, A R (Ed.) (1986). *The Boxworth Project: 1985 Annual Report*. ADAS, Tolworth. 114 pp.

Hardy, A R (1986). 'The Boxworth Project—a progress report.' *Proceedings of the 1986 British Crop Protection Conference, Pests and Diseases, vol. 3*. pp. 1215–1224.

Hardy, A R (1986). 'Boxworth examines wildlife spray patterns.' *Farmers' Weekly*, 29 August, pp. 42–43.

Hart, A D M (1990). 'The assessment of pesticide hazards to birds: the problem of variable effects.' *Ibis*, **132**, 192–204.

Hart, A D M, Fletcher, M R, Greig-Smith, P W, Hardy, A R, Jones, S A & Thompson, H M (1990). 'Le Projet Boxworth: effets de regimes de pesticides contradictoires sur les oiseaux de terre arable.' In: *Relations entre les traitements phytosanitaires et la reproduction des animaux*, Annales de l'ANPP. No. 2. pp. 225–232.

Jarvis, R H (1987). 'The Boxworth Project.' *Farmers' Weekly* **23**, 5 June pp. 18–19.

Jarvis, R H (1988). 'The Boxworth Project'. In: Harding, D J L (Ed.) *Britain since Silent Spring*. Institute of Biology, London. pp. 46–55.

Jarvis, R H (1988). 'The Boxworth Project: Crop protection practices and profitability.' *Aspects of Applied Biology* **17**, 37–45.

Jarvis, R H (1989). 'The Boxworth Project.' *Boxworth Experimental Husbandry Farm Review 1989*. ADAS, Cambridge, pp. 1–3.

Johnson, I P, Flowerdew, J R & Hare, R (1991). 'Effects of broadcasting and of drilling methiocarb molluscicide pellets on field populations of wood mice, *Apodemus sylvaticus*.' *Bulletin of Environmental Contamination and Toxicology* **46**, 84–91.

MAFF (1989). *The Boxworth Project and subsequent studies—Topic Review*. MAFF, London. 135 pp.

Marshall, E J P (1985). 'Weed distributions associated with cereal field edges—some preliminary observations.' *Aspects of Applied Biology* **9**, 49–58.

Marshall, E J P (1985). 'Field and field edge floras under different herbicide regimes at the Boxworth EHF—Initial studies.' *Proceedings of the 1985 British Crop Protection Conference—Weeds, vol. 3*. pp. 999–1006.

Marshall, E J P (1986). *Studies of the flora in arable field margins*. Long Ashton Research Station, Weed Research Division, Technical Report No. 96, pp. 1–33.

Marshall, E J P (1987). 'Drawing a bead on the need for weed control.' *Farmers' Weekly* **23**, 5 June: 19–20.

Marshall, E J P (1987). 'Using decision thresholds for the control of grass and broad-leaved weeds at Boxworth EHF.' *Proceedings of the 1987 British Crop Protection Conference—Weeds, vol. 3*. pp. 1059–1066.

Marshall, E J P (1987). 'Herbicide effects on the flora of arable field boundaries.' *Proceedings of the 1987 British Crop Protection Conference—Weeds, vol. 3*. pp. 291–298.

Marshall, E J P (1988). 'Field-scale estimates of grass weed populations in arable land.' *Weed Research* **28**, 191–198.

Marshall, E J P (1988). 'The dispersal of plants from field boundaries.' In: Park, J R (Ed.). *Environmental Management in Agriculture: European Perspectives*. Belhaven Press, London. pp. 136–143.

Marshall, E J P (1988). 'The ecology and management of field margin floras in England.' *Outlook on Agriculture* **17**, 178–182.

Marshall, E J P (1989). 'Distribution patterns of plants associated with arable field edges.' *Journal of Applied Ecology* **26**, 247–257.

Marshall, E J P & Smith, B D (1987). 'Field margin flora and fauna: interaction with agriculture.' In: Way, J M & Greig-Smith, P W (Eds). *Field Margins*. British Crop Protection Council, Croydon. pp. 23–34.

Stanley, P I & Hardy, A R (1984). 'The environmental implications of current pesticide usage on cereals.' In: Jenkins, D (Ed.) *Agriculture and the Environment*. Institute of Terrestrial Ecology, Cambridge. Symposium No. 13. pp. 66–72.

Tarrant, K A, Johnson, I P, Flowerdew, J R & Greig-Smith, P W. (1990). 'Effects of pesticide applications on small mammals in arable fields, and the recovery of their populations.' *Proceedings of the 1990 Brighton Crop Protection Conference—Pests & Diseases, Vol. 1.* pp. 173–182.

Tarrant, K A, Thompson, H M & Hardy, A R (1992) 'Biochemical and histological effects of the aphicide demeton-S-methyl on house sparrows under field conditions.' *Bulletin of Environmental Contamination and Toxicology*, 48, 360–366.

Thompson, H M (1991). 'Serum 'B' esterases as indicators of exposure to pesticides.' In: Mineau P. (Ed.) *Cholinesterase-inhibiting insecticides: impacts on wildlife and the environment*. Elsevier pp. 110–125.

Thompson, H M, Walker, C H & Hardy, A R (1988). 'Esterases as indicators of avian exposure to insecticides.' In: Greaves, M P, Greig-Smith, P W & Smith, B D (Eds). *Field Methods for the Study of Environmental Effects of Pesticides*. Monograph No. 40, BCPC, Croydon. pp. 39–45.

Thompson, H M, Walker, C H & Hardy, A R (1989). 'The use of avian serum 'B' esterases in monitoring exposure to organophosphorus insecticides.' In: Somerville, L & Walker, C H (Eds). *Effects of Pesticides on Terrestrial Wildlife*. Taylor & Francis, London. pp. 348–349.

Vickerman, G P & Burn, A J (1987). 'How invertebrates prey on pests and feed wildlife.' *Farmers' Weekly* **23**, 5 June: 20–22.

Vickerman, G P (1988). 'Farm scale evaluation of the long-term effects of different pesticide regimes on the arthropod fauna of winter wheat.' In: Greaves, M P, Greig-Smith, P W & Smith, B D (Eds). *Field Methods for the Study of Environmental Effects of Pesticides*. Monograph No. 40, BCPC, Croydon. pp. 127–135.

APPENDIX VI

Participants in the Boxworth Project

Steering Group

Chairmen: P Wiggell (1981–85),
M J Griffin (1986–1990)

Secretaries: A R Hardy (1981–86),
P W Greig-Smith (1986–90)

Members: S A Evans, R A Lelliott,
H J Wilcox, P I Stanley,
H J Gould, D J Yarham,
R H Jarvis, A Roberts,
P Skeels, M Hancock,
G W Cussans, J E King,
J C Sherlock, J MacLeod.

Staff at Boxworth EHF

R H Jarvis, P Bowerman, J S Rule,
M J Strickland, J T Clapp, A J Roberts,
P J Skeels, N E Tinkley, N F Cross,
L H Webb, S M Smith, L F Wright,
P R Blundell, R F Church, S D Dry,
D J Eavis, M J Eavis, E A Geburtig,
E G Girdlestone, L P Lister, H C Shelton,
J E Wilkins, S J Newell.

Pest Monitoring

R Gair, M Hancock, S M Roberts,
H M Maher, A J Sherwood, E M Balderstone,
G A Talbot, S Corless, P J Saunders, plus
students and casual staff.

Botanical Studies

E J P Marshall, G W Cussans, B J Wilson,
O Leedham, G Bird, H Oakes, C Steveni,
Y Jones, E Spearing, C Munkley, L Dugdale,
M Dowlen, S R Moss, K J Wright,
P A Phipps, J E Birnie, D Bailey,

D R Tottman, A M Blair, T D Martin,
P M Steer, J Barker, S J Barrett, M Betty,
R D Cousens, A Croxford, D G Cussans,
J Cussans, R A P Denner, A Duggal,
S Hilger, B van Hille, C J Marshall,
N K Scott, B Thurley, B Wilson,
S L Wooliams.

Plant Pathology

D J Yarham, B V Symonds, D Ellerton,
B McKeown, N Giltrap.

Pesticide Residues

D Eagle, P M Brown, P N Kendrick

Invertebrate Studies

A J Burn, M Bryan, G P Vickerman, M Free,
P Halfpenny, D Farrington, S Ameer,
U Banda Ekanayeke, J Mauremootoo,
S D Wratten, T Coaker.

Soil Fauna

G K Frampton, J Smith, L Reeves,
H Shortridge, D Harratt, H Dickens,
P Williamson, F Wimpress, S D Langton.

Mammals

R Hare, I P Johnson, J R Flowerdew,
K A Tarrant

Birds

M R Fletcher, S A Jones, A D M Hart,
K A Tarrant, J K Blakey, J Norman, P Burns,

S Bunyan, D Mossidegi, A Norman,
J Linnell, R Lansdown, D Phillips, K Hayson,
D Whitehorne, A Ratcliffe, R Bowen,
H M Thompson, P W Greig-Smith,
A R Hardy, R J Fairall, K Miah, D J Flynn,
J Collins, G K Frampton, J Mauremootoo,
G P Vickerman, G E Westlake, L Vick

Photography

R Page, R Blakeman

Preparation of Reports

M Beams, J Rapley, M Spittle, K Farrow,
J M Crow

INDEX

spurge, see *Euphorbia* spp
standardisation (Project design), 10–11, 18
Staphylinidae, 87, 98–101, 104, 112, 114, 154, 176, see
    also beetles, predatory
starling, see *Sturnus vulgaris*
statistical tests
    agricultural trials, 7
    Chi-squared, 150–151, 154–155
    cluster analysis, 77
    Fisher exact, 151
    multivariate analysis, 70
    regression analysis, 134, 188
Steering Group
    function and responsibilities, 17
    membership, 2
Stellaria media, 25, 31, 68–72, 77
Strix aluco, 163
Sturnus vulgaris, 160–171, 194–197, 199, 203
Supervised programme
    birds, 160–174, 175–199, 203
    continuation study, 214
    crop choice, 11, 42
    crop walking, 22–23, 58
    design, 7–8, 11, 15–16
    diseases, 4, 37–46
    husbandry, 13
    insect pests, 19–24, 83–108, 112–131
    invertebrate (crop), 82–108, 110–131, 207
    monitoring/management, 16, 22–23, 37, 200–201,
        210–211
    programme, 2, 6, 200, 208–211, 214
    residues, 62–67
    small mammals, 145, 149–156
    soil fauna, 138–143
    weeds, 3, 25–36, 68–81, 205
    yield, 15, 50–59, 209–210
Sylvia communis, 160
Symphyla, 103
synergism, carbendazim and prochloraz, 40
Syrphidae, 96, 103, see also flies, predatory

Tachydromyia arrogans, 85, 98, see also Empididae,
    flies, predatory
Tachyporus spp, 87–88, 98, 104, see also Staphylinidae
take-all, see *Gaeumannomyces graminis*
TALISMAN, 214–215
territorial behaviour, birds, 160–163
thistle, creeping, see *Cirsium arvense*
thousand grain weight, 50, 52
threshold, 1
    disease, 15, 37–39, 48
    pests, 15, 19–23, 48, 83–85, 122, 150
    weeds, 15, 48

weeds, broad-leaved, 26, 31–35
weeds, grass, 26–31
thrips, see *Limothrips* spp
Thysanoptera, see *Limothrips* spp
timing
    agricultural operations, 10
    sowing, 6, 10, 35
translocation of pesticide, 64
trapping methods, 144–145
    rabbits, 157
    small mammals, 144–152
Trechus quadristriatus, 100–101, 105–107, 115, 124,
    see also Carabidae
tree sparrow, see *Passer montanus*
tri-allate, 25, 29, 31, 65, see also herbicide programme
triadimenol, 142, see also fungicide programme
    with fuberidazole, 46
triazophos, see also insecticide programme
    effects on hedgerow invertebrates, 60–65
    effects on starlings, 194–197
    persistence, 60–62
    use, 13, 84, 93, 105, 107, 130, 142
triplets of fields, 10, 88, 134
Troglodytes troglodytes, 160–163
Turdus merula, 160, 213
Turdus philomelos, 160

Urtica dioica, 79

Veronica spp, 25, 31, 72
vertical distribution of invertebrates, 105–108, 140–142,
    206

weasel, see *Mustela nivalis*
weather, effects on
    bird census visits, 161
    disease, 44
    husbandry, 10–11, 18, 19
    invertebrate populations, 82, 105, 107, 108, 207, 209
    pesticide use, 13, 18, 37, 42, 85
    pests, 21, 85–87
    small mammal trapping, 145
    soil fauna, 141
    weed control, 32
    yield, 57
weed assessment
    control programme, 25, Appendix I
    methods, 26–36, 77, 119, 205
    spray decisions, 26–35
    survey, 68–81
    thresholds, 15, 26–31
Weed Research Organisation, 2
wheat bulb fly, see *Delia coarctata*

Printed in the United Kingdom for HMSO
Dd293426 6/92 C11 G531 10170